Fluvial Processes in the Pleistocene of Northern Europe

Wei Chu

BAR International Series 2797

2016

First Published in 2016 by
British Archaeological Reports Ltd
United Kingdom

BAR International Series 2797

Fluvial Processes in the Pleistocene of Northern Europe

ISBN: 978 1 4073 1461 7

Printed in England

All BAR titles are available from:

British Archaeological Reports Ltd
Oxford
United Kingdom
Phone +44 (0)1865 310431
Fax +44 (0)1865 316916
Email: info@barpublishing.com
www.barpublishing.com

Contents

Abstract

This book explores the effects of fluvial processes on Pleistocene lithic artifacts and assemblages in northern Europe. Lithic artifacts are commonly found within fluvial sediments indicating that they have been subject to moving flow. Their incorporation into the bedload may result in changes to morphology and spatial context, thereby obscuring site interpretations. This has been well-documented among numerous sites worldwide.

This study reports on three experiments aimed at understanding the effects of fluvial processes on lithic artifacts. In the fluvial-field experiment, 456 painted and measured replica lithic artifacts were placed in nine different 'scatters' along the River Glenderamackin (UK), where river flow and animal activity were monitored for 18 months. Artifacts were visited at four different times and each time their locations, orientations, and physical conditions were recorded. The results showed that artifact movement depends on lithic morphology and discard location (in the river channel or on the banks), with vegetation and burial playing critical roles in inhibiting transport. Artifact transport also resulted in characteristic edge micro-flaking.

In the tumbling barrel experiment, 188 replica lithic artifacts were subjected to an environment simulating the effects of coarse-grained fluvial abrasion. The results showed that artifact lithic abrasion is affected by time, the amount, size, and lithology of the abrasive sediment as well as the artifact's lithology and shape.

The last experiment reports on aspects of fine-grained fluvial abrasion. In this study, 27 artifacts were marked with nine microscopic observation areas and abraded in an annular flume with three different sediment sizes for up to 168 hours. When analyzed with a scanning electron microscope, the results showed that time, sediment size and the artifact's entrainment mode have varying effects on artifact edge micro-flaking, surface texture, and ridge widths.

The implications of these experiments are discussed along with previous experiments and applied to the sites of Happisburgh 1 (UK), Happisburgh 3 (UK), Pakefield (UK), and Caours (FR), questioning the formation of these sites and indicating that they were subject to post-depositional processes.

List of abbreviations

%originallength	Percentage of original measured length
%originalweight	Percentage of original measured weight
%originalwidth	Percentage of original measured width
AAR	Amino Acid Racemization
ANOVA	Analysis of variance
AOVcircular	Single criterion Analysis of variance for circular data
BOR	Binary Ordinal Regression
df	Degree of freedom
DGPS	Differential Global Positioning System
EDX	X-ray spectrometer
ESR	Electron Spin Resonance
GIS	Geographic Information System
INRAP	*l'Institut National de Recherches Archéologiques Préventives*
K-S test	Kolmogorov-Smirnov test
mAOD	meters Above Ordnance Datum
NGR	National Grid Reference
OLR	Ordinal Logistic Regression
OSGB1936	Ordnance Survey Great Britain 1936
OSL	Optically Stimulated Luminescence
p	*p*-value
QGIS	Quantum Geographic Information System
RFID	Radio-Frequency Identification
SD	Standard Deviation
SEM	Scanning Electron Microscope

Acknowledgements

I am grateful to the many people who have supported and facilitated this project. I would like to thank Dr Robert Hosfield and Prof. Martin Bell, who have guided me throughout this work. Thanks also go to Prof. John Allen and Dr Stuart Black for discussions and advice. I would also like to thank Prof. Mark White and Dr Nick Branch for carefully reading the work and providing valuable comments which have helped to clarify and improve the work. My thanks also are to the University of Reading that generously funded this research as part of my PhD and the University of Cologne which allowed me time to prepare this manuscript.

For the fluvial field experiments, my gratitude goes to Dr Jeroen De Reu, the Environment Agency of Northwest Penrith, Mr Dave Thornley for his help with the DGPS, and Mrs M.E. Hutton and family for the use of their land.

The tumbling barrel experiments would not have been possible without support from the Lithic Studies Society and the John Wymer Bursary, Dr Radu Iovita, and Dr Andy Howard for furnishing materials and expertise. Thank you.

Dr Charlie Thompson was a friend and critical partner in performing the flume experiments. Thanks also go to Dr Suzanne MacLachlan at the British Ocean Sediment Core Research Facility (BOSCORF) for use of the SEM for the flume experiments.

Un grand merci à Jean-Luc Locht and the kind staff of INRAP Picardie. I am also indebted to Ms Gisselle Garcia at the American Museum of Natural History, and to Dr Nick Ashton and the staff at the British Museum for allowing me access to the archaeological collections.

I am fortunate to have a number of people who love me enough to have put up with the inevitable pangs of writing this book. I cannot thank my family, particularly my mother and siblings, and close friends enough for their support. Very special thanks are reserved for Sophie, who, by virtue of proximity, had the most to put up with.

Chapter 1

Introduction

[The antiquarian] has, from a few detached facts, to fill up a living picture; so to identify himself with the past, as to describe, and follow, as though an eye-witness, the different changes which have at various periods so greatly modified the surface of the earth. (Prestwich 1857: 6)

1.1. Background and aims

Lithic artifacts are a main source of information for reconstructing the geographical origins and movements, behavior, and diets of ancient hominins (Binford 2002; Braun *et al.* 2006; Henry *et al.* 2004; Keeley 1980; Schick 1991; Stout 2011). Their morphology, location, and spatial relationships provide archaeologists with key datasets to test hypotheses.

Since the early 1970s, archaeologists have recognized that the unquestioning interpretation of the lithic artifact record leads to erroneous conclusions as sites have been disturbed by natural processes such as animal trampling, weathering, or geomorphic activity (Binford 1979; 1983; Schiffer 1983; 1985; Villa 1982). The study of these processes is known as *taphonomy* (Efremov 1940).

The following examples, chosen from many, illustrate what taphonomy is and how it has affected archaeological site interpretations:

- When the open-air site of Terra Amata, Nice, France (200–300kya) was excavated, several distinct sedimentary layers with conjoinable flint pieces, postholes, and hearths were found (de Lumley 1969; 1975; Villa 1977; 1978). These were interpreted as the remnants of campsites with distinct living areas. However, later research showed considerable vertical migration of artifacts between the layers, possibly the result of trampling, bioturbation, and sediment wet-dry cycles, ultimately rendering the original conclusions untenable (Villa 1977; 1982; Villa and Courtin 1983).
- At Cagny-l'Epinette, Picardy, France, researchers originally suspected that a layer (I1) represented a single occupation surface or living floor, based on artifacts found in thin layers of fine fluvial sediments in discrete 'activity' areas (Tuffreau *et al.* 1986; 1995). However, based on the artifacts' spatial distribution, orientation, and size comparisons to the natural clast background, a reanalysis concluded that they had likely been subject to erosion and entrainment by a palaeochannel (Dibble *et al.* 1997).

Though Terra Amata and Cagny-l'Epinette are both classic examples of Palaeolithic archaeology modified by natural processes, the latter highlights the importance of *fluvial processes* whereby artifacts are entrained within ancient river systems. In recent years, the effect of fluvial processes on the lithic artifact record has been extensively investigated both in the field and through laboratory experiments. Still, no clear patterns for these effects have yet been established because the literature reports important discrepancies in characteristic behavior of lithic artifacts in fluvial settings (Hosfield and Chambers 2005c; Schick 1993). In addition, there is a lack of experimental data concerning the impact of the sedimentary parameters involved (Basell *et al.* 2011; Hosfield and Chambers 2005c; Kaufmann *et al.* 2011; Thompson *et al.* 2011).

Among these are:

1. The effect of river discharge on artifact entrainment (Hosfield 2011).
2. The effect of sediment size and lithology on artifact abrasion (Hosfield 2011).
3. The effect of artifact lithology, class, and entrainment mode on abrasion (Chambers 2003; Hosfield and Chambers 2005c).

What emerges from the literature is that fluvial processes are complex processes that are only partially understood.

Understanding the ways in which artifacts interact with fluvial systems will lead to fewer incorrect interpretations of archaeological sites and to more accurate analyses. To gain such an understanding, it is important to:

1. Understand how lithic artifacts behave in different fluvial settings.
2. Recognize the effect of fluvial transport and deposition on lithic artifact morphology.
3. Identify the effect of fine-grained fluvial transport on artifact surfaces.

The primary aim of this study is to assess the effect of fluvial action on the spatial and morphological, especially abrasive, modifications of Palaeolithic artifacts. The following specific research questions form its point of departure:

1. How do fluvial erosion, transport, and deposition affect the spatial composition and appearance of Palaeolithic assemblages?
2. How does coarse-grained fluvial abrasion affect Palaeolithic artifact size/shape and physical appearance?

3. How does fine-grained fluvial abrasion affect Palaeolithic artifact morphology?

1.2. Book structure

To address these questions, this book first presents a review of lithic artifact taphonomy (Chapter 2; Figure 1), with an overview of site formation models focused on fluvial processes. This review is followed by an overview of experimental approaches used to understand lithic fluvial taphonomy (e.g., Hosfield and Chambers 2005a; Isaac 1967; Petraglia and Nash 1987; Schick 1986). Avenues for further investigation are then outlined and the aims of this study are defined. After documenting the methodological approaches used in this study and justifying the methods and techniques employed (Chapter 3; Figure 1), three experimental regimens, each corresponding to the main goals, are discussed in succession. The results of a fluvial experiment that explores the spatial movement and modification of replica Palaeolithic artifacts in the Glenderamackin River (UK) are reported in Chapter 4 (Figure 1). Chapter 5 presents the results of a tumbling barrel experiment targeted at understanding morphological changes to lithic artifacts in coarse-grained fluvial environments. Chapter 6 reports the results of an annular flume experiment that assesses changes to artifact morphology in fine-grained deposits and low-energy fluvial settings (Figure 1). The results obtained in these three experiments are compared to previous experiments and applied to artifacts collected at key fluvial Palaeolithic sites in northern Europe (Chapter 7). The study concludes with a robust summary of the research and draws out new avenues for future research.

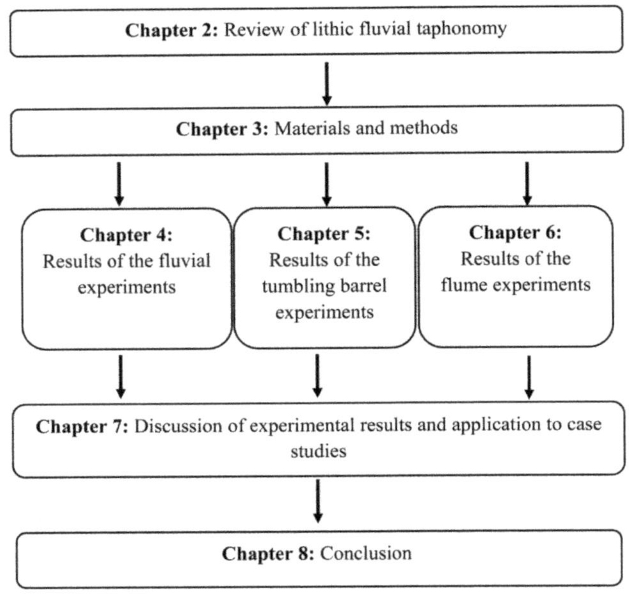

Figure 1: Outline of this study.

Chapter 2

Fluvial processes and lithic artifacts

2.1. Introduction

> *The discovery of flint implements wrought by the hand of man, in what are certainly undisturbed beds of gravel, sand, and clay, both on the continent and in this country—tends to show that... in this region of the globe, at least, its surface has undergone far greater vicissitudes since man's creation than has hitherto been imagined. (Evans 1862: 1)*

This chapter addresses the sedimentological context of Palaeolithic artifacts in northern Europe and reviews the current understanding of how assemblages are formed. First, key geo-archaeological contexts of artifacts will be examined. Next, the reader's attention will be turned to river terrace formation and the characteristic preservation of archaeological sites in fluvial deposits. The latter is important because understanding 'how' and 'why' river terraces form is crucial to interpreting many early Palaeolithic assemblages. The issue of lithic taphonomy will then be discussed—it is through this discipline that archaeologists have come to fully appreciate the complexity of interpreting ancient archaeological remains.

The next part of this background chapter reviews experiments relating to how rivers affect archaeological sites. The first section (2.3) deals with artifact transport. It begins by reviewing clast transport in rivers and then outlines previous archaeological experiments pertinent to artifact transport in rivers. The next section (2.4) deals with the abrasion of lithic artifacts in fluvial systems. Like the previous section, it will begin by explaining what is known about clast abrasion and then move on to experiments that have sought to clarify how artifacts abrade. The third section (2.5) continues to explore lithic abrasion, focusing on more subtle microscopic alterations in fine-grained sediments. The final section (2.6) links this to the wider context of fluvial taphonomy and outlines necessary further work.

2.2. Geo-archaeological contexts

Palaeolithic artifacts are commonly found within Pleistocene sediments (Mishra *et al.* 2007; Wymer 1995; 1999). For archaeologists, understanding how these ancient sediments were laid down is an important goal because the conditions that deposited them may have dramatically and repeatedly changed during the Quaternary (Wymer 1999). Usually, this goal is achieved using a uniformitarian approach that compares key morphological traits of the sediments to those observed because of present day depositional processes. Understanding how these sediments form allows researchers to partially disentangle artifacts from their taphonomic histories and contextualize them within appropriate spatial and temporal frameworks (Hiscock 1985).

Sediments are commonly categorized according to the degree of energy that deposited them. As greater force is required to move larger particles, sediments can be classified by the largest clasts that have been transported. Traditionally, these distinctions are distilled into a binary classification system and called *fine-grained* or *coarse-grained* deposits (e.g., Roebroeks and Van Kolfschoten 1994), generally based on the particle size of the surrounding sediments. When artifacts are deposited within a fine-grain environment, they are assumed to be minimally transported (in the order of cm and m). Conversely, when they are deposited in a coarse-grained environment, it is assumed that they were heavily transported (upwards 10 of m). Though such assumptions are problematic because they do not account for the variable size of the artifacts, for this introduction, they will be used tentatively, keeping in mind the complexities of depositional processes.

Among fine-grained deposits, *Aeolian* (windblown) sediments are the most common. Aeolian sediments are deposited by ambient winds as dune sands or fine-grained silts or clays that, when loosely cemented by calcium carbonate ($CaCO_3$), are known as loess (Wymer 1995). Loess formations blanket areas of hundreds of square kilometers across the earth and can be tens of meters thick. In northern Europe, loess, derived by deflation from the unvegetated floodplains of glacio-fluvial runoff, are important for Palaeolithic research in the southern United Kingdom, northern France and parts of Germany and the Low Countries such as at Veltwezelt-Hezerwater (Bringmans 2006) and Maastricht-Belvédère (Roebroeks *et al.* 1997). These sites provide excellent preservation conditions for archaeological materials and are commonly datable by luminescence (thermoluminescence, optically stimulated luminescence) and biostratigraphy (Vallin *et al.* 2001).

Fine-grained deposits are also associated with low-energy alluvial channels and lacustrine sediments formed at the margins of palaeolakes, channels, or saltmarshes, where dry land was covered with fine muds or silts as the water bodies expanded or changed their course. Examples are found at Hoxne, the Boxgrove Slindon silts, and Neumark Nord (Ashton *et al.* 2006; Mania *et al.* 1990; Pitts and Roberts 2000: 110). Artifacts recovered in these deposits usually maintain high degrees of spatial integrity, having moved little more than a few centimeters from their original locations. The preservation of these sites is important to archaeologists as it produces high-quality refits and artifacts that are often suitable for micro-wear analysis (Keeley 1980).

Coarse-grained deposits are the result of powerful Pleistocene water or ice actions that form a broader and ubiquitous context for Palaeolithic artifacts (Hosfield 2001; Wymer 1995). The most common of these are the river terraces of southern Britain and northern Europe that fossilize the remains of ancient powerful fluvial systems usually containing Palaeolithic artifacts. To a lesser extent, Palaeolithic artifacts in northern Europe are also preserved within *diamictons*, accumulations of unconsolidated and poorly sorted debris. Glaciers that have either pushed, or 'conveyer-belted' sediments off an existing landscape may have formed many of these deposits (Hart 1999). Though the characteristic transport and/or taphonomic processes of glaciers is not fully understood, it is likely that they are highly stochastic and probably vastly deleterious of both faunal and lithic remains (Saville 1997).

While a plethora of other sedimentary contexts exist and are examined elsewhere (Wymer 1999), the focus of this book is on fluvial systems. While fluvial systems are akin to glacial deposits in that they can be difficult to source (cf., Lewis 1998), they offer benefits to our understanding of the Palaeolithic record. Indeed, they are far higher-yielding of artifacts than any other Quaternary sediments and, as broad time/space-averaged sequences, less dependent on local environments that function as sedimentary traps (Hosfield 2001: 80; Shott 2008).

Ancient fluvial sequences are important analytical units for palaeoanthropological research as they regularly contain multiple hominin and faunal fossil and/or stone tools or débitage that reveal long-term trends. Fluvial sequences offer artifact-rich alternatives to cave and abri (rock shelter) sites, which represent only one mode of hominin landscape exploitation. These datasets are therefore essential to our understanding of prehistoric behavioral adaptations, in addition to the probable interrelationships between environmental constraints and hominin taxa evolution.

However, there is variation in the preservation and resolution of fluvial sediments as the following examples of European sites and landscapes show:

- *Soucy 5 level II*: Acheulean industries are found *in situ* in association with the remains of large mammals (Lhomme 2007). Artifacts were found within two main layers: an upper fine sediment fill of several channel systems and a lower flint pebble layer (Figure 2). Flint concentrations correspond to biface knapping locations where 80% of the lithic remains can be linked to the operational sequence of production suggesting a high level of archaeological resolution (Lhomme 2007).
- *Clacton (golf course site)*: a high proportion of unrolled artifacts, retouched flakes, cores, and mammalian faunal

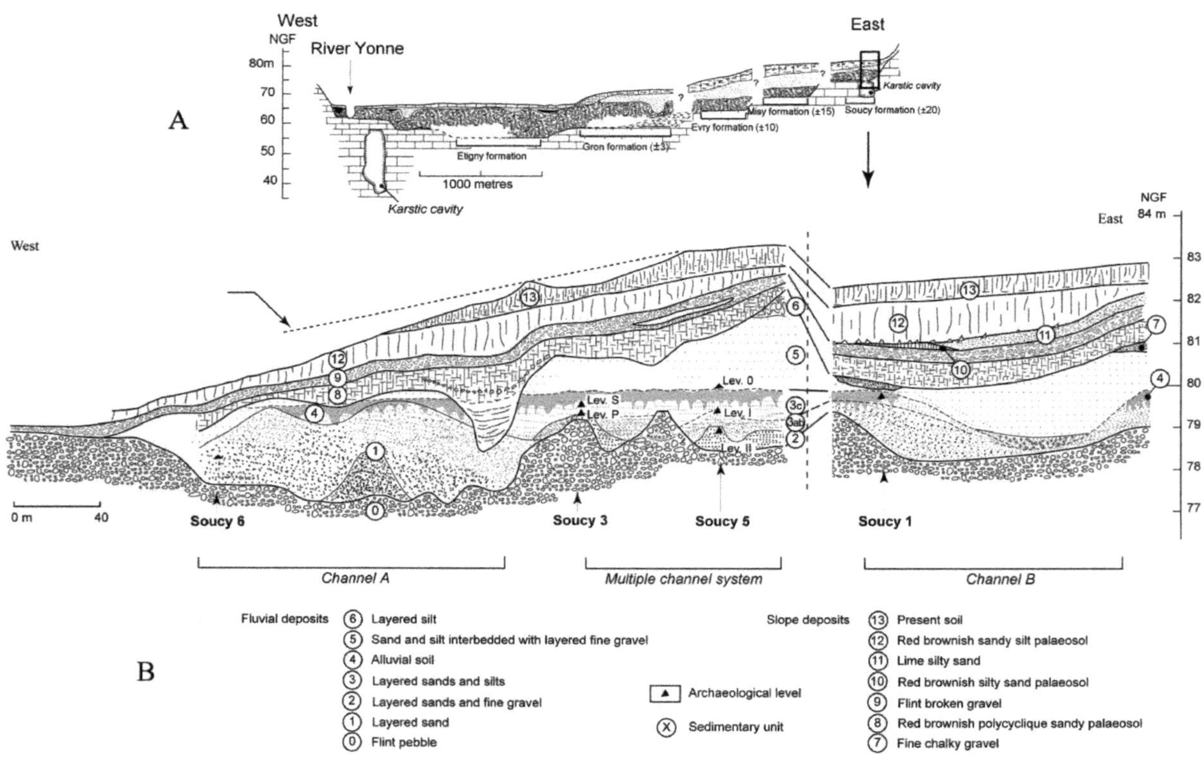

Figure 2: A. Cross-section of the Yonne Valley at Soucy. B. Transect of the alluvial plain at Soucy. Integrated stratigraphy and position of the Lower Palaeolithic settlements (Reproduced from Lhomme 2007; figure 2, Cambridge University Press, used with permission).

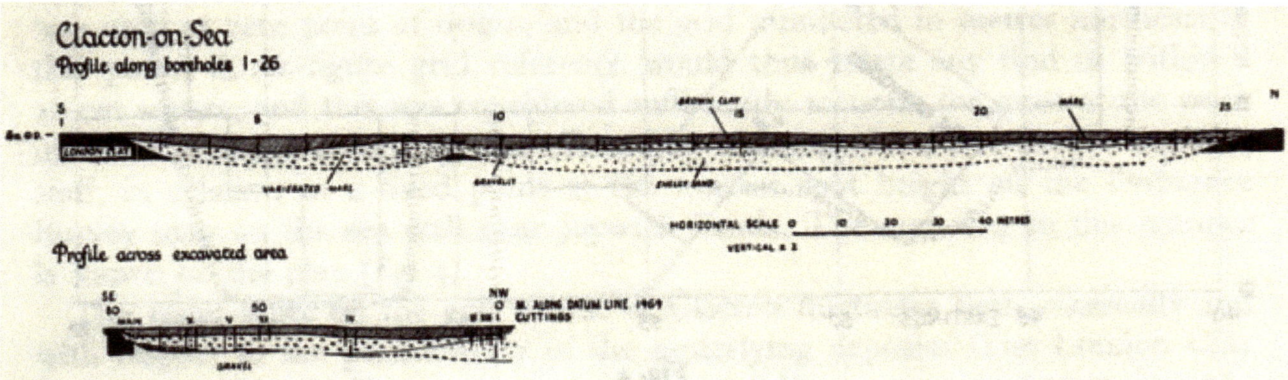

Figure 3: Sketch profiles of Clacton on Sea (golf course site) based on sections revealed by boreholes and excavation (Singer *et al.* 1973, Cambridge University Press, used with permission).

remains were found within three major deposits of gravels, marls, and clays (Singer *et al.* 1973: 17; Figure 3). The concentration of the artifacts near the top of the gravel, along with the refitted artifacts and micro-wear studies suggest that the site is preserved *in situ* (Hosfield and Chambers 2005c; McNabb 2007: 69).

- Elveden and Barnham: handaxes, flake tools, flakes, and cores were found with vertebrate and ostracod remains recovered from a late glacial/early interglacial channel, cut into a series of glacial deposits and infilled with sands, silts, and clays (Ashton 1998; Ashton *et al.* 1994; 2005; Figure 4). Multiple assemblages were recovered with a range of preservation from 'fresh' to 'rolled' to undisturbed, suggesting a variation of spatial integrities from *in situ* to more disturbed (Ashton 1998; Ashton *et al.* 1994; 2005).
- Lynford: *in situ* mammoth remains and associated Mousterian stone tools and débitage are sealed within a dark organic fill of a palaeochannel, approximately 21m long by 12m wide, representing a former meander cut-off or oxbow lake environment (Boismier *et al.* 2003; Gamble and Boismier 2012; Schreve 2006; Smith 2012). Faunal remains were likely deposited under still water conditions during the infilling of the

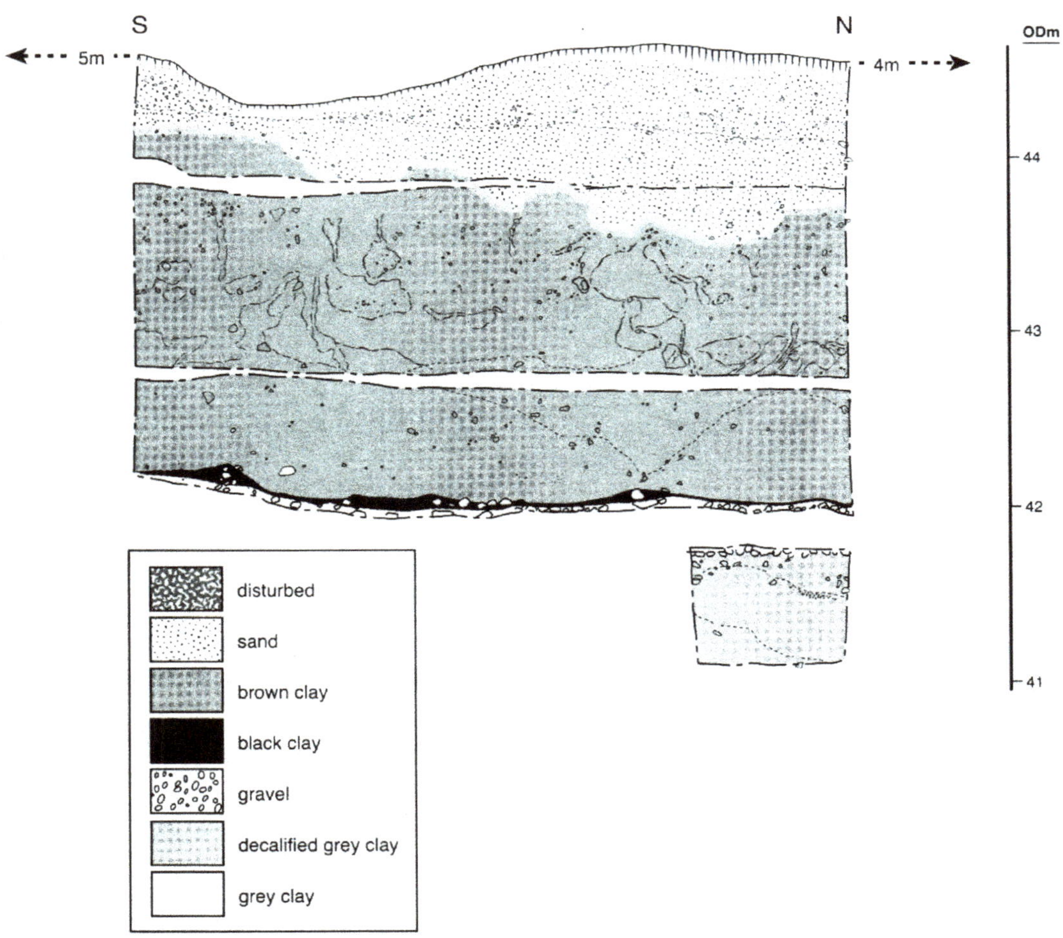

Figure 4: Elveden Area I section (Ashton *et al.* 2005; fig 11, Cambridge University Press, used with permission).

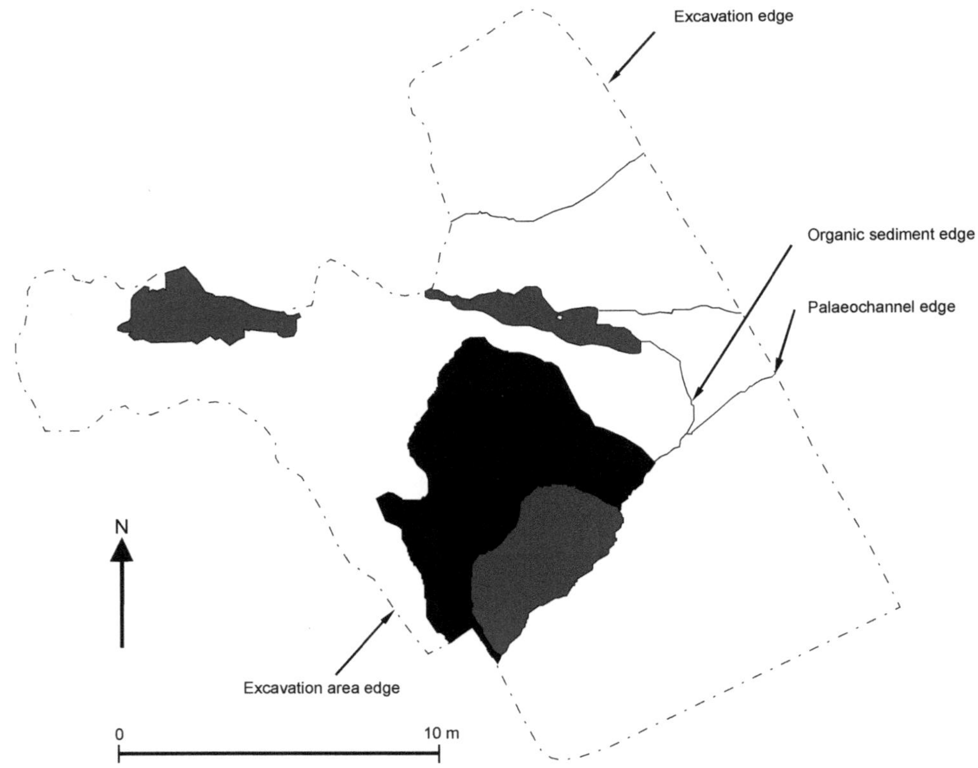

Figure 5: Plan of organic deposits at Lynford; shaded areas represent bank collapse and sediment gravity flow (Redrawn after Smith 2012; fig. 2 after Schreve 2006).

main palaeochannel though others may have lain on the adjacent land surface for multiple decades before being incorporated in the channel through debris flows, bank collapse or over-bank flooding (Schreve 2006; Figure 5). There is no direct evidence of association between the lithic and faunal remains, however the fresh condition of many handaxes (*n*=47), the non-natural/carnivore mortality profile of the faunal remains, the unusually high number of pathologies, and the clear absence of major meat-bearing long bones suggesting that the lithics also maintain a relatively high level of spatial integrity (Schreve 2006). More recent analyses of the lithic and faunal remains have suggested that natural processes such as surface runoff, mass movement, and animal interaction may have modified the lithic and faunal remains from strictly *in situ* assemblages (Smith 2012; White 2012).

- Hackney: vertebrate, plant, insect, and mollusk remains occur within a thick unit of organic sands and silts occupying a channel near the confluence of the River Thames and the River Lea (Figure 6). Likely, these represent a short time interval of deposition and a relatively high faunal resolution. Sands and gravels superimposed with organic sediments suggest floodplain deposition of a braided river under cool or cold climatic conditions (Green *et al.* 2006).
- Solent River: many highly abraded handaxes and cores, without a substantial amount of faunal remains due to sediment chemistry, are distributed over a large terrain (Figure 7). Most of the archaeological record derives from the fluvial sands and gravels of the former Solent River and its principal tributaries, the Rivers Frome, Stour, Avon, and Test (Ashton and Hosfield 2010; Hosfield 2001).

In the past, Palaeolithic archaeologists have tended to focus excavation efforts on sites of unabraded and conjoinable lithic materials found in high concentration. These indicators are taken as a sign that artifacts are in primary, or near primary context (Wymer 1995). It is argued that, when artifacts are found *in situ*, they can inform us about stone tool reduction, subsistence strategies, and behavioral patterns (Binford 2002; Schiffer 1983). While this may be true, focus on primary sites has created a bias towards environments that act as natural sedimentary traps, which, in turn, affect the way we understand the archaeological record and place an overwhelming focus on 'brief moments in time' (Roe 1980). Such a narrow focus on primary context sites fails to represent the true diversity of hominin occupation. Indeed, it is necessary to extend the reach of archaeological research to secondary sites to answer questions related to hominin prehistory that operate at different scales (Gamble 1996; 1999: 114).

Artifacts with highly abraded surfaces are also found in secondary or 'geological' contexts that have broad spatial associations. These artifacts constitute some 95% of the evidence for Lower Palaeolithic activity in Europe and, as a result, form an important part of the Palaeolithic record (Wymer 1995). They also make up most of the museum assemblages, which, because of their often-uncertain contexts and abraded condition, have been

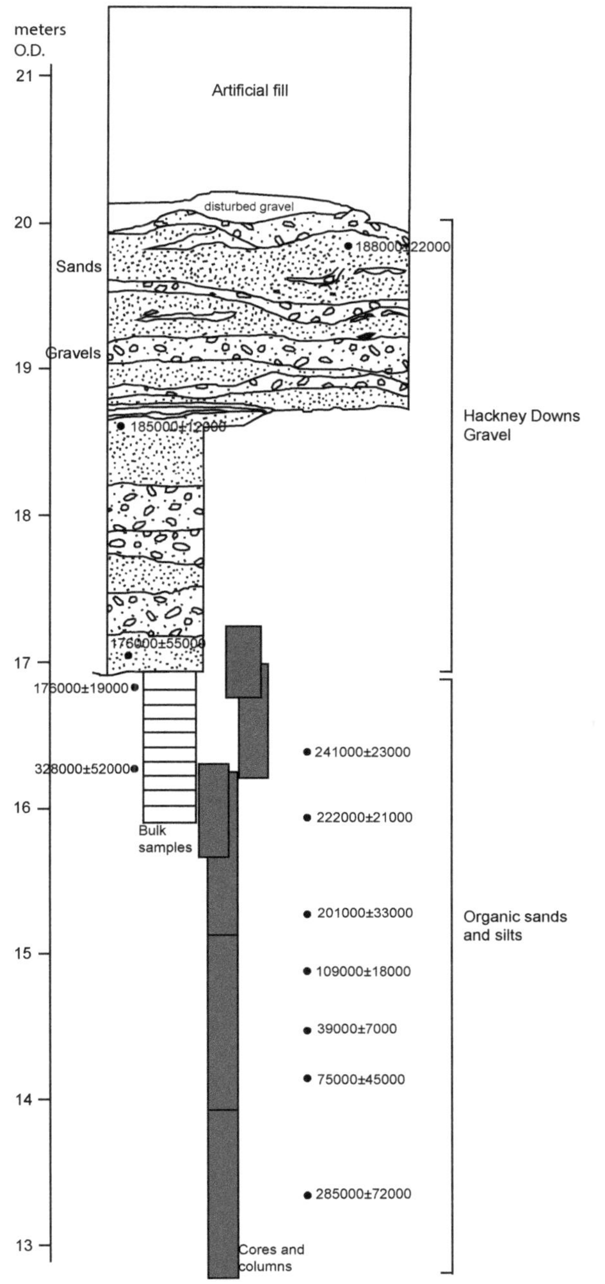

Figure 6: Sampling scheme, Nightingale Estate, Hackney. Dates determined by OSL (Redrawn after Green *et al.* 2006; figure 3).

largely overlooked, especially since the shift from typo-chronological schemes of the early twentieth century (e.g., De Sonneville-Bordes and Piveteau 1960).

The derived nature of secondary context artifacts, however, needs not be a fatal factor in their potential to offer important information for Palaeolithic research. In fact, while they provide a more complicated taphonomic history, these artifacts through their very quantity provide robust demonstrations of human presence and activity over time-averaged scales from which patterns of hominin dispersal and ecological factors, as well as hominin technological responses, can be determined (Ashton and Hosfield 2010; Ashton and Lewis 2002; 2012).

2.2.1. River terraces and northern European climate

The northern European sedimentary record presents a unique opportunity for Pleistocene fluvial research. Many taphonomic complications associated with documenting the ancient fluvial systems are alleviated by the fact that, in areas unaffected by Pleistocene ice sheets, large Quaternary fluvial deposits are well preserved (Gibbard and Lewin 2009). Intense quarry exploration (Roebroeks 1996) of the gravel beds of rivers near urbanized areas (i.e., the Thames and the Somme), from the 19th century onwards, has exposed large lithofacies allowing the vertical and horizontal mapping of defined stratigraphic units (Wooldridge 1939; Wymer 1999).

As a result, northern European Pleistocene fluvial deposits are relatively well understood and well correlated. Research has uncovered a pattern of ubiquitous valley *staircases*, whose steps form large-scale aggradational river terraces (Antoine *et al.* 2007; Bridgland 2000; 2010; Bridgland *et al.* 2006). It is well understood by the dates given from Optically Stimulated Luminescence (OSL), Electron Spin Resonance (ESR) and Amino Acid Racemization (AAR; e.g., Bahain *et al.* 2007; Lauer *et al.* 2010; Penkman *et al.* 2013; Schreve 1998) that these terraces are the remnants of ancient floodplains that formed during the Quaternary, during successive incisions into bedrock below the bed of the contemporary floodplain.

Researchers have long hypothesized that the key driving force behind these erosional incisions was a response to sea-level fluctuations of the Pleistocene (Wymer 1968; Zeuner 1945). During the last 2 million years, the history of the earth has been marked by cold *glacials* or ice ages, punctuated by shorter *interglacials*, which were characterized by intervals of increasing global average temperatures lasting thousands of years (Lisiecki and Raymo 2005). These phases are commonly attributed to Milankovitch cycles, a result of the earth's eccentricity cycles (Roe 2006) and are documented by a host of geological (i.e., rock scouring/scratching, valley cutting, drumlins, tills), chemical (marine isotope sediments, ice core samples), and biological (biostratigraphy) phenomena (Lowe and Walker 1997). The amplification of climatic oscillations in the last 700 kya resulted in the regular occurrence of the above-mentioned staircase formations, whereby the cutting of new floodplains was formed under cold/warm transitional conditions (Bridgland and Westaway 2008). The refinement of timescales in recent decades has offered the possibility of examining the geomorphological effect of river activity at a single cycle level, illustrating that incision is most likely expressed at times of climatic change (Vandenberghe 1993; 1995). Successive stages of terrace development have been linked to river architecture, vegetation and climate: the Middle Thames Valley is often used as a standard sequence for the late Middle Pleistocene (Bridgland 1996).

It is now understood that climate cyclically affects long-term uplift, which is the essential requirement of river

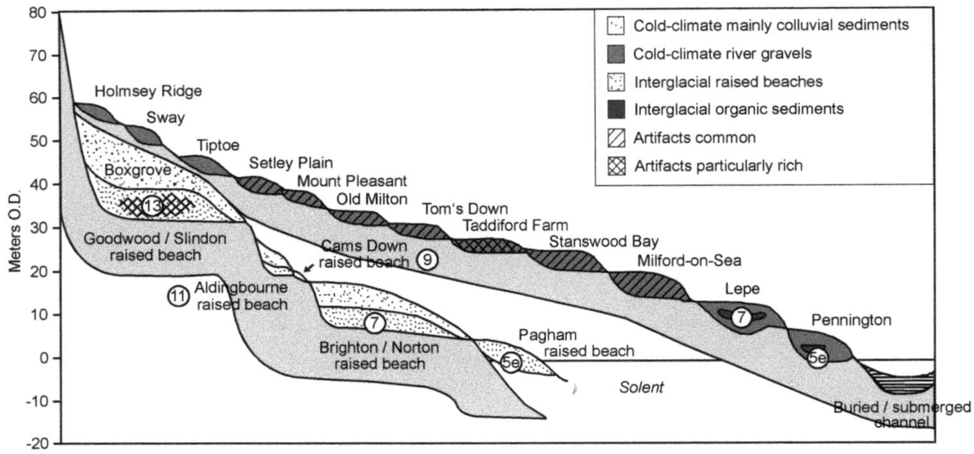

Figure 7: Idealized and superimposed north–south sections through the staircase of raised beaches bordering the Sussex coast and the terraces of the River Solent (Redrawn after Bridgland 2000; fig. 5).

incision (Maddy *et al.* 2000; Maddy and Bridgland 2000; Westaway *et al.* 2002). In glacial and interglacial periods, when erosion is at a minimum due to seasonal melting and low water flow, sediments are deposited offshore on the wide European continental shelf where the Upper Crust is thick. During climatic transitions, as erosion increases because of pleniglacial runoff and harsher inclement weather, the pressure gradient from inside the earth is increased inland, creating uplift as the weight of the overlying eroding sediment is washed away.

Uplift is also a result of the cyclical loss of the weight of ice-sheets, which allows the Upper Crust to expand, a process known as *denudation isostasy*. This process can be observed in the Lower Rhine area (see Figure 8; Bridgland 2000), where the southern central part of the North Sea sedimentary basin has exhibited continuing well-documented subsidence while upstream, the Rhine has an extensive flight of terraces (Brunnacker *et al.* 1982). Between these two zones is a discrete region with virtually no Quaternary fluvial sediments preservation. This represents a *hinge area* between the subsiding basin and the continental interior, which has been rising during the Quaternary. As Bridgland (2000: 1297) notes, 'The slope of the Rhine catchment towards the North Sea basin has illustrated a clear convergence of the Rhine terraces downstream to the hinge area, beyond which there is a divergence of the buried sediments beneath the Netherlands.'

Climate drives terrace formation in other non-geological ways (Gibbard and Lewin 2002). It has a major influence on soil, vegetation, and water supply, which in turn influence the erosive potential of catchments in uplifting areas. As climate persistently drops below freezing temperature, riparian landscapes are heavily influenced. Water becomes frozen in glaciers, and its supply is vastly decreased. Furthermore, muds and plants, likely sparse herbaceous vegetation (Gibbard and Lewin 2002), freeze in the periglacial zone, impacting the overbank matrix and the potential sediment supply to the rivers. By contrast, glacial maximums and full interglacials seem to limit the colluvial

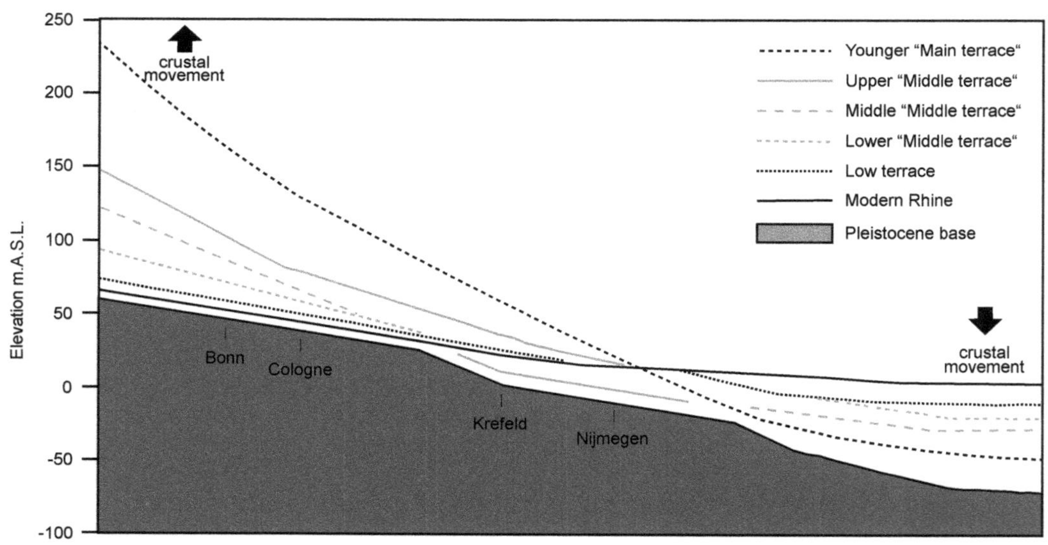

Figure 8: Long profiles of the River Rhine's sediment terraces along with crustal movement. The figure shows their downstream convergence and eventual disappearance beneath later sediments in the subsiding area of the Netherlands (Redrawn after Bridgland 2000; figure 6; from Brunnacker *et al.* 1982).

movement of sediments as they are either anchored by ice or plants (Gibbard and Lewin 2002).

A variety of other influences can also affect terrace formation. A summation of these influences, suggested by Bridgland (2000) for the Lower Thames, can be considered as a putative model (see Figure 9):

- During the cold/warm transition, receding glaciers, rising land, and increased water supply from melting permafrost result in higher glacio-fluvial activity, which triggers rapid incision through the old flood plain (Phase 1). The cool climate does not support riparian vegetation, which would otherwise act as an erosion buffer by reducing sediment supply on the floodplain. This produces a terrace out of the old riverbed and redeposits material at the bottom of the new channel and its floodplain (Phase 2).
- During the interglacial, erosion and coarse sediment deposition subside as the network of roots created by vegetation stabilizes slopes. Deposition is therefore limited to fine silts and sands (Phase 3). Reworking is still a factor, but it is limited to sediments within the channel and pre-existing downstream fossil sediments.
- As the climate oscillates back towards a cooler phase (i.e., warm/cold transition), riparian vegetation once again wanes and slopes are weakened by the increasing influence of freeze-thaw actions (Phase 4). Thus, erosion again deletes large parts of the Phase 2 aggradation and widens the floodplain though this depends on the underlying bedrock (Bridgland 1985). Over time, the subsiding of the river's competence results in the main depositional event of a terrace cycle (Phase 5).
- As the glacial part of the cycle takes full effect, water and sediments become frozen and deposition reaches a minimum, a so-called glacial lockdown (Phase 6).

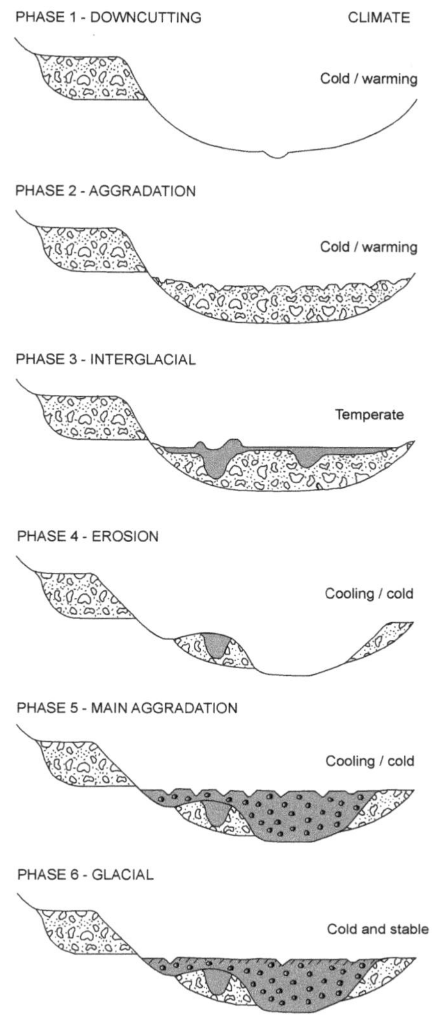

Figure 9: The Bridgland model of river terrace development and the climatic conditions under which they were formed (Redrawn after Bridgland 2000; fig. 1).

The result of this depositional schema is the characteristic 'sandwich' form of each terrace, which has an interglacial fine-grained deposit imbedded within two coarse-grained deposits associated with major climatic transitions. Because northwestern European rivers share the same wide continental shelf and former pleniglacial environment, some have asserted that the Lower Thames model is the best narrative of terrace formation for other rivers in the United Kingdom, Europe, and even worldwide (e.g., Van Huissteden *et al.* 2001; Bridgland 2000; Bridgland and Westaway 2008). Evidence suggests, however, that river terrace formation is greatly affected by local factors even within the same river system.

Sedimentological records from other well-documented Pleistocene river systems illustrate that there are often significant challenges to overcome in correlating gravel beds. Proximal systems to the Thames, such as the Solent in southern Hampshire and Dorset, exhibit nearly twice as many post-Anglian terraces as the Thames, while others, such as the River Medway, exhibit considerable gaps in terrace preservation (Bridgland 1996). Though these gaps may reflect post-formation preservation differences, they may also relate to the underlying bedrock as well as the historical mapping approaches and conventions used by the British Geological Survey (Bridgland 1985; Brown *et al.* 2009).

The linkage of specific parts of the glacial/interglacial cycle to terrace formation has permitted attempts to correlate terraces to the better-understood marine isotope record (Bridgland 1994; 1996; Bridgland *et al.* 2001). The ability to do this accurately has improved because of revised geomorphological surveys and advancements in OSL, ESR and AAR dating (Bahain *et al.* 2007; Lauer *et al.* 2010; Penkman *et al.* 2013). The result is in synchrony with the climatic scales, with few unequivocal exceptions (e.g., the Middle and Upper Thames Valley (Bridgland 1994), the Neckar, and the Wipper (Bridgland *et al.* 2006).

2.2.2. Fluvial site formation processes

A necessary first step to interpreting fluvially deposited artifacts involves constructing geological frameworks of the sedimentary processes, stratigraphic relationships, and depositional controls. Characteristics of ancient fluvial systems and their deposits are commonly inferred from their association with one of three categories of modern rivers and interpreted according to channel patterns and the nature of their overbank, or floodplain sediments (Boggs 1995; Leopold and Wolman 1957; Nanson and Croke 1992).

Two common categories of modern rivers are the braided river, characterized by multiple interconnected networks of channels separated by sediment bars; and the meandering river, a single channel with point bars and well-developed floodplain (Ashmore 2003; Figure 10). The meandering river is similar to a sinuous river, which has the same floodplains pattern but with less curvature and smaller point bars (Hicken 2003; Figure 10).

A third modern river type is the anastomosing system, which is marked by a multiple avulsions that make diverging and reconverging channels in a web-like pattern (Figure 10). Most fluvial systems are related to low annual hydrological regimes, except for the braided rivers, which are typified by a highly fluctuating hydrograph (Boggs 1995; Nanson and Croke 1992).

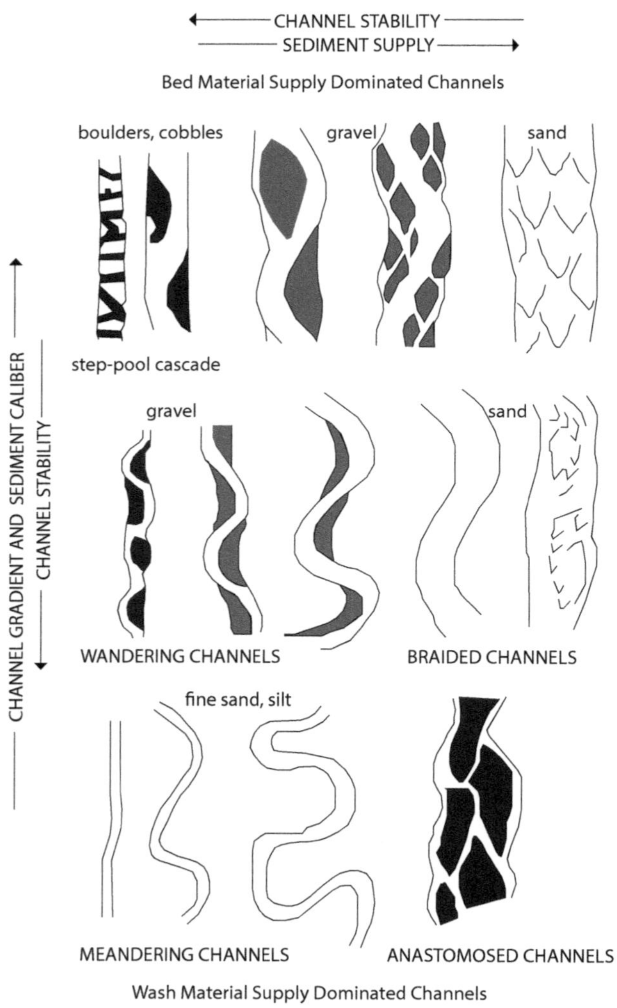

Figure 10: Common river channel patterns and their relation to channel stability, sediment supply, channel gradient, channel stability, and sediment caliber (Redrawn after Church 1992).

Identifying ancient meandering systems by their sediments is relatively straightforward, particularly when point bar deposits are present. Recognizing the sedimentary archive of braided and anastomosing systems is usually more problematic (Willis 1993). This largely stems from the complexity of local geometries and individual processes associated with rivers and a lack of detailed geological models to demonstrate the potential array of channel and bar forms generated by ancient systems (Allen 1985). A host of complex variables can affect our understanding of fluvial sediment; they can broadly be categorized as follows:

- Establishing frameworks for palaeorivers is complicated by the tendency of different types of fluvial systems to generate indistinguishable sedimentary sequences, and the potential for any given system to accumulate an array of disparate sequences (Allen 1985; Best and Bristow 1993).
- Pleistocene fluvial sequences are often fragmentary, tending towards disjunctive deposits separated by erosional surfaces, indicative of deleted large gaps in the geological sequence (Bridgland and Maddy 2002; Brown and Gathogo 2002).
- Poor lateral/vertical exposure resulting from modern sedimentary activity and the nature of quarrying renders informative sections inaccessible.
- Interpreting ancient floodplains requires *a priori* knowledge of a landscape's paleogeography and an understanding of the active channel's location (Nanson and Croke 1992).
- Extant river genetics may not be analogous to those of the Pleistocene. Extensive changes to the landscape during the Bronze Age deforestation have likely left fluvial behavior in the Holocene different from even that of the Pleistocene interglacials (Brown *et al.* 2001).

Though these variables confound interpretations of palaeorivers, it is rare that they are all found in concert. Moreover, multiple sampling locations can also ameliorate our understanding of these deposits.

2.2.3. Lithic taphonomy

> *From the artifact's standpoint the environment is filled with hostile forces. (Schiffer 1987: 143)*

The following section will begin by defining taphonomy and then will explore its impact on our interpretation of archaeological sites. After that, it will discuss how taphonomy directly relates to stone tool assemblages and review the field of experimental taphonomy and some key experiments that are directly related to lithic assemblage formation.

What is taphonomy?

In its original sense, taphonomy (*taphos*-burial, *nomos*-law) refers to the decay of paleontological material as it moves from the biosphere to the lithosphere (Efremov 1940: 88). There are conflicting views on what taphonomy means today as some authors propose a strict interpretation of the term (Lyman 2010) while others argue a more flexible use, closer to definitions of natural formation processes (Schiffer 1983). It is safe to say though that, when archaeologists refer to taphonomy, the term (incorrectly or not) generally is applied to non-anthropogenic processes responsible for the formation of the archaeological site (Ashton 1998; Bernatchez 2010; Enloe 2012; Lyman 1994; Vallin *et al.* 2001; Schiffer 1972).

Taphonomy is a critical epistemological concern because it provides us with contextual information about how the site and the surrounding area have changed since an artifact's original deposition. Much of our current understanding of this was formed through the dialogue between Michael Schiffer's *formation-process theory* and Lewis Binford's *middle-range theory*. Binford (1981) argued that archaeological sites *could* represent a 'moment frozen in time' (so-called the Pompeii premise; Ascher 1961) while Schiffer argued that archaeological sites were necessarily disturbed by innumerable biological, geological and even cultural agents (Schiffer 1985). In spite of the subtle and changing differences in these theories, a broad consensus was that understanding site formation processes helps archaeologists to better interpret archaeological sites (see Kelly 2011 for discussion).

Indeed, there are numerous examples of how an understanding of taphonomy has impacted our reconstruction of hominin behavior, and many, if not most, show that what was originally interpreted as reflecting hominin behavior is, in fact, partially the result of complex natural processes in the past (Dibble *et al.* 1997; Nowell and d'Errico 2007; Villa and Courtin 1983).

Research has shown that taphonomic processes affect sites in different ways (Behrensmeyer *et al.* 2000):

- Spatial fidelity: Sediment movement modifies artifacts' spatial distributions.
- Temporal resolution: The deposition rate of archaeological assemblages is higher than the sedimentological deposition rate—in archaeological terms, this is often called a *palimpsest*. By superimposing different artifact structures created at different times, it becomes difficult to extract a clear narrative of the 'history' of past activities.
- Compositional fidelity: Refers to the differential preservation of biological structures because of diagenesis. Organisms, or parts of organisms, disintegrate at different rates. This can lead to a biased understanding of the original habitat.
- Completeness of time series: It is well known that the geological and sedimentological record is episodic and therefore discontinuous at all scales, suggesting that an accurate record of time is also necessarily segmented. Additionally, sedimentological processes can erode as well as aggregate, which means that segments representing specific portions of time in the sedimentological record can also be deleted.
- Consistency in preservation over geologic time: A major supposition of paleontologists is that the preservation rate of organisms decreases *en masse* over time. However evolutionary changes to organisms' morphology or behavior—so called mega biases—can improve their preservation rates leading to a positive bias in the record (Martin 1999: 309). A robust example of this would be the development of exoskeletons before the Cambrian revolution, leading to better preservation of organisms compared to their ancestors.
- Human bias: Lastly, the study of taphonomy depends on the recovery of artifacts. It is understood that innumerable collection biases, both intentional and unintentional, exist among the exploration done by scientists. For example, Roebroeks and Corbey (1999) document important 'double standards' in the way that we collect and identify both Middle and Upper Palaeolithic assemblages to fulfill preconceived notions of *Archaic* and *Modern* hominins.

What is flaked stone taphonomy?

The interpretation of lithic assemblages offers many of the same epistemological post-depositional problems as that of biological remains, including variable sedimentation rates (creating palimpsests) and diagenesis (Hiscock 2002; 1985). Strictly speaking however, taphonomy only refers to the post-burial processes of biological material (Lyman 2010). As a result, Eren has proposed the term *flaked-stone taphonomy*, as 'a subfield identifying and analyzing the processes affecting the appearance and context of lithic artifacts subsequent to their cultural use lives' (Eren *et al.* 2011). Though a distinction is perhaps not entirely warranted, it is true that the concerns of lithic preservation differ from those of biological material in two primary ways:

1. Lithic artifacts have different spatial preservation pathways (Hiscock 1985). Stone artifacts occupy a narrower range of size and material diversity (especially when they are from the same assemblages) when compared to the range of biological organisms. Even biomineralized components like bone have variable surfaces, densities, and shapes occurring within the same organisms. This has ramifications for the way in which their spatial fidelity is affected by natural forces before deposition. Additionally, stone artifacts have a different set of taphonomic agents acting upon them. For instance, they are not subject to carnivore caching, scavenging, or modification but they are prone to patina formation and weathering.
2. Lithic artifacts *generally* maintain a higher degree of compositional fidelity due to the different diagenetic pathways. Though it is assumed that lithic artifacts are better preserved than biological materials, which are subject to organic decomposition, lithics almost invariably undergo mechanical and chemical changes.[1]

[1] Lithics formed only one component of prehistoric technology. In addition to the suite of taphonomic processes affecting lithics, the

Lithic taphonomy

The last three decades have produced an impressive body of experimental work to help understand the taphonomic preservation (*sensu stricto*) of Palaeolithic sites (e.g., Aslan and Behrensmeyer 1996; Behrensmeyer 1990; Behrensmeyer *et al.* 1991; Boaz and Behrensmeyer 1976; Gifford and Behrensmeyer 1977). In addition, a suite of lithic experiments has been conducted with the aim of trying to understand how assemblages and artifacts are modified. A major (if often ignored) conclusion of these experiments is that taphonomic studies are a necessary component of understanding *any* lithic assemblage.

Figure 11 shows a list compiled by the author (after Appendix 1 from Eren *et al.* 2010) of nine experimentally determined natural processes and their characteristic modifications to archaeological sites or lithic artifacts. While this table is by no means exhaustive, it demonstrates that common natural processes (sediment consolidation, heat/frost action, plowing/curation, trampling, wind abrasion, sweeping/dumping, soil movement, chemical weathering, and fluvial transport) that can affect biological material, can also affect the preservation of lithic residues. Furthermore, it demonstrates that these natural processes can have similar if not identical effects to each other (equifinality), which can be confused with hominin agency (as is the case of micro-flaking and spatial distribution; Hosfield and Chambers 2005c).

The issue of lithic taphonomy is complicated by the many characteristic diagenetic markers that are easily confused with anthropogenic modification. If the main goal of lithic analysis is to extract meaningful behavioral interpretations from the spatial or formal variation of lithic artifacts, then a critical antecedent is the recognition and disentanglement of natural modifications from hominin agency (Schiffer 1983; Shott 1998). This may seem to be an impossible task because (i) analytical or inferential equifinality may exist and (ii) hominin intent is an inference, not a certainty (Eren *et al.* 2010). One promising way of approaching the disentanglement process is by using/performing experiments. In the context of this book, fluvial field and laboratory experiments were chosen.

2.3. Artifact transport and field-based experiments

> *Geology is, in fact, but an elder brother of archaeology, and it is therefore by no means surprising to find that the one may occasionally lend the other brotherly assistance. (Evans 1862: 1)*

Previous sections have detailed the effects of Pleistocene glacio-fluvial activity on archaeological site preservation. The aim of this section is to elaborate on the effects of fluvial processes on archaeological sites, specifically by discussing how they transport lithic artifacts and affect assemblage spatial arrangement. This section begins by exploring the current knowledge of clast transport from a hydrological point of view. Key experiments aimed at understanding lithic artifact transport will then be reviewed. The results of these experiments are then synthesized to explore what is still unknown about Palaeolithic artifact fluvial transport. The conclusion focuses on formally stating outstanding questions and outlines how they can be answered.

2.3.1. Sediment loads and transport paths

The available literature on non-cohesive sediments (such as sand and gravel) and their transport provides a good starting point to understanding artifact fluvial transport. Fluvial transport, or *entrainment,* deals with the movement of grains, or *particles* within a river system (Bridge 2003). A long sequence of experimental work has demonstrated that there is a critical size and entrainment value for grain transport to occur, as illustrated by the Hjulström-Sundborg Curve (Figure 12; Sundborg 1956). Once entrained, particles follow transport patterns that can be estimated by theoretical and experimental methods (Church and Hassan 1992; Wiberg and Smith 1985; 1989; Yang and Lim 2003). However, such estimates are rarely accurate due to the wide range of variables, including sediment composition (e.g., size, density, shape; Soulsby 1997), and the geometric and hydraulic properties of the stream channel (i.e., stream depth, width and length; slope, velocity, discharge; Simons and Şentürk 1992: 565; Wilcock 2001).

Both theoretical and experimental methods suppose that once a particle is mobile, one of three modes of transport is possible: (1) *bedload*—initially, particles roll or slide along in constant contact with the bed; (2) *saltation*—as the flow velocity increases, the particles begin to move in a series of ballistic hops; or (3) *suspension*—the particle is carried, fully suspended in the fluid (Van Rijn 1984). In a given sediment body, all three modes can occur simultaneously, and the predominant mode depends on flow conditions and grain size (Bridge 2003). This can make such distinctions arbitrary at times because the distinction between the bedload and suspended mode of transport is not discernible in nature (i.e., there is no definition of how long particle 'hops' must be before the transport mode is considered suspended; Engelund and Fredsøe 1976; Yang and Lim 2003).

Though definitions vary (Bagnold 1980; Einstein 1950), bedload transport is generally taken to be the lowest layer of moving particles. In the bedload, sand and larger particles travel randomly by traction transport characterized by rolling and sliding along the bed (Soulsby 1997). The specific movement is affected by a variety of factors such as grain size (usually b-axis; but see Church (2003) for discussion), mass, surrounding matrix, and entrainment fluid viscosity (Nichols 2009: 45). Particle movement in bedload is also highly sensitive to many seasonal and local

hominin extrasomatic toolkit included organic components (e.g., Chu 2009; Thieme 1997). Also important is that we may not know to what extent the lithic record is biased towards harder conchoidally fracturing lithics. A good example is the recent identification of heat-treated silcrete as a raw material (Brown *et al.* 2009).

Figure 11: A non-exhaustive summary of taphonomic processes affecting lithic artifact morphology.

	Sediment consolidation	Heat and frost	Plowing and curation	Trampling	Wind abrasion	Sweeping/ Dumping	Soil movement (i.e., solifluction/ gelifluction)	Chemical/ weathering	Fluvial transport
Edge damage/ micro-flaking	Eren *et al.* 2011; Warren 1914; Warren 1905		Mallouf 1982	Eren *et al.* 2010; Flenniken and Haggarty 1979; Lopinot and Ray 2007; McBrearty *et al.* 1998; Nielsen 1991; Pargeter 2011; Pargeter and Bradfield 2012			Dennell 2004; Stapert 1976		Chambers 2003; Grosman *et al.* 2011; Grosman *et al.* 2013; Harding *et al.* 1987; Hosfield and Chambers 2005a; Hosfield and Chambers 2005b
Polish/gloss/ surface				Shea and Klenck 1993	Stapert 1976		Levi Sala 1986	Goodwin 1960; Howard 1999a; Stapert 1976	Howard 1999b; Howard 2002)
Patina							Sheppard and Pavlish 1992	Ackerman 1964; Clark and Purdy 1979; Goodwin 1960; Honea 1964; Rottländer 1975; Stapert 1976; Van Nest 1985	
Scratching			Moir 1914 (implied)				Stapert 1976		
Impact scars/ herzian cones							Stapert 1976		Hosfield and Chambers 2005c
Cracking		Stapert 1976							
Staining							Borrazzo 2006; 2007		
Ridge widening									Burroni *et al.* 2002; Chambers 2003; Shackley 1974; 1975; 1978
Changes to site patterning			Bryan *et al.* 2011; Odell and Cowan 1987	Gifford-Gonzalez *et al.* 1985; Nielsen 1991		Speth 2006; Vaquero *et al.* 2001	Bertran *et al.* 2010; Brink 1977; Bowers *et al.* 1983; Cahen and Moeyersons 1977; Esdale *et al.* 2001; Hilton 2003; Johnson and Hansen 1974; Johnson *et al.* 1977; Moeyersons 1978; Viklander 1998		Harding *et al.* 1987; Hosfield and Chambers 2005b; 2005a; 2004; Shackley 1974; 1975; 1978

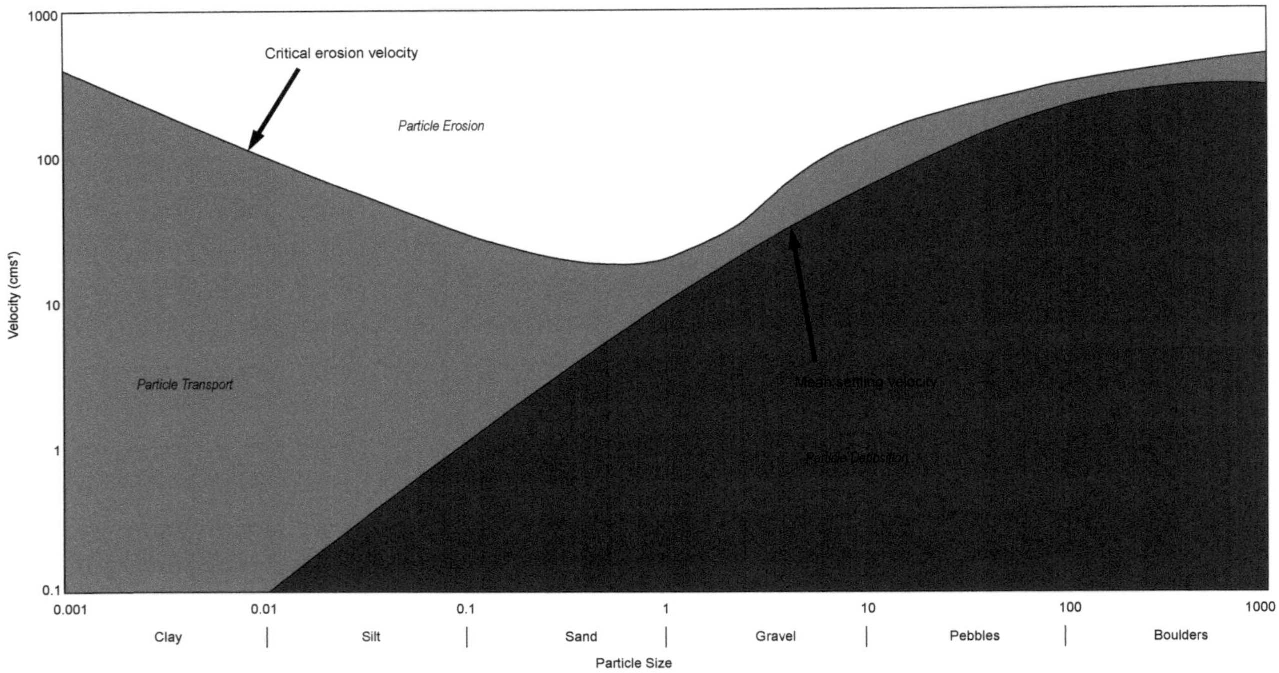

Figure 12: Hjulstrom's curve modeling the relationship between particle size and fluid velocities where erosion, transport and deposition will occur. Note: increased flow velocity is required to entrain clays and silts as a result of van der Waals forces (Redrawn from Hjulström 1935; then modified by Sundborg 1956).

factors such as river discharge, channel morphology, and bed conditions (Hassan and Church 2001).

Moreover, bedload texture can have an impact on the way in which particles are entrained (Boggs 1995: 33–34). Smooth beds tend to have small, thin zones between the bed and the river where viscous forces dominate and water follows a laminar flow (Boggs 1995: 42). These channel beds generally behave as Bingham plastics,[2] meaning that the channel bed is normally non-viscous but, when bed shear exceeds the critical yield stress or Reynolds's number (the critical force required to move clasts), they switch from a laminar flow pattern to a turbulent flow pattern where erosion commences (Hogg 2003). As a result, bedload transport distance is not a direct function of time, but of local hydrological conditions and local morphology (Schmidt and Ergenzinger 1992). Turbulent flow is the interaction of non-linear forces that create highly turbulent areas also known as *eddies* (Komar 2003). These random motions cause collisions with other particles or *creep*, which results in particle dislodgement and entrainment. This phenomenon is particularly apparent in gravel-bedded rivers where the profile is extremely rough thereby producing more turbulent flow (Komar 2003).

Saltation occurs as the bed shear velocity increases further. Particles are influenced by Bernoulli's effect[3] and move along the bed in regular jumps. Particle saltation transport paths can be predicted according to size and shape by theoretical and practical models (Bagnold 1980; Lee *et al.* 2000). Experimental work has illustrated that there is a critical size and entrainment value for saltation to occur, as illustrated by the Hjulström Curve (Figure 12). As particles increase in size, they require more force to be entrained. However, for particles below approximately 5mm in diameter, they also require more force to be entrained because of cohesion due to van der Waal forces[4] (Chattopadhyay and Chattopadhyay 2003). Though saltation is a typical method of entrainment during seasonal freshets in pleniglacial conditions, it is unlikely that larger clasts such as handaxes would be entrained in that manner (Nichols 2009: 49).

Particles become part of the suspended load when the value of the bed-shear velocity exceeds the forces of gravity. Particles are lifted out of the bedload to a level where upward turbulent forces exceed the submerged weight of the particles (Van Rijn 1984). In extant rivers, this happens only to smaller particles such as sands or silts; however, in pleniglacial conditions, slightly larger particles may have been transported as well (Boggs 1995).

2.3.2. Reviewing artifact transport experiments

For three decades, archaeologists have tried to answer the question whether lithic artifacts in a river behave as normal clasts (Chambers 2004; Hosfield and Chambers

[2] a material that behaves as a rigid body at low stresses but flows as a viscous fluid at high stress (Bingham 1916).

[3] for an inviscid flow, an increase in the speed of the fluid occurs simultaneously with a decrease in pressure or a decrease in the fluid's potential energy.

[4] the totality of intermolecular forces.

2004; Petraglia and Nash 1987) or whether their unique size and shape (Bertran *et al.* 2012; Schick 1986: 43) and anthropogenic insertion point (Bertran *et al.* 2012) substantially modify their behavior. Given the complex and unpredictable nature of modeling particle transport, as well as archaeologists' focus on particles (vs. geomorphologists' focus on transport phenomena), studies have attempted to experimentally model characteristic lithic fluvial behavior.

Experimental archaeology, in the laboratory or in the field, is an important instrument in the Quaternary researcher's investigatory toolkit. Though the years, it has played a key role in developing the understanding of taphonomic processes that are difficult to understand solely from sedimentological investigations of the site (Macklin 1995). Experimental replication of taphonomic processes in a laboratory offers a window into the archaeological record if one accepts the concept of uniformitarianism (the present is the key to the past; Hutton 1788). Experimental replication of taphonomic processes can help archaeologists form tentative hypotheses of disjunctive archaeological assemblages. Experimentation has the added benefit that variables can be controlled so that the influence of one cause can be isolated and magnified. In addition, it can replicate real world scenarios through field studies in which the full range of unforeseen variables can be realized (even if not necessarily controlled; Coles 1966-67). Such *in vivo* experimental archaeology offers the only way to test inferential linkages between artifact assemblages and taphonomic processes.

In the following section, both the European and non-European experimental record are reviewed in a chronological fashion, identifying the advancements which they have made in our understanding of fluvial taphonomic processes. The section will present an overview of the current state of knowledge, while identifying key weaknesses as well as areas of significant research potential.

Isaac (1967)

Experiments targeting the behavior of lithic artifacts in rivers were initiated by Glynn Isaac (1967: 191–198). Isaac tried to explain the inverse relationship between bifaces (n=100) and smaller flakes in five locations at the Lower Palaeolithic site of Olorgesailie, Kenya. While Binford (1977) had suggested that these differences were a result of hominin curation, Isaac (1967) argued that they were due to hydrological sorting.

To test this hypothesis, Isaac (1967) placed replica lithic artifacts akin to those found at Olorgesailie at five different locations along the braided course of an ephemeral stream channel (*laga*) near the site. Lithic artifacts were laid out in rectangular grid plots 30–50cm apart in no preferred orientation. Two years later, Isaac (1967) observed that in some instances lithic artifacts were buried undisturbed within fine fluvial silts while in others flakes were separated from bifaces both in sand and hard bed rivers. In sandbeds, Isaac (1967) also noticed many upstream tilting bifaces as a result of scouring, and a tendency for bifaces to be arranged transverse to the current after transport.

While these observations failed to replicate the recorded patterns at the Olorgesailie sites (Isaac 1967: 82), they did suggest that hydrological processes have an effect on site formation, highlighting the need for further investigations.

Schick (1986, 1987, 1993)

Isaac's (1967) work was expanded upon by Schick in similar ephemeral streams around Lake Turkana, Kenya (Schick 1986; 1987; 1993). Schick's four-year project provided a data-rich alternative approach to Isaac's (1967) study, as it analyzed the movement of over 10,000 artifacts from 35 different simulated archaeological sites in a variety of alluvial environments (delta flood basins, alluvial flood basins, banks/levees) and channels (islands, bars, scours). Sites generally consisted of numerous flakes ($55<n<5000$) within a square meter along with a few cobbles and cores (usually less than 10). Most were knapped *in situ* though animal bones were added to some sites as well.

Schick (1986) observed dramatic changes to site composition based on artifact classes and maximum length dimensions. Fluvially altered sites were typified by a winnowing of smaller artifacts (<2cm) from the site while larger artifacts were characteristically buried in up to 26cm of sediment (Schick 1986: 244). Based on the results, the author suggested a proportionally small amount of small flakes compared to large ones at archaeological sites was an indicator of fluvial rearrangement. However, the author also argued for the potential of differential preservation of archaeological materials within different hydrological environments. The author found that sites which had not been heavily disturbed were either 'in elevated situations above the associated channel, as on a channel island or bank, or else in more distal parts of a flood basin… while more dispersed sites were for the most part within active channels but also included some that were on bars or banks close to a channel' (Schick 1986: 61).

Schick (1986)

Schick (1986) elaborated on the Koobi Fora experiments with a flume (54ft (16m) long x 2ft (0.6m) wide; water depth >25cm; slope gradient >0.5%), with an aim to observing fluvial forces on archaeological materials in real time. During the course of the experiments, Schick (1986) observed artifact entrainment in a variety of conditions: flow velocity (1ft/sec (0.30m/sec), 1.4ft/sec (0.43m/sec), 2.5ft/sec (0.76m/sec)), artifact morphology (size and shape; *unspecified*) and channel substrate (bare or sandy).

Similarly to the real world experiment, Schick (1986) observed that:

1. Smaller artifacts move first and further (Schick 1986: 58). However, sudden surges transport artifacts *en masse* and deposit them downstream (Schick 1986: 58). Sudden flow surges affected a large range of materials nearly simultaneously and initiated travel through the flume; assemblage composition was partially maintained when they were deposited.
2. Artifact morphology affects transport (Schick 1986: 50). Schick found that in higher stream velocities, increasing thickness, sphericity among cobbles, flakes, and cores increased their transport distances.
3. Artifacts shifted their orientation under current flow to a position of maximum stability (Schick 1986: 51). Artifacts with a long-axis oriented themselves parallel to the flow direction. This happened because of transport, but artifacts could also be oriented in place (Schick 1986: 53).
4. Substrate affects particle transport (Schick 1986: 53). Sandy bed cover impeded artifact movement by either stopping their transport or temporarily covering artifacts as sand pockets moved downstream. Additionally, Schick noted that scouring hindered entrainment by 'digging' artifacts into the sediments (Schick 1986: 55).
5. Particle interaction hinders erosion, enhancing artifact redeposition (Schick 1986: 58). Schick noted that when artifacts came close to each other, they stopped moving, producing small-scale clusters of artifacts.

Schick concluded from these findings that artifact transport was a complex process that required further exploration (Schick 1986: 58).

Petraglia and Nash (1987)

Schick's (1986) conclusions were quickly challenged by Petraglia and Nash (1987), who emphasized a need for greater methodological rigor in experimental design. Eight 'stations' with replica assemblages identical to each other, were monitored at semi-annual intervals for 2 years in different alluvial environments around the ephemeral Jemez River (New Mexico) floodplain (Figure 13).

At the end of the two-year experiment, Petraglia and Nash (1987) found artifact movement was statistically significantly different between sites, but there was no association between the contexts (Figure 13).

Additionally, they found a weak relationship between fluvial transport distance and weight. While larger artifacts (>15g) generally stayed put, smaller artifacts showed no significant trend—some stayed indiscriminately *in situ,* while others were eroded and transported. In most cases, this was because some smaller artifacts were protected from the stream flow by sediment burial. Simply put, fluvial action did not necessarily transport smaller artifacts away from the site because only a few small pieces were transported; rather it only slightly diminished their numbers in relation to larger artifacts. Petraglia and Nash (1987) suggested that fluvial impact on archaeological sites was significant and that it was reliant on the 'tempo, magnitude, and duration of hydrological events' (Petraglia and Nash 1987:126).

Harding et al. (1987)

Although the work of Petraglia and Nash (1987) covered a greater variety of local settings than previous experiments, it still largely focused on emulating depositional conditions of ephemeral streams within East Africa. In contrast, Harding *et al.* endeavored to explore the behavior of derived terrace gravel assemblages in Britain in a one-year experiment. To do so, they conducted experiments in a more analogous high-gradient (mean annual flood of $90m^3s^{-1}$; Harding *et al.* 1987) gravel-bedded river in Wales (Afon Ystwyth). In addition to using the standard measurements (a-, b-, c-axis and weight) Harding *et al.* classified artifacts into lanceolates, cordiforms, and ovates according to Bordes (1961). Polishes and edges were examined after transport to see how their wear compared to usewear. Unlike in previous experiments, however, monitoring periods were dictated by extreme hydrological events that were recorded by a local gauging station rather than by time as in previous experiments.

The authors demonstrated that handaxe weight and local environments (i.e., gravel bars and 'areas of slacker water' (Harding *et al.* 1987: 118)) did have an impact on the transport of artifacts and that their weight loss was proportional to their distance traveled. Though no empirical relationships were established, this was in broad accordance with the development of polish and micro-wear when evaluated microscopically.

Figure 13: Petraglia and Nash's 1987 experimental scatter (station) locations.

Station	Context
Station A	Control plot – flat clearing of Aeolian sediments
Station B	A hillslope bounded by ephemeral streams
Station C	A large ephemeral stream (4.5m wide, 4m deep)
Station D	A sandy lag deposition in a small ephemeral stream (85cm wide, 20cm deep)
Station E	The intersection of the Jemez River floodplain and an alluvial fan of an ephemeral stream.
Station F	Within a small ephemeral stream channel (1m wide, 60cm deep)
Station G	A gently sloping anthill
Station H	On the Jemez River floodplain

Dennell (2004)

During their work at Pabbi Hills, Hurcombe and Dennell studied clast transport in the Soan Valley of Pakistan. 148 maroon stones were placed in a hillside stream gully. One year later, four pebbles were recovered—three of which appeared to have been trapped within irregularities in the streambed. The extreme nature of the monsoons in the area and the flash flooding that resulted from them suggested to the authors that the stones were likely transported as a result of floods rather than being buried.

Hosfield and Chambers (2004; 2005a; 2005b)

Harding *et al.*'s study was followed up by Hosfield and Chambers (2004; 2005a; 2005b), who focused on the transportation and transformation of replica Lower and Middle Palaeolithic artifacts at Llanilar (NGR SN 628754) and Grogwynian Reach, Llanafan (NGR SN 709719; Figure 14). During the course of the three-year experiment, 563 Palaeolithic flint flakes and 83 flint and greensand chert handaxes were deposited in 13 artificial assemblages for an average time of approximately 9 months (though some assemblages were monitored for upwards of 1 year and a half). Assemblages were measured according to their primary axes, weighed, and painted to aid in their recovery. They were subsequently placed along various locations of a semi-stable floodplain and point bar complex.

Hosfield and Chambers (2005c) noted no apparent relationship between artifact size (using weight as an index of size) and transport distances, a phenomenon they attributed to the frequent burial and 'trapping' of artifacts within the gravel bed. Similar to other authors (Lenoble 2005b; Petraglia and Nash 1987; Schick 1986), Hosfield and Chambers (2005c) noted that flake scatters travelled largely intact initially (after approximately 2 months), but were then widely dispersed. Like Petraglia and Nash (1987), they concluded that this might have been the result of variant fluvial episodes of different magnitude. Transported handaxes similarly responded to flow velocity, however the authors noted that local channel morphologies also played a role.

Figure 14: The Llanafan (Grogwynian Reach) site, Afon Ystwyth, mid-Wales, UK (Hosfield and Chambers 2004; figure 1; October 2002, Cambridge University Press, used with permission).

In accordance with Harding *et al.*, the authors (2004; Chambers 2004) observed that of the 20 recovered bifaces, the widening of artifact ridges (arêtes) developed rapidly but rarely exceeded 100μm (0.1mm) over distances of up to 300m. By contrast, laboratory-abraded handaxes (Shackley 1974) showed values as large 150μm after 100 hours. Also noted were percussion cones on most artifacts, irrespective of transportation distances. Handaxe abrasion development and related damage (e.g., edge micro-flaking) developed during phases of partial burial, as well as during periods of active fluvial transport. The relationships of the transport distances, at least for handaxes, led to the conclusion that the transport of archaeological artifacts was largely stochastic as evidenced by the range of transport responses by nearly identical artifacts in the same areas (Chambers 2004).

Lenoble (2005)

Recent work by Lenoble (2005) has targeted the effects of fluvial transport in more upland and overland flow regimes. In one particular experiment (Experiment 7), the author examined the destruction of a 37-flake assemblage in a 50cm ephemeral channel in Fumel, France (Figure 15). After 26 months, the site was excavated and the following conclusions were made:

- Part of the assemblage was left intact while episodic floods transported another part downstream.
- Torrential flows transported artifacts regardless of size.
- Morphological threshold traps formed secondary concentrations downstream.

Lenoble's (2005) conclusions were largely in line with those

of Schick (1986) who also suggested that in ephemeral

Figure 15: View of Lenoble's site 7 (Lenoble 2005; figure 59: 82, British Archaeological Reports, used with permission).

streams, assemblages were transported downstream *en masse*, giving the appearance of an *in situ* site.

2.3.3. Summary and aims of the field-based experiments

The presented field-based experiments highlight six key findings about the behavior of archaeological artifacts in fluvial environments:

1. Fluvial entrainment can obscure artifact discard location (Bunn *et al.* 1980; Hosfield and Chambers 2005c; Petraglia and Nash 1987; Schick 1987).
2. Artifact transport is related to river velocity (Hosfield and Chambers 2005c; Petraglia and Nash 1987).
3. Artifact transport is related to the size and shape of the artifact (Hosfield and Chambers 2005a).
4. Local fluvial environments influence artifact entrainment and deposition (Lenoble 2005; Petraglia and Nash 1987; Schick 1986).
5. Fluvial entrainment can morphologically modify archaeological artifacts (Harding *et al.* 1987; Hosfield and Chambers 2005c).
6. Fluvial entrainment affects the orientation bias of artifacts' long axis (Schick 1986).

In spite of this rich experimental record, five main points remain controversial or require further study:

1. How do fluvial locations affect transport, entrainment, and deposition in meandering gravel-bedded rivers?
2. What is the best artifact metric predictor of transport, if any?
3. What is the relationship between transport and artifact morphological modification?
4. What is the relationship of stream power to artifact transport?
5. Under what fluvial conditions is orientation and dip of artifacts altered? Do they orient *in situ* or only if they are transported or both?

In previous studies, these factors are focused upon because they are often visible traits within the archaeological record. Chapter 4 presents a fluvial experiment that answers these questions.

2.4. Coarse-grained abrasion (tumbling barrel experiments)

> *Where they have undergone wear in the gravel, they, like the ordinary flints, have their edges blunted and are irregularly broken, and they are not infrequently truncated at the point. (Prestwich 1860: 295) In reference to early Palaeolithic finds from Hoxne and Abbeville.*

As clasts/particles are entrained within a fluvial system, they experience countless collisions with each other eventually resulting in surface attrition that affects their superficial condition and morphology (Charlton 2008: 96). Fortunately, particle abrasion is relatively well understood as a result of decades of debates over the role of downstream fining, even though this complex topic still has many unknowns (Frings 2008). This is further complicated by discrepancies in experimental techniques.

When these particles are artifacts, fluvial abrasion presents an interpretive problem, particularly for archaeologists interested in making inferences based on form (i.e., systematics; e.g., Cardillo 2010; Eren and Lycett 2012; Iovita and McPherron 2011). In the past, researchers have attempted to use abrasion to their advantage, as a proxy for relative degrees of transport (Ashton 1998; Petraglia and Potts 1994; Shea 1999; Wymer 1968). However, artifact abrasion mechanisms are poorly understood, making these inferences broad if not outright false.

Surprisingly little is known about fluvial abrasion of flint artifacts in spite of the excavation of vast Quaternary fluvial landscapes in Europe and a rich history of water-based bone abrasion research (Aslan and Behrensmeyer 1996; Behrensmeyer 1990; Boaz and Behrensmeyer 1976; Fernández-Jalvo and Andrews 2003; Gaudzinski-Windheuser *et al.* 2010; Pante and Blumenschine 2010; Shipman 1981; Thompson *et al.* 2011). Hazeldine Warren ventured that the dearth of research on this topic was because the study had fallen between the 'two stools' of archaeology and geology, both relying on the other for elaboration (O'Connor 2003).

In the following subsections, clast abrasion will be reviewed before discussing artifact abrasion. The remaining unknown facts about artifact abrasion will then be pointed out before concluding with the aims and objectives for the tumbling barrel experiment in Chapter 5.

2.4.1. Clast abrasion

To understand the effects that fluvial processes have on flaked lithic tool morphology, it is important to briefly review the underlying mechanisms of abrasion of fluvial particles. When sediments are subjected to moving bodies of water, bed erosion and subsequent entrainment of its constituent particles occur (Charlton 2008: 10). The results are their selective transport and repeated stochastic collisions with one another as well as with stationary objects within the system (Charlton 2008: 10).

For more than a century, scientists have studied the evolution of bedload characteristics during fluvial transport and found that interclast interactions result in their gradual attrition. This is commonly called *rounding* (Krumbein 1941; Kuenen 1956; Wentworth 1919). Though numerous schemes exist to describe rounding (Hayakawa and Oguchi 2005), it is generally measured by its medium axis length (width). Over time, rounding can alter the size, shape, and surface texture of these particles (Krumbein 1941).

Abrasion in fluvial systems is commonly described as the splitting, crushing, chipping, cracking, and grinding of clasts (Charlton 2008: 97; Kuenen 1956: 351). From a tribological (the study of wear) perspective, these forms

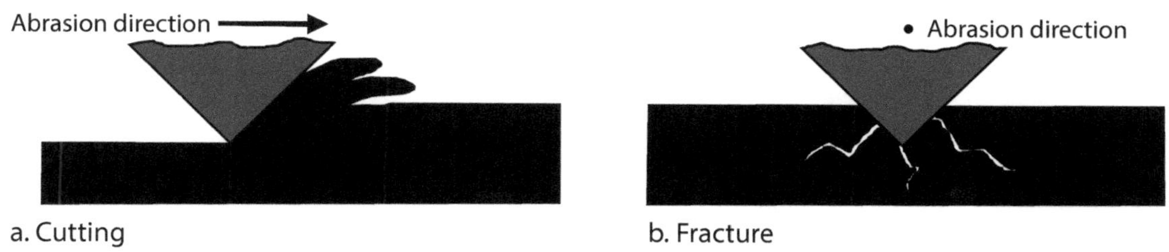

Figure 16: Cutting wear and fracture wear (Redrawn after Stachowiak and Batchelor 2005; fig. 11.1).

of abrasion result from two distinctive mechanisms of abrasive wear: *cutting wear* (grinding) whereby hard asperities deform softer, ductile surfaces through micro-cutting; and *fracture wear* (cracking, crushing, chipping, splitting) whereby cleavage of brittle solids propagates as a result of applied tensile stress (Figure 16; Stachowiak and Batchelor 2005: 484).

Given all the relevant parameters, fluvial abrasion is a systematic process. Generalized models indicate that particles initially abrade rapidly through fracture wear and later, more asymptotically through persisting cutting wear, as a result of which they approach their *ultimate shape* (Krumbein 1941; Rayleigh 1942). The continued disconnect between theoretical, laboratory, and field results, however, suggests that in practice, there are simply too many variables involved to develop accurate predictive models (Lewin and Brewer 2002). As a result, targeted field and laboratory studies are our best way of understanding particle abrasion.

However, this research itself is fraught with problems. To date, it has remained largely qualitative in nature, often due to inadequate laboratory equipment and setups (Lewin and Brewer 2002). Nevertheless, it still suggests many primary parameters that influence the intensity of abrasion.

Detailed below and outlined in Figure 17 is a non-exhaustive list of the most common factors influencing abrasion:

1. Transport distance increases interclast interaction, leading to higher abrasion rates: Sternberg's law (1875) indicates, 'The wear of a grain is proportional to its submerged weight (W) and the distance (L) it has travelled.' Mathematically, the statement is expressed as:
$$W=W_0\, e^{-\alpha L}$$
where α is the weight loss rate coefficient, and is the starting weight. However, abrasion is complex both spatially and temporally and cannot be predicted by raw travel-distance alone (Lewin and Brewer 2002). This suggests that artifact abrasion is proportional to transport distances, but is modified by other factors.
2. Lithology is also a prime regulator of abrasion within a fluvial system: All things being equal, hard particles produce higher wear rates than soft particles (Attal and Lavé 2009; Frings 2008; Šolc *et al.* 2012). This is confirmed by multiple laboratory experiments showing particle durability (hardness) and grain structure (i.e., brittleness) affect abrasion rates at logarithmic scales depending on the lithology (Abbott and Peterson 1978; Attal and Lavé 2009; Morris and Williams 1999). Recent studies even suggest that variable lithological properties can influence particle shape more than transport mechanisms (Brook and Lukas 2012). In polyclastic sediments, this is exacerbated as the natural abrasion rates of harder particles are temporarily suppressed while softer particles abrade more rapidly (Abbott and Peterson 1978). This would suggest that for instance, flint artifacts would abrade at a slower rate in riverbeds mostly containing softer lithic (e.g., quartzite) materials rather than harder materials (this is further discussed in Section 7.3.6)
3. Particle size affects abrasion rates: Early experiments found that, while overall abrasion among larger particles is greater, the percentage of reduction for all particle sizes is constant over the distance traveled (Frings 2008; Sternberg 1875). Generally, larger particles produce higher rates of abrasion but it is unclear if this is the result of particle size or the higher kinetic energy required to maintain the same velocity as smaller particles. Attal and Lavé (2009) found, in independent tests of different clast sizes, that size had a positive but weak effect on abrasion. In polyclastic samples, size plays a more significant role, as the attrition rate of a given particle depends more on its size, relative to the median size of the mixture, than on its overall mass. Smaller particle abrasion is accelerated as a result of increased fracture wear (Attal and Lavé 2009; Wentworth 1919). One result of this is that, when these particles are sand-sized (but no smaller), they can diminish the abrasion rate of larger particles as a result of cushioning impacts (Kodama 1994b; Kuenen 1956). This implies that flint artifacts abrade more rapidly and severely as particle size increases (see Section 7.3.5 for discussion).
4. Particle velocity affects abrasion: At low speeds, impact stresses can be insufficient to incur any deformation. Beyond the critical velocity at which abrasion commences, as particle velocity increases, abrasion theoretically increases quadratically (Frings 2008). Among brittle materials, this tends to be expressed primarily by subsurface cracking (Stachowiak and Batchelor 2005: 509). However, in experimental applications, this tends not to be observed. Kuenen's (1956) experiments showed that abrasion rates did not always increase along with particle velocity. Lewin and

Figure 17: Experimentally derived factors influencing the fluvial abrasion of particles in gravel-bedded rivers (nonexhaustive; based on Frings 2008).

Factor	Experimental findings	Reference
Transport distance	• Longer transport means more rounding	Krumbein 1941; Sternberg 1875; Wentworth 1919
Particle lithology	• Fragmentation depends on lithology • Lithology controls abrasion rates and fragment production	Attal and Lavé 2009; Daubrée 1879; Kodama 1994b; Kuenen 1956; Morris and Williams 1999; Rayleigh 1942
Particle size	• Larger clasts abrade quicker than smaller clasts • Particle abrasion rate decreases with increasing grain size for poorly sorted material	Daubrée 1879; Wentworth 1919
Particle velocity	• Mass loss by attrition scales with particle velocity squared • The influence of velocity on fragment production rate is unclear, due to the high variability of fragment production rates • High traveling velocities tend to promote fragmentation	Attal and Lavé 2009; Bonney 1888; Daubrée 1879; Kuenen 1956
Particle angularity	• The greater the roundness, the smaller the percentage loss in weight per unit of time or distance rolling	Daubrée 1879; Krumbein 1941; Kuenen 1956; Rayleigh 1942
Weathering	• Weathering accelerates abrasion processes	Bradley 1970; Jones and Humphrey 1997; Kuenen 1956
Amount of moving particles	• Increased sediment increases abrasion, though Attal says this is a direct function and that lower sediment levels tend to promote fragment production	Attal and Lavé 2009; Wentworth 1919

Brewer (2002) found only a low dependency of abrasion rates on flow velocity. This is possibly explained by increasing particle velocity, which raises the percent of saltating particles, counteracting the higher impact energy (Attal and Lavé 2009; Frings 2008). This would suggest that artifacts entrained in high velocity channels would experience much higher abrasion than those entrained in low velocity channels.

5. Particle angularity affects abrasion: The sharpness of a particle accelerates its abrasive wear as acute-angled protrusions are prone to chipping, resulting in an abrupt loss of mass (Daubrée 1879; Gölz *et al.* 1995; Krumbein 1941). Angular particles also have a larger surface area per volume than rounded particles, which promotes mass loss through grinding and cracking. Lastly, angular particles experience more fluid drag than rounded particles, resulting in a higher particle velocity for angular particles (Frings 2008). The extent of this influence is unclear and may depend on the exact shape of the particle. Kuenen (1956) found shape influence to be negligible. Lewin and Brewer (2002), however, found shapes differed, with cubes losing most weight in tumbling barrel experiments but least in annular flume experiments, a finding that can possibly be attributed to the ability of clasts to 'free-fall' in a tumbling barrel. This would suggest that artifact classes with higher angularity (i.e., flakes) abrade more rapidly than those with obtuse angles (i.e., cores; see Section 7.3.7 for discussion).
6. Weathering accelerates particle abrasion: A myriad of chemical, mechanical and/or biological processes may increase the susceptibility to abrasion of particles (Kuenen 1956; Lewin and Brewer 2002). Bradley (1970) argued that abrasion only removes the weathered skin of sediment particles when they are transported during a flood. This implies that small particles have a greater percent diameter reduction than coarse particles, because the thickness of the weathered skin is equal, but the particles are smaller (Jones and Humphrey 1997). This would suggest that weathered artifacts (i.e., patinated) in equivalent conditions, would abrade faster than unweathered artifacts (see Section 7.3.7 for discussion).
7. Increased sediment supply (volume of sediment load) affects abrasion: Increasing particle density changes the way in which particles move and collide (Lewin and Brewer 2002) and this is likely to affect the nature of abrasion. Whether the abrasion rate increases or decreases with increasing sediment load remains unclear, because tumbling barrel experiments give inconsistent results (Gölz *et al.* 1995; Lewin and Brewer 2002), possibly due to differences in experimental methods (revolution frequency, barrel diameter, water depth, experimental duration). More recent work suggests that abrasion is a direct function of sediment mass and that lower sediment levels tend to promote fragment production as opposed to abrasion (Attal and Lavé 2009). This would suggest that artifacts abraded in a fluvial system with greater sediment entrainment would yield higher abrasion rates (see Section 7.3.4 for discussion).

2.4.2. Reviewing coarse-grained abrasion experiments

The sedimentological/hydrological literature provides a good starting point for the understanding of fluvial abrasion of artifacts but it is worth noting that there are four key differences between the two:

1. Although flaked lithic artifacts are subjected to the same downstream sorting and fining processes as the surrounding bedload, once they enter a system, their initial deposition into a fluvial system is artificial through hominin discard. As a result, they have not been subjected to the same selective sorting and fining processes as the surrounding bedload and may be distinct in size. This can drastically affect their behavior

in a fluid medium (Bertran *et al.* 2012; Schick 1986: 43).

2. Lithic artifacts are *a priori* shaped differently from fluvial gravels or else they would not be identified as such. The nature of flaked lithic artifacts architecture is dictated by the application of mechanical impact to reach desired fracture planes, not by the preferred orientation of the material or by other geochemical processes such as frost cracking. Used often for cutting, flaked lithic tools have a high proportion of acute angles relative to river pebbles whose shapes tend to be rounded and shaped by stochastic processes. This suggests they both (a) abrade differently from fluvial particles (Hosfield and Chambers 2005c) and (b) have different transport mechanisms (Bertran *et al.* 2012).
3. The lithology of fluvially derived Palaeolithic flaked lithic assemblages is often different from the Pleistocene gravels from which they are recovered. Although flaked lithic artifacts are commonly manufactured on local river cobbles (Ashton *et al.* 2006), ancient hominins selectively exploited raw materials that conchoidally fractured for flaked lithic tool production. As a result, flaked lithic assemblages are comprised of a lithology biased towards harder, more brittle cryptocrystalline materials when compared to the heterogeneous constituents of a palaeochannel that are often composed of multiple lithologies (i.e., Allen and Gibbard 1993; Bridgland 1994; Gibbard 1985; Gibbard 1994).
4. Sedimentological literature deals with erosional and depositional processes at larger scales and often-longer terms than are prevailing in the immediate circumstances of archaeological site preservation.

Information relating specifically to artifact abrasion is therefore vital to understanding the *signatures* of fluvial deposition for the Palaeolithic. If fluvial depositional characteristics can be related to characteristic artifact abrasion, then we can begin to form more accurate hypotheses of original spatio-temporal trends by looking at both Palaeolithic artifacts and their Pleistocene sediments. In the rest of this section, specific experiments relevant to large-scale fluvial abrasion are reviewed.

Boule (1905)

Marcelin Boule (1905) carried out the earliest experiments in Prehistoric archaeology dealing with fluvial abrasion. Ironically, the goal was to disprove artifact authenticity. In the first *coup* to the Eoliths hypothesis, Boule (1905) found natural concoidal fracturing in fluvial contexts. Boule investigated the remnants of an industrial cement mixer in the small town of Guerville near Mantes (northern France). The mixer was a large tank that rotated 16 times per minute for a period of 29 hours. Inside was a mixture of clay, chalk, and flint cobbles (Boule 1905). Boule suggested that the speed and motion of this particle mixer would create a situation analogous to entrainment in a river like the Rhône. During the 29 hours, the flint cobbles were subjected to what Boule described as 'thousands of natural shocks that, from the sound of the rattling, is similar to the noise of a torrent (*Translation mine*)'. Boule observed that most of the cobbles became rolled identically to clasts found in palaeogravels but also found that there were many retouched samples. Boule's work was, subsequently, supported by Warren (1905), who suggested, 'Water-abrasion is capable of producing a contusion of surfaces, a chipping of edges by battering and a considerable amount of 'free-flaking'.'

Hosfield (1999; 2000)

Hosfield (1999; 2000) aimed to establish a localized reference set for abraded handaxes in the Hampshire Basin. A single handaxe was tumbled for 50 hours in 3kg of River Test (UK) gravels (from 1–95mm in a-axis) in a cement mixer with padded paddles. The ridge widths, standard dimensions (Roe 1981), and weight were recorded after 1, 2, 3, 4, 5, 10, 15, 20, 25, 30, 35, 40, 45, and 50 hours (Hosfield 1999: 114–116). Contrary to Shackley's (1975) findings (next section), Hosfield found a strong positive linear relationship between the running time of the experiment and the degree of ridge widening on the artifact (*correlation coefficient*=0.990, *R*2=97.94%; Figure 18; Hosfield 1999: 115). Nevertheless, he observed that the average rate of attrition was 39µm hr^{-1} for the first 35 hours, but for the last 15 hours, dropped to 10µm hr^{-1}. The average ridge width after 50 hours was 1567µm, nearly 10 times larger than the largest ridges abraded by Shackley (1975), implying that the extended running time was warranted but may also reflect a difference in the experimental apparatus.

Hosfield used weight as a proxy for mass loss and found that, after 50 hours, the 170g handaxe had lost 26.47% of its original weight, though part of this may have been due to the effects of the metal cement mixer they were tumbled in. Similar to ridge width, weight loss was inversely correlated to time (*correlation coefficient*=0.934, *R*2=87.19%; Figure 19; Hosfield 1999). Still Hosfield noted that the rate of weight loss after the first 5 hours slowed from 4.4g hr^{-1} to 0.51g hr^{-1}. He ascribed this to edge damage and breakage '[occurring] rapidly during the early stages of abrasion before subsequent stabilization, possibly reflecting the vulnerability of the freshly knapped bifaces edge.' (Hosfield 1999: 115).

Hosfield made no note of the length and width rates of attrition, but did observe that the shape classification changed (Roe 1981): L_1/L (where L=artifact length; L_1=distance from the butt to the maximum breadth). The initial handaxe started as an ovate (L_1/L=0.397) but after 30 hours, became classified as a cleaver (L_1/L=0.570). This confirmed Hosfield's findings that mass was lost mainly from the margins of the handaxe and cautioned that shape analysis should not be readily performed on heavily abraded assemblages (Hosfield 1999: 115). However, the extreme rate of abrasion was possibly due to the hardness of the equipment used by Hosfield (2013 *pers. comm*).

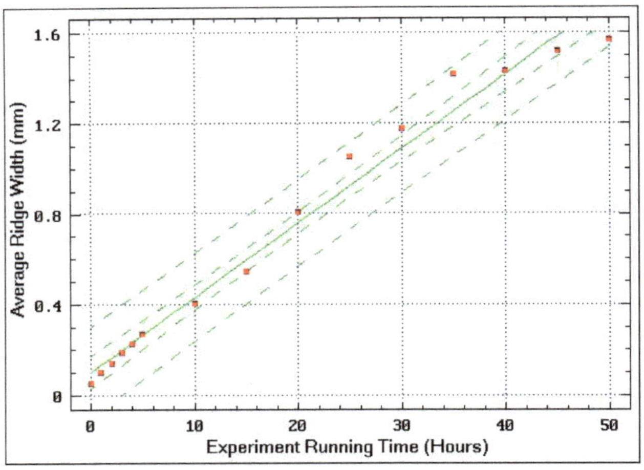

Figure 18: Graph of ridge width and time for 15 points on an experimentally abraded handaxe. ***Correlation coefficient*=0.990, *R2*=97.94% (Hosfield 1999; figure 6.1, British Archaeological Reports, used with permission, courtesy Rob Hosfield).**

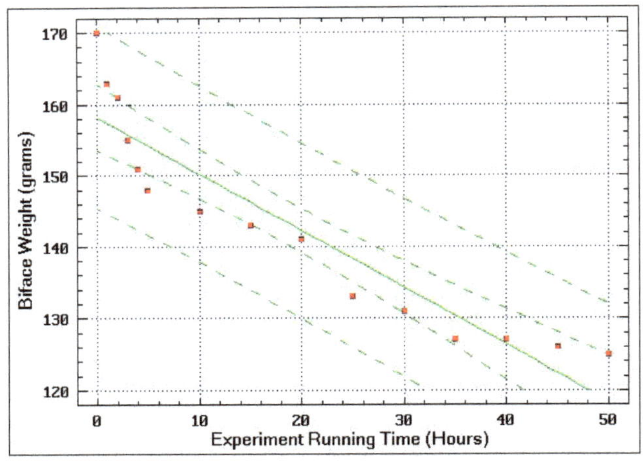

Figure 19: Handaxe weight versus experiment running time. ***Correlation coefficient*=0.934, *R2*=87.19% (Hosfield 1999; figure 6.4, British Archaeological Reports, used with permission, courtesy Rob Hosfield).**

Chambers (2003; 2004)

Chambers' work on the Solent River system expanded on previous work dealing with the abrasion of handaxes. The author saw a need for new experiments that focused not only on the mechanisms underlying the rounding of ridges, but examined the effect of fluvial entrainment on fragile handaxe edges. Critically, Chambers outlined the danger of subjectively assessing macroscopic handaxe ridge widths. In a practical study, the author found that 21 volunteers incorrectly assessed Shackley's microscopic handaxe abrasion categories (1–6) 86% of the time.

The main thrust of Chambers argument was that ridge widths measured at different points in six 'zones' along the handaxe surface provided more detailed artifact transportation histories. To assess fluvial damage, a series of flume experiments were undertaken to assess the relationship between artifact damage and time. The flume was a gravel-lined 12m tilting recirculating flume. The static 'bedload' of this experiment may have made replicated an abnormal bedload. Four replica handaxes of Greensand chert and Chalk flint were knapped into lenticular and plano-convex forms (one of each).

Each handaxe was transported a minimum of 250m; the handaxe was stopped after every 10m for the first 100m, then every 25m for the last 150m (16 intermediate steps). Lenticular forms were then transported a further 750m and stopped every 50m until they reached 1000m (31 intermediate steps). During each time-step, the dominant mode of transport, ridge wear, and edge damage was recorded. The data gathered are provided in Figure 20.

Chambers observed the following:

- Raw material affects ridge abrasion rates: finer-grained flint had consistently narrower ridges in spite of similar attrition trends.
- Artifact morphology affects transportation mode: plano-convex bifaces prefer sliding: lenticular prefer rolling/ saltating.

Figure 20: Summary of artifact transportation trends and the damage documented during flume experiments (modified from Chambers 2004; table 4).

Biface No.	Form	Raw material	No. recording events	Dominant mode of transportation	Face A ridge damage	Face B ridge damage	Other damage
F#1	Lenticular	Flint	30	Saltation	Highly heterogeneous	Highly heterogeneous	Substantial edge micro-flaking; single 'large' flake removal; scratches
					Range of both faces highly comparable		
Ch#2	Lenticular	Greensand chert	30	Saltation	Highly heterogeneous	Highly heterogeneous	Substantial edge micro-flaking
					Range of both faces highly comparable		
F#9	Plano-convex	Flint	15	Sliding	Moderately homogenous	Highly heterogeneous	Very limited edge micro-flaking
					Range of both faces differs		
Ch#1	Plano-convex	Greensand chert	15	Sliding	Highly homogenous	Highly heterogeneous	Very limited edge micro-flaking
					Rage of both faces differs substantially		

- Transportation mode affects artifact damage type: saltation exhibits more edge micro-flaking than sliding.

Chambers suggested that these better-reconstructed histories would be able to inform local reconstructions of Palaeolithic assemblages in Britain.

Grosman et al. (2011; 2013)

Attempting to understand fluvial modification to handaxes at the Gesher Benot Ya'kov (GBY) site in Israel, with the aim of building a regional model of handaxe morphological distribution, Grosman *et al.* (2011; 2013) heeded the warning of Hosfield (2003) and initiated experiments exploring the gross morphological changes to handaxe abrasion. A total of eight handaxes were manufactured and abraded with fifteen basalt cobbles with an average weight of 1062g (min=250g, max=2920g) in a 100 liter cylindrical barrel by manual rotation. The rolling of the barrel was interrupted after 5, 10, 20, 40, 60, 100, and 200 complete rolls. Before and after each abrasion, handaxes were digitized using a high precision 3D structured light scanner with an accuracy of 0.2mm per pixel.

Following the tumbling, volume was calculated from the generated point clouds. Point clouds of each abrasion step were matched using homologous points in each scan. Point clouds were then compressed to produce 2D contours of the handaxes. Commercial and purpose-designed software then evaluated: (1) the localized change along the handaxe perimeter; (2) the artifact asymmetry (Saragusti *et al.* 2005); and (3) the level of perimeter undulation (roughness).

Grosman *et al.* (2011; 2013) found the following:

1. After tumbling, artifacts lost approximately 10% of their original volume (Figure 21A).
2. Damage was mostly sustained in the tip (Figure 21B).
3. Mean asymmetry increased through rolling (Figure 21C).
4. Roughness increased as damage progressed (Figure 21D).

2.4.3. Summary and aims of the tumbling experiments

The impact of coarse-grained fluvial abrasion on Palaeolithic artifacts is of considerable archaeological interest, for open-air contexts. However, there is limited knowledge concerning the characteristics of fluvial abrasion as well as the factors involved in their expression. Previous studies demonstrating that coarse-grained fluvial abrasion can affect overall morphology have focused mainly on extremely large sediment size categories (usually cobbles and boulders; Wentworth 1922). These studies have not fully explored other sedimentary factors that might affect fluvial abrasion.

The literature detailed in this section suggests that fluvial processes are dynamic and significant and affect artifacts in the following ways:

1. Coarse-grained fluvial entrainment can lead to significant morphological changes and condition.
2. Coarse-grained fluvial entrainment heavily affects the preservation of artifact surfaces (Chambers 2005).
3. Lithology, particle size, particle velocity, particle angularity, weathering, the amount of moving particles, transport distance, and bed rugosity are among the prime regulators of abrasion (Figure 17).

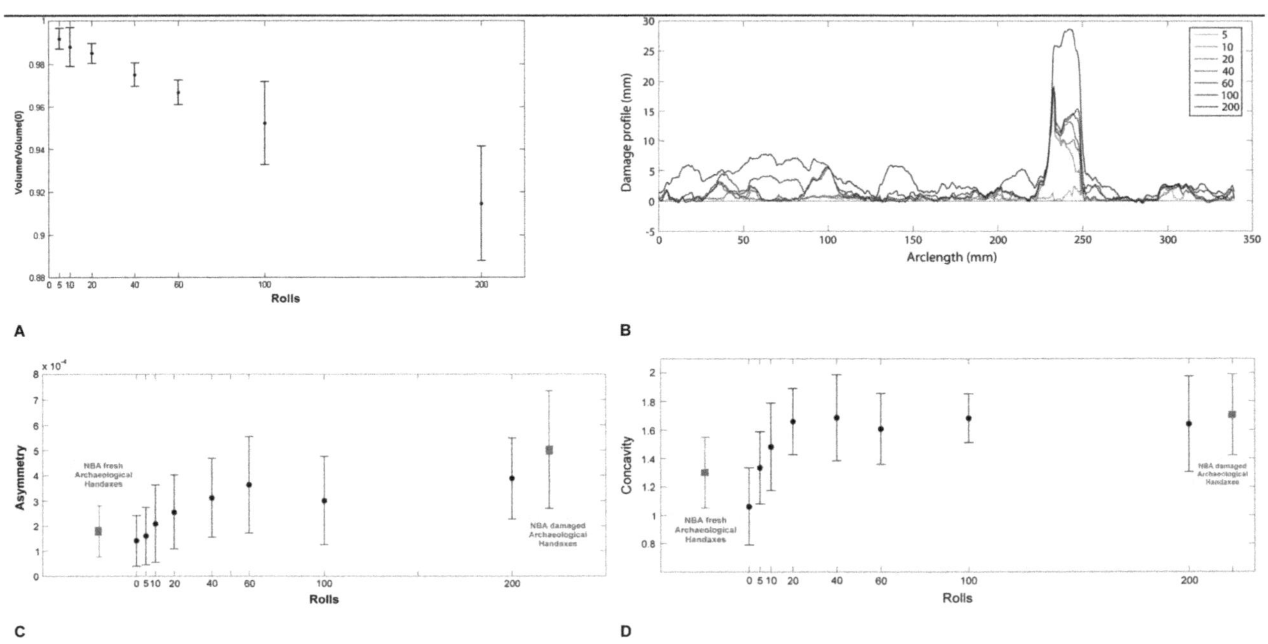

Figure 21: Handaxe abrasion graphs showing changes over time to (A) volume (B) profile (C) asymmetry and (D) concavity (Reprinted from Grosman *et al.* 2011, with permission from Elsevier).

This leads to my principle hypothesis: if above stated factors regulate fluvial abrasion of gravels, then they may also alter the rate and manner of abrasion of Palaeolithic artifacts.

Information relating specifically to artifact abrasion is vital to understanding the *signatures* of fluvial transportation and deposition for Palaeolithic assemblages. If fluvial processes can be related to characteristic artifact abrasion, then we can begin to form more accurate hypotheses of original spatio-temporal trends and assemblage homogeneity/ heterogeneity. This will be therefore tested in the tumbling experiment outlined in Section 3.2.

2.5. Fine-grained abrasion (flume experiments)

Fluvial entrainment affects artifacts microscopically as well as macroscopically. This is fortunate because macroscopic abrasion (e.g., tip breakage, well-rounded ridges) only happens in high-energetic sedimentary situations and can be confused with post-depositional processes such as trampling, weathering, or freeze-thaw action (e.g., Bertran *et al.* 2010; Eren *et al.* 2010; Lenoble 2005; Levi Sala 1986; Lopinot and Ray 2007; Miller *et al.* 2009; Paddayya and Petraglia 1993; Stapert 1976; Tringham *et al.* 1974). The result is that isolated macroscopically unmodified artifacts recovered in finer sand-gravel caliber sediments can be mistaken for being *in situ* even though they may have been transported kilometers from their discard location (Hosfield and Chambers 2005a; Shackley 1975; Shackley 1974).

The following sections introduce the key extant literature on microscopic fine-grained abrasion and in the process, introduce Chapter 6, which explores whether Scanning Electron Microscope (SEM) analysis of flaked lithic artifacts can be used to identify fluvial abrasion in sand-gravel sized sediments.

2.5.1. Reviewing fine-grained abrasion experiments

The following discussion details the most important list of published experimental work relevant to fine-grained abrasion of lithic artifacts. Publications are introduced chronologically.

Tringham (1974)

To better distinguish anthropogenic edge damage from fluvial processes, Tringham *et al.* (1974) placed five flint flakes in a plastic container along with 'water, sand and stones,' and agitated them for an hour. Observations of the modifications to the edges were made afterwards:

> *Scars were distributed at random along the entire perimeter of the flake, with no localization of scarring as was observed with those flakes used by human beings. The orientation of the scars was random with no standardization of scar size or shape on each flake. Both of the features are unlike the damage caused by deliberate usage in which pressure is uniform and comes from one direction, producing scars which have the same orientation and, within a narrow range, the same shape and size. (Tringham et al. 1974: 49)*

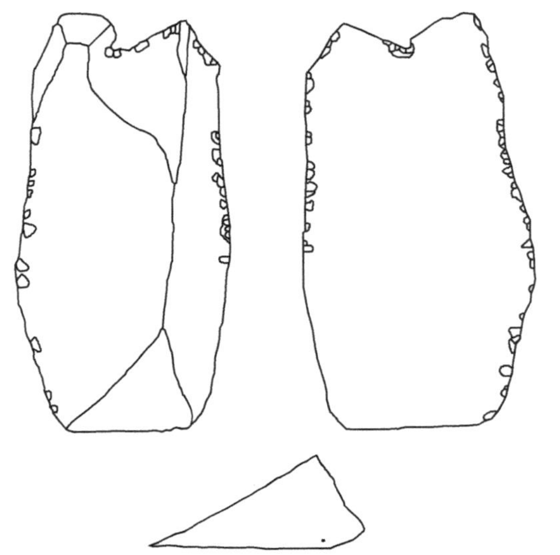

Figure 22: A flake damaged by agitation in water with sand and stones (Redrawn after Tringham *et al.* 1974).

Tringham *et al.* compared the flint flakes used in this experiment to other artifacts that had been experimentally trampled or used to work various materials. The authors noted that the edge damage incurred in the plastic container was distinct and unlike any of the other wear patterns observed. They concluded that water-born flint artifact edge damage was distinctly identifiable from micro-wear both in type and location (Figure 22).

Shackley (1974; 1975)

To test the hypothesis that the widths of flint ridges correspond to artifact transport distances, Shackley experimentally 'transported' six replica flint handaxes in a large plastic barrel at 25rpm with sand and clay sediments. At 0.25, 0.5, 0.75, 1, 5, 10, 20, 40, 60, 80, and 100 hours, ridges were re-measured. The experiments showed that, irrespective of the shape of the handaxe and the volume of the water in the barrel, ridge widths increased rapidly with time and sediment size (up to between 70µm and 150µm; Shackley 1975: 46). Although it is not explicitly noted, the data suggest an asymptotic relationship between ridge width and abrasion time (Figure 23). This implies that heavily abraded ridges can continue to be transported without showing further appreciable damage.

Additionally, Shackley noted ridge abrasion stages (Shackley 1975: 47) starting with 'braided' stress cracks that later became smooth hairline cracks. When artifacts were abraded with larger sediments (pebbles), 'braided' cracks were replaced by 'pitting' (Shackley 1975: 47), largely in line with the cutting wear (surface abrasion) and deformation wear (formation of crack networks) patterns observed among brittle materials (Bayer 2004: 35). The outer margin of the handaxe was also chipped (micro-flaking). This micro-flaking developed rapidly and may be

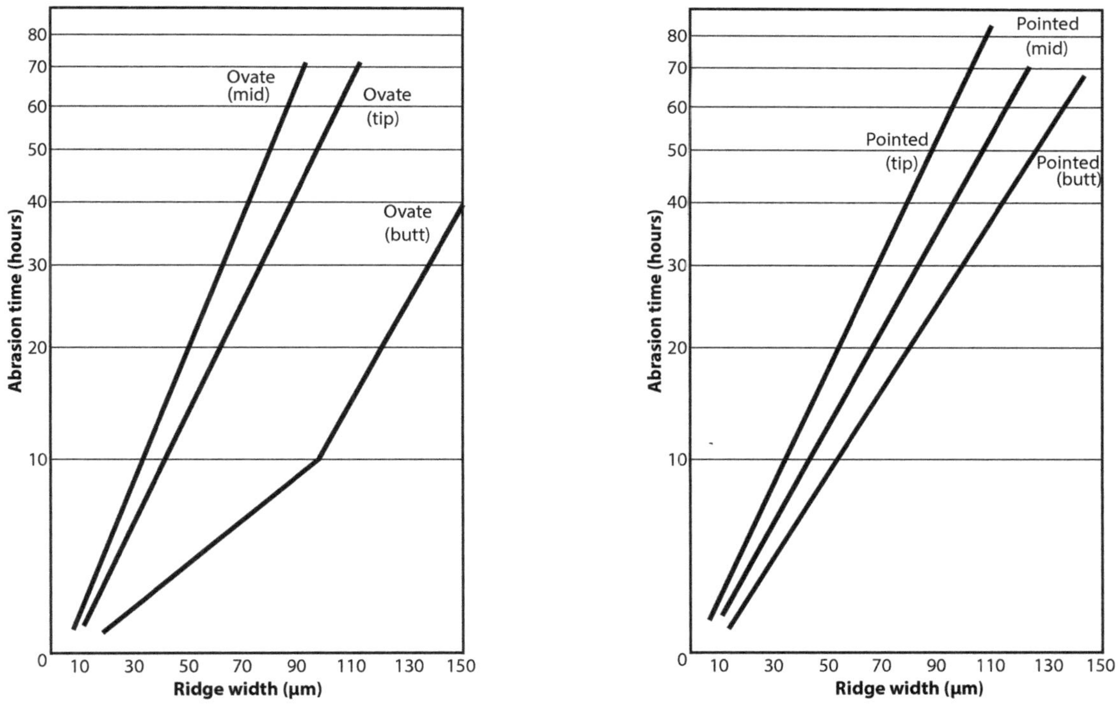

Figure 23: Ridge width v. time for experimentally abraded handaxes (Redrawn after Shackley 1975; figure 17). Note the scale of the y-axis is logarithmic.

attributed in part to the tumbling barrel (Shackley 1975: 47).

Burroni et al. (2002)

Burroni *et al.*'s (2002) tumbling barrel experiment reproduced Shackley's as part of a larger study of the *stigma* or surface features of flint artifacts. Additionally, the authors wanted to test the influence of artifact lithology and lower water content. Chert flakes were placed in an 800ml tumbling barrel at 40rpm for 32 hours and stopped at 2, 4, 8, 16, 27, and 32 hours to measure the ridge width, weight, and edge damage. They found that flint surfaces could sustain movement and interaction with the sediment without showing any signs of modification. However, surface cracking was sometimes observed and striations were common, depending on the hardness, shape, and velocity of the grit size. After time, edges became 'micro-flaked' and surfaces became smoothed over and glossy (river patina; Howard 2002; 1999). Burroni *et al.* (2002)

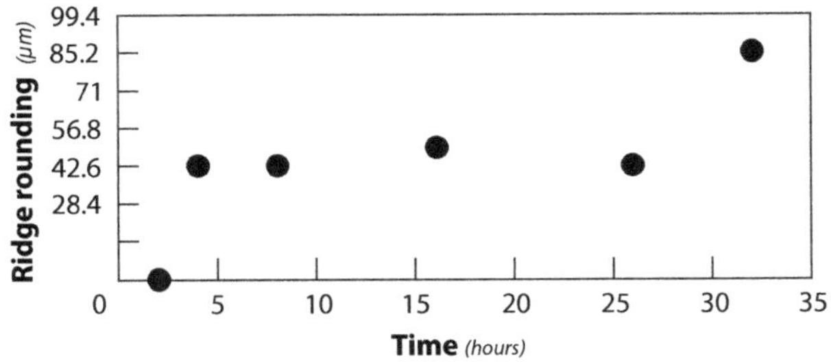

Figure 24: Chert flake ridge width over time in a tumbling barrel with sand and water (Redrawn after Burroni *et al.* 2002).

also observed ridge-widening patterns, though they were different from those of Shackley. Ridge widening was faster in the beginning and final stages of abrasion, a switch attributed to a change in wear mechanisms (Figure 24). Both water volume and lithological grain size were associated with increasing ridge width development.

In contrast to Shackley's experiments, Burroni *et al.*'s (2002) experiments were shorter in scale, but illustrated similar abrasion rates. The different trends may be due to differences in lithic material as Burroni *et al.* used stones that were more varied in the micro-structure of their grain (2002).

Lenoble (2005)

To observe the effect of natural abrasion on flint artifacts in overland slope runoff, Lenoble (2005: 107) put 20 moderately sized flint flakes (2–3cm long) in 1 liter rotating glass bottles with sand and water for 24 hours at 30rpm, checking them at regular 6 hour intervals. Flakes became glossier and lustrous, and at high (1000x) and low (25–35x) magnification, lithic surfaces were characterized by widened ridges, blunted edges, and surface ablation recognizable at four different levels: (1) rough and matte, (2) slightly glossy, (3) glossy, and (4) lustrous (Figure 25).

While he observed that these conditions increased over time, they did not increase regularly. Between 6 and 12 hours, the changes to condition were small. However, between 12 and 18 hours, and 18 and 24 hours, he found that the differences were large. Lenoble proposed that this suggested a nonlinear surface abrasion pattern of artifacts comparable to that found by Burroni *et al.* (2002).

Figure 25: Different stages of mechanical abrasion from Lenoble's tumbling barrel experiments at both the macroscopic and microscopic level (Lenoble 2005; table 39; after Plisson 1985; *translation mine*).

Score	Naked eye	Under 25 to 35x binocular microscopy
1	Rough and matte surface due to micro-relief fracturing.	Low reflectivity; mainly due to the facets that form breaks in the quartz grains.
2	Slightly glossy, 'soapy' appearance. The surface fracture relief is still fresh.	A reflectivity of the quartz grains is seen in the micro-relief. The reflection of the asperities is smaller than grains of quartz. The delineation of the ridge is not initiated by wear, which is discontinuous.
3	Glossy surface, ridges, and edges are blunt.	Strongly reflective flat areas. Quartz grains are still discernible. The ridges form shiny lines.
4	Surface noticeably lustrous; edges and ridges are blunt to very blunt.	A patterned and uniformly reflective surface. Crystal faces are no longer visible. Edges are faceted and shiny.

2.5.2. Summary and aims of the flume experiments

Ridge widths, surface conditions, and micro-flaking have typically been used to reconstruct the depositional history of lithic artifacts in alluvial deposits. These factors, either measured objectively or, in the case of ridge widths, quantitatively, can provide a baseline from which to understand their transport history, potentially providing information on the time they were entrained, the mode of their transportation, and the type of abrasive in which they were transported. A known source of uncertainty limiting the use of these factors is a clear understanding of the effect of variations in time, sediment size, and the artifact entrainment mode. Although a recent study suggested a possible involvement of these factors in bone abrasion (Thompson *et al.* 2011), the impact of these factors in the case of lithic artifacts remains obscure and requires further study.

An experimental regimen was devised to test the following hypotheses:

1. To see if microscopic signs of fluvial surface abrasion was observable on lithic artifacts transported under low-energy sedimentary conditions.
2. To understand if variations in flow regimes and sediment sizes, and transport modes might differentially affect surface abrasion, ridge width, and edge micro-flaking (see Figure 38).

2.6. Summary and experimental research aims

It has long been known that fluvial systems exert a wide range of influences on archaeological assemblages. Since the 1960s, researchers have been interested in verifying and quantifying the exact nature of this relationship. Experimental work was developed to address assemblage formation and a steady stream of subsequent experiments has succeeded in elaborating the knowledge we can extrapolate from these experiments (Figure 26).

Experimental research into fluvial taphonomic processes has had different foci since it began. Broadly, this research can be understood within the framework of three different, though often overlapping categories:

1. Understanding the characteristic formation and modification of artifact assemblages.
2. Quantifying characteristic damage of individual artifacts for comparative means.

Figure 26: Current methods to evaluate archaeological site modification by fluvial processes.

	Size distribution	Fabric analysis	Edge damage	Ridge abrasion	Surface texture (polish)
Application	• Dense archaeological assemblages in relatively *in situ* alluvial contexts	• Dense archaeological assemblages in relatively *in situ* alluvial contexts	• Individual artifacts and assemblages • Not expressed on artifacts without acute edges (i.e., cores)	• Individual artifacts and assemblages • Small to medium assemblages • Recognizable on most flaked artifact types (except *éclats entame*)	• All flaked-lithic artifacts
Advantages	• Provides detailed site history	• Able to distinguish fluvial rearrangement from other taphonomic processes	• Direct evidence of artifact transport as a result of clast-artifact interaction. • Applicable to all cryptocrystalline artifact types. • Unobscured by artifact patination	• Easily identifiable • Easily quantifiable (ridge width measurements)	• Extremely sensitive to artifact taphonomic processes
Limitations	• Requires large numbers of artifacts excavated with modern techniques • Dependent on a certain degree of site preservation	• Requires large numbers of artifacts excavated with modern techniques. • Dependent on a certain degree of site preservation	• Can be indistinguishable from anthropogenic retouch or other taphonomic processes • Can render artifacts unidentifiable	• Underlying mechanisms poorly understood (may be more than just fluvial transport) • Asymptotic process– only a finite amount of abrasion can be expressed • Possibly affected by artifact patination	• Can require expensive microscopic equipment • Time consuming • Can be obscured by chemical changes to artifact • Initial taphonomic information may be erased by subsequent modifications
Formation conditions	• Low energy alluvial settings	• Low energy alluvial settings	• High energy gravel-bedded rivers • Transport within bedload-sliding or saltating	• Widely associated with high-energy gravel assemblages, but possibly a result of *in situ* abrasion	

3. Relating characteristic damage incurred by artifacts to their degree of transport.

2.6.1. Areas for further research

Despite much experimental research that has been done on fluvial assemblages and their taphonomic processes, further work is required to fully appreciate the factors involved in these processes. The need to expand the body of experimental data that address the reworking and derivation of artifacts is clear and the main lacunae in the research can be summarized as follows:

1. Many of the experimental studies into high-energy fluvial deposits have neglected the wide variety of factors involved in transporting clasts. As geographers have shown, river architecture and behavior play a main role in determining the distances over, and processes by, which artifacts are entrained and thus may have an important influence on clast transport. A main shortcoming of previous experiments is that they have been skewed to ephemeral braided systems that represent only one subset of the various types of rivers that exist. In addition, as shown above, river architecture and channel morphology (bed porosity, depth, and discard location), aspects that are observable in Pleistocene sections, can influence clast transport. Local environmental factors are also known to affect clast behavior. Experimental research on clast transport has been mostly carried out in the African Rift Valley. This is problematic both spatially and temporally because recent studies have illustrated that regional characteristics can highly influence artifact transport and assemblage formation (Texier *et al.* 1998). For example, the periglacial locale of Palaeolithic Britain and northern Europe exhibited strong annual nival regimes and would have had a different hydrography than the African Savannah during Pleistocene times.
2. Notwithstanding new methods devised in recent years to assess artifact abrasion, there are still problems: There is still no clear understanding of how ridge widths relate to other characteristics of artifact abrasion such as polish or edge wear damage. A main difficulty has been to understand the ways in which different artifacts abrade. Though authors have recognized that the degradation of ridge width and formation of polish is related in part to the raw material durability, a true quantitative process to account for this variable has not yet been adequately explored (Burroni *et al.* 2002).
3. Experimental research has had a distinct bias for larger artifacts, partly because the goal of much research has been to explain the spatial origins of handaxes due to their dominance in the Lower Palaeolithic record, partly also because of the pragmatics of recovering smaller artifacts, especially in field experiments. Though important patterns of structural integrity and dispersal of flake scatters are known, the damage and modification of flakes during fluvial transportation and spatial and vertical patterns of flake dispersal in gravel-bed river environments are still widely understudied (Hosfield and Chambers 2005c). As a result, our understanding of assemblage formation is still largely skewed away from smaller artifacts such as flakes or chips, which can be informative to the archaeological record in locally derived/lightly disturbed assemblages.
4. While artifacts entered fluvial systems through scouring of Pleistocene landscapes and the erosion of extinct river systems, it is also recognized that terrestrial processes (Lenoble *et al.* 2008) such as soil attrition, trampling (e.g., Clacton golf-course site), frost creep, and mass-wasting have also affected riparian landscapes where artifacts would have been abundant (Ashton *et al.* 2006). Little is understood about the time lag between the manufacture and the discard of artifacts and their incorporation into fluvial systems, however, if the Bridgland model is correct, it is safe to assume that this time lag may have been considerable. Radiocarbon estimates of Holocene palaeochannels indicate that the rates of lateral migration can be low for extended periods of time and that deposits may survive intact for upwards of five thousand years (Lewin *et al.* 2005; Macklin *et al.* 2006). As a result, long periods of terrestrial exposure could have had an influence on the characteristic breakage patterns and abrasions of artifacts and assemblage formation.

2.6.2. Aims of the study

A main finding of the experimental record is that fluvial processes are a multivariate process for which key variables have remained understudied. This work will help to understand fluvial processes by answering the following questions:

1. What is the role of rivers in the dispersal and modification of artifacts? How do local river features affect the ways in which artifacts are transported and deposited in meandering river channels?
2. How are artifacts abraded in a northern European river context? How are different lithic artifact classes affected? How do fluvial factors such as sediment load impact abrasion?
3. How are assemblages formed? What is the vertical dispersal of transported artifacts within gravel deposits such as bars within these northern European river contexts?

It is clear that experimental archaeology focused on fluvial modification of artifacts remains in the early stages, particularly when it comes to understanding the various factors that influence site formation and artifact modification. To address these gaps, a new set of experiments will be undertaken drawing upon research from both the archaeological and earth science disciplines. The goal will be to use statistical modeling to predict patterns in artifact representation, dispersal, and modification.

Chapter 3

Materials and methods

3.1. Introduction

This chapter outlines the materials and methods for the field-based experiments (Chapter 4), the tumbling experiments (Chapter 5), and the flume experiments (Chapter 6). These experiments test hypotheses of changes to artifact morphology and spatial displacement as a result of actual or laboratory simulated fluvial processes.

3.2. Field-based experiments

The field experiments were conducted between February 2011 and July 2012, using replica artifacts placed in the River Glenderamackin in Threlkeld, Cumbria, United Kingdom (NGR NY3124). These artifacts were periodically monitored and changes to location, dip and strike, and morphology were recorded to test their relationships to artifact dimensions, fluvial discharge (maximum and average), animal trampling, burial, artifact trapping, and initial point of discard.

3.2.1. Sample preparation for fluvial experimentation

Replica artifacts used in the fluvial experiment were knapped into Mode 1-style hard hammer flakes and cores (Clark 1977) by the author from East Anglian chalk flint. Maximum length, maximum width, maximum thickness, and midpoint thickness were recorded to the nearest 0.01mm using digital vernier calipers, as described in Figure 27. Weight was recorded to the nearest 0.1g using a Fisher Scientific SG-2001 digital scale.

Here a field experiment was chosen as a way to directly observe the effects of fluvial systems on replica archaeological assemblages over time. The experimental tracing of gravel movement has long been practiced by researchers (Black *et al.* 2007). Tracer technologies fall into three predominant groups:

1. Visual tracers such as painted rocks and exotic lithologies account for the bulk of tracer experiments undertaken to date (Dornbusch *et al.* 2002; Matthews 1983).
2. Passive tracers can be detected remotely but cannot be tracked during events. They include radioactive, aluminum, magnetic, or electronic tracing technologies (Allan *et al.* 2006; Lamarre 2005; Lamarre and Roy 2008; Schmidt and Ergenzinger 1992 ; Schneider *et al.* 2010).
3. Active tracers, which include acoustic or radio-tagged gravels that transmit signals that can be monitored in real time (Bray *et al.* 1996; Lee *et al.* 2000).

It was decided to use painted visual tracer techniques to track the artifacts because it is an easy, cheap method of recording artifacts well-suited to the short duration of the project. Furthermore, attempts by the author to insert passive tracers (RFID chips) were hampered by (1) the artifact dimensions, which were often too small to fit tags, and (2) the practical time investment of drilling out artifacts and inserting tags (>1 h).

Artifacts were therefore aerosol painted with a single uniform layer of bright orange matte finish on all surfaces and marked with unique numerical identifiers on dorsal and ventral faces with an acrylic paint pen (Figure 28).

Between manufacture and insertion, replica artifacts were stored in plastic containers with individual artifacts separated with plastic bubble wrap. Artifacts were also visually checked upon arrival at the experimental site to insure that they did not incur damage during transport.

3.2.2. Fluvial site: the River Glenderamackin

Experiments were carried out on a 1.4km section of the River Glenderamackin in northwestern England at Threlkeld, Cumbria, between the Threlkeld gauging station (national grid reference (NGR) 332231, 524807) and the Threlkeld Bridge (NGR 331424, 524656; Figure 29). Permissions to conduct the experiment at this location were secured from the Environment Agency's North Cumbria division (Mr Andy Gowans) and the local landowner, Setmabanning Caravan Park (Mrs M.E. Hutton). The Cumbria Historic Environment Service was also alerted to the project (Mr Mark Brennand) and a publicly available website with

Figure 27: Descriptions of artifact measurements (Andrefsky 2005; Marshall *et al.* 2002).

Weight (g)	Measured to the closest 0.1 gram.
Maximum length (mm)	Length from the distal end of the flake as a line perpendicular to the flake length line at the most remote point on the distal end.
Maximum width (mm)	Measured perpendicular to length. Note that this is not necessarily the widest point, but rather the width of the fitted rectangle.
Maximum thickness (mm)	Measured across the thickest portion of the artifact. Note that this is not the thickest point (see Edge view thickness at L1), but that of the fitted rectangle.

Figure 28: Two artifacts used in the fluvial experiment after their recovery.

photographs of the replica artifacts was maintained for the duration of the project:

The River Glenderamackin runs westward and drains the southern side of Blencathra Mountain. The catchment geology is dominated by the Skiddaw Slates (metamorphosed Ordovician, marine turbiditic and hemi-pelagic greywacke, arenite, siltstone and mudstones), with small areas of the Borrowdale Volcanic Group, a dominantly extrusive volcanic association, in the southern watershed (Hatfield and Maher 2008).

The valley floor is filled with Holocene alluvium and the riverbed is largely comprised of smooth oblate, bladed, and prolate gravels of local lithology, of up to 15cm in a-axis. The river has a steep gradient (approximately 5.6m per 1000m for the project area derived from Ordnance Survey Digital Elevation Model) and, in parts, the northern banks are channelized with concrete gabions, though most of the river is characterized by active bar development and the transport of bed materials. The banks of the river are grassy alluviums that slump from localized erosion creating a staircase-like (between 0.5 and 1m per step) profile in some parts (Figure 30).

The surrounding catchment terrain is characterized by high-relief rural moorland with a maximum height of 868m above ordnance datum (mAOD) and a minimum valley floor of 125mAOD. The project area was largely unaffected by foot traffic as the land was privately owned and cordoned off by fences and barbed wire. The valley is largely agricultural, grazed by cattle and sheep, though during this experiment the project area was left fallow except for a brief period in between April 2011 and September 2011.

This project area was chosen for five reasons:

1. The frequent mobility of the large clasts (>15cm) within the gravel-bedded system (A. Gowans 2010 *pers. comm.*).
2. The topography of the river's catchment makes it an extremely flashy river, prone to very rapid rises and falls in level depending on rainfall. Daily discharges from the Threlkeld gauging station between January 1, 2009 and January 1, 2012 showed a mean discharge of $2.749 \pm 4.617 m^3/s$ with a minimum discharge of $0.06 m^3/s$ and a maximum output of $59.04 m^3/s$.
3. The water visibility was ideal for tracer recovery.
4. There are no known examples of Lower Palaeolithic archaeology in the area so there was little chance of contaminating the archaeological record (A. Gowans 2010 *pers. comm.*). Flint is an exotic lithology in the River Glenderamackin which prevented confusion between experimental artifacts and river clasts. This setting is not analogous, however, to archaeological contexts, in which flint artifacts are transported with a partially flint bedload. However, studies have reported that the river clasts in the River Glenderamackin (mostly

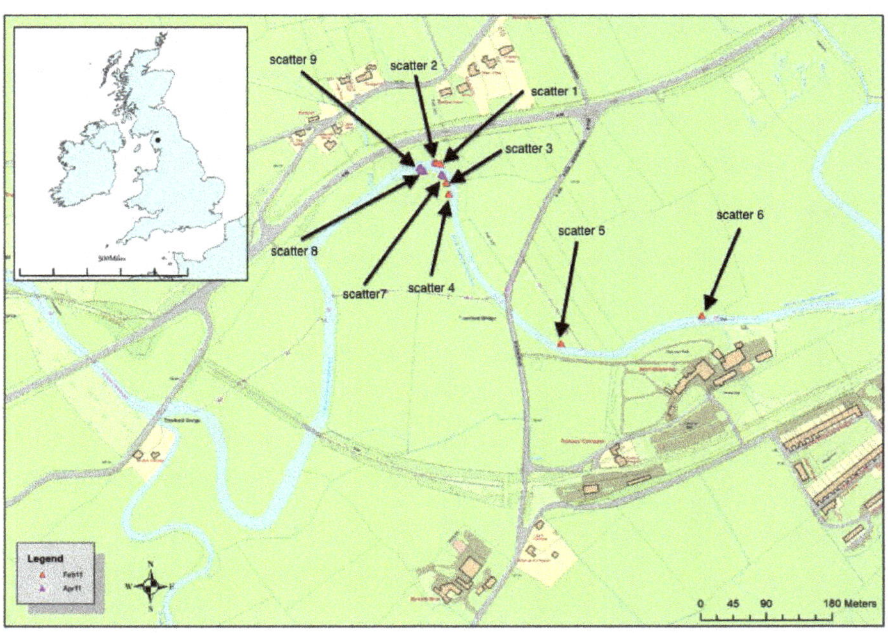

Figure 29: The project area of the field-based experiments showing the initial insertion points of artifacts in February and April 2011. Note: the river flows from east to west.

Figure 30: River Glenderamackin at the northern most point of the project area in April 2011 (photographic viewpoint: OSGB 1936: 331847.8, 525024.4). Note active erosion of the banks and stepped slumping of the southern bank. The point gravel bar visible was the location of Scatters 7–9.

Greywacke) would abrade the experimental artifacts similarly to flint bedloads, differing less than 1% in clast abrasion when transported downstream under 1km of transport (Morris and Williams 1999; Figure 31).

5. The gauging station's reading and its online telemetry system were ideal for remotely monitoring the river.

The river flow statistics

River discharges were gauged by the Environment Agency at the Threlkeld gauging station, a 13m wide concrete weir that measures river levels with a stilling well and floats (Figure 32; NFRA ref. 75007). The weir provided digital

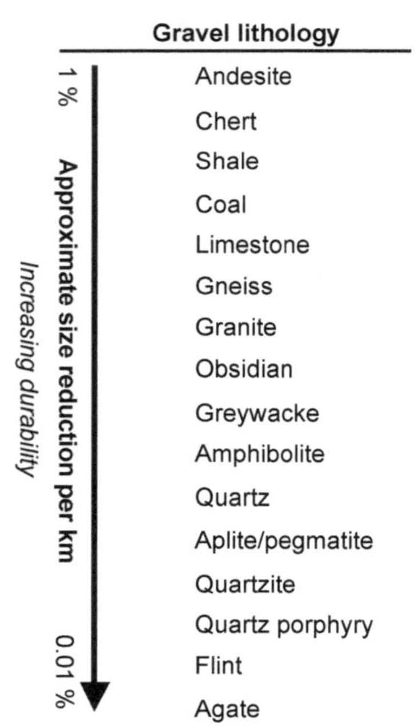

Figure 31: Relative abrasion rates from experimental studies (Redrawn after Frings 2008; fig. 4 after Morris and Williams 1999).

river height levels every 15 minutes telemetrically to the EA website:

http://www.environmentagency.gov.uk/homeandleisure/floods/riverlevels/default.aspx

This website was regularly monitored to see when the river flooded. At the conclusion of the experiment, Ms. Helen Reynolds of the Environment Agency provided spreadsheet readouts of average daily flow means, maximums, and minimums, from January 2011 to August 2012.

The artifact trampling patterns on the fluvial experiment site

During a 4-week period in August/September 2011, animals grazed parts of the project area. Seven adult cows were allowed access to land containing Scatters 2 and 3 while approximately 10 sheep grazed an area that included Scatters 4 and 5. A series of fences and barbed wire restricted the movements of the animals in these areas allowing us the opportunity to assess the effects of different sized (Categories 3 and 4; Brain 1983; Smith *et al.* 2003) ungulate trampling on fluvial assemblage rearrangement and modification.

3.2.3. Fluvial experiment protocol

In February 2011, 300 replica artifacts were divided into six scatters and placed at various parts of the river distributed across the project area. A further three scatters, totaling 156 replicas, were placed during a following visit in April 2012 (Figure 33). Scatters were approximately 2x1m in extent, composed of 50 randomly selected artifacts spaced 15–20cm apart in 5 rows of 10 artifacts. In alternating rows, artifacts were oriented parallel or perpendicular to the stream flow with either the dorsal or ventral face

Figure 32: Weir and gauging station at Threlkeld in April 2011 (NRFA reference 75007).

Figure 33: Descriptions of artifact scatters. (see Figure 29 for locations).

Scatter	Description
1	Edge of a low-lying northern bank of the channel at the beginning of an outside bend of the river
2	Edge of a low-lying northern bank upstream from Scatter 1
3	Edge of a low-lying southern bank at the beginning of a river bend
4	Edge of a low-lying southern bank on a straight part of the channel
5	Edge of a low-lying southern bank on a straight part of the channel
6	Northern bank of the river on a straight part of the channel
7	Beginning inside bend on a depositional gravel bar
8	End of an inside bend on a depositional gravel bar
9	End of an inside bend of a depositional gravel bar after Scatter 8

(unsystematically) resting on the ground's surface (Figure 34). Artifacts were positioned in this unnatural manner so as to avoid complicated artifact-artifact interactions and provide comparable positions between assemblages.

Artifact location was recorded with a Leica GS09 Differential Global Positioning System (DGPS) real-time rover working in tandem with the 'SmartNet' network of reference stations accurate to 1cm. Artifact location was recorded in the British National Grid system. Spatial data was analyzed using Quantum Geographic Information System (GIS) 1.8 (Lisboa) and ArcGIS 10.x.

The project site was visited during low flow periods in February, April, and September 2011, and July 2012 (Figure 35). During monitoring visits, the length of the riverbed from the weir to the Threlkeld Bridge was walked to recover artifacts with a minimum search time of 6 hours during each visit (Figure 29). The area underneath Townfield Bridge was not surveyed because water depths always exceeded 2m.

When found, artifact locations were recorded with the DGPS and dip and strike were recorded as described in Ragan (2009). Dip and strike measurements were taken with an Apple iPhone 4 running iOS 5.x and later iOS6.x using the app GeoID 1.6x (Engineering Geology and GIS Lab., SNU). This technology has been demonstrated to measure strike, dip, and location with the same accuracy as traditional methods such as a Brunton compass (Haupt *et al.* 2009). GeoID also recorded the longitude and latitude of each artifact (to the nearest 11m). Artifacts were noted if they were trapped between clasts or other artifacts, had sustained edge damage (either transverse snapping or micro-flaking; *sensu* Hosfield and Chambers 2005c), or were buried either partially or completely in alluvial sediments.

At the final visit, 97% of the recovered artifacts were buried under an 8cm thick layer of fine alluvial sediment (6.5Y 2.3/2.4) and grass and an approximate 4 square meter

Figure 34: Scatter 4 initial placement in February 2011.

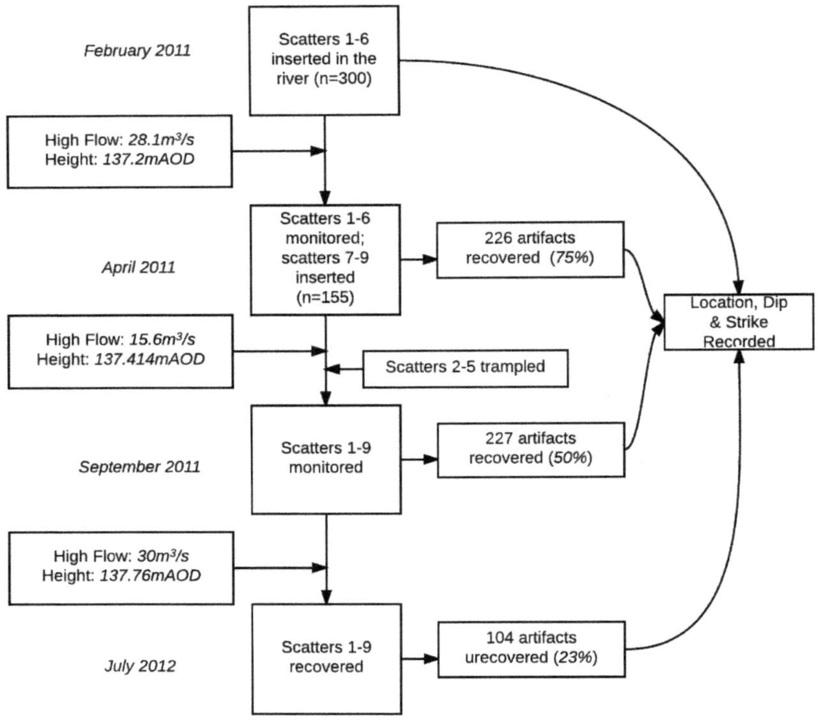

Figure 35: Flowchart of the protocol of the fluvial experiment.

area covering their original location was excavated with a folding military shovel and trowel.

3.2.4. Data post-processing procedures

Artifact dip and strike values, locations, and photographs were exported to an SQL database and then imported into ArcGIS 10x. Matching equivalent artifact location points at each monitoring period and converting these to distances created a vector file of artifact travel distances. Where artifacts had traveled through bends in the river, a vector file of the stream flow was created to simulate the most parsimonious artifact travel path. Data were coded as listed in Figure 36 and all analyses were performed with SPSS 20.0.

3.3. Tumbling barrel experiments

This section presents the materials and methods of the tumbling barrel experiments presented in Chapter 5. In this experiment, replica Palaeolithic artifacts were placed in a tumbling barrel with Pleistocene sediments, to simulate high-energy fluvial abrasion and test the hypothesis that gravel size, gravel volume, gravel lithology, artifact lithology, artifact class, and time impact upon the morphology and abrasion condition of artifacts.

Figure 36: Experimental design of the fluvial experiment showing the variables.

Dependent variables				**Independent variables**										**Controlled variable**
				Between groups factor									**Within groups factor (repeated measure)**	
Transport distance	*Height change*	*Edge damage*	*Dip and strike*	*Maximum length*	*Maximum width*	*Maximum thickness*	*Weight*	*River discharge (M and Max*	*Trampling*	*Trapping*	*Burial*	*Scatter*	*Time*	*Artifact lithology*
Measured (1.0cm)	Measured (1.0cm)	Micro-flaking (>2mm in maximum diameter)	Measured (1.0°)	Measured (.01mm)	Measured (.01mm)	Measured (.01mm)	Measured (.1g)	Measure (.001m³/s)	None	No	No	1	Apr 2011	Flint
									Large animal (size 4)	Yes	Partially	2	Sep 2011	
									Medium animal (size 3)		Yes	3	Jul 2012	
												4		
												5		
												6		
												7		
												8		
												9		

Figure 37: An example of a handaxe (K*) and a flake (1yen) used in the tumbling barrel experiment.

3.3.1. Sample preparation

To test the effects of abrasion on lithic tool morphology, a total of 170 replica handaxes and unretouched flakes were manufactured (Figure 37). These lithic tool forms are types commonly associated with Early and Middle Pleistocene archaeological finds.

The handaxes used in this experiment consisted of 10 cordiforms manufactured by John Lord on raw materials sourced from the Lynford gravels in Norfolk, UK. These handaxes were previously used in a fallow deer butchery experiment that was recorded on digital video, so their life history is well-documented (Machin *et al.* 2007).

The 160 hard-hammer flakes were manufactured from four nodules of East Anglian chalk flint by the author and students at the University of Reading during an introductory flintknapping course. Flakes were randomly chosen from the leftover debris if they had a defined bulb of percussion, and were at least 2cm in maximum length.

The metrics taken were the maximum length, maximum width, and maximum thickness; these were recorded using vernier calipers (Figure 27). During the course of the experiment, abrasion was recorded using a four category ranking scheme from 'fresh' to 'heavily rolled' similar to that used by Ashton (1998), Ashton and Hosfield (2010), and others (Figure 38).

Artificial artifact patination

One set of artifacts was artificially patinated to test the influence of patination on artifact abrasion. Artifacts were placed in a 30% solution of Sodium Hydroxide (NaOH) and water (H_2O) following the protocol of Rottländer (1975). Artifacts and the solution were stored in a tightly sealed low-density polyethylene container for 3 months, during which time a 2mm thick white patina was formed around the surface of the artifacts (Figure 39).

3.3.2. The tumbling barrel

The main goal of the experiments was to examine the combined effect of time and sediment on artifact morphology. Accordingly, an apparatus was needed that could contain gravel, water, and artifacts to simulate the mechanical effects of fluvial transport in a Pleistocene river on flaked lithic artifacts.

The author designed a tumbling barrel, depicted in Figure 40, that was constructed by Richard Tegg and Paul Lock of the University of Reading's School of Human and Environmental Sciences. The barrel consists of a cylindrical polyethylene drum (84cm height, 49cm diameter) whose axis is traversed by a metal rod. Both ends of the rod are affixed to ball-bearing hubs mounted to a wooden base (112 x 79 x 77cm). Two (2.5cm thick, 0.5cm high) 23-cm-long plastic vanes are fastened to the inner surface, parallel to the rotational axis of the drum, so that contained particles are carried upward on the rotating vane and dropped back onto artifacts at the bottom of the barrel. This mechanism roughly simulates the natural sediment motion of field experiments (Jones and Humphrey 1997; Kodama *et al.* 1992; Lewin and Brewer 2002).

The middle circumference of the drum has a rectangular window (20cm/20cm) providing access to the sediments

Figure 38: Comparison of the abrasion indexes.

Abrasion index	Descriptions	Ashton (1998; Ashton and Hosfield 2009)	Shackley (1975)	Wymer (1968; plate *xi*)
1	Little to no traces of abrasion; no more than a few discontinuous 'nicks'	Fresh	Mint, Very fresh	Mint, Sharp
2	Some bifacial edge damage (<50% of total edge)	Slightly rolled	Fresh, Slight abrasion	Slightly rolled
3	Considerable bifacial edge damage (>50%) and ridge rounding	Rolled	Abraded	Rolled
4	Total removal of original edge and rounding of ridges	Heavily rolled	Heavily abraded, Very heavily abraded	Very rolled

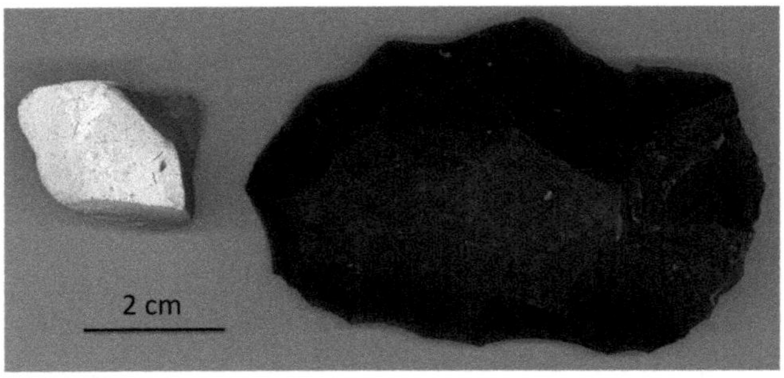

Figure 39: Example of two flint artifacts manufactured from the same flint nodule. The left has been patinated and broken open with a hammer.

and artifacts. It is fastened with two locking metal straps along the outer circumference of the drum. The entire mechanism is enclosed with a wooden and plastic cover with dimensions of 97 x 79 x 44cm.

The axis of the drum is affixed to an aluminum gear and turned with a belt driven 1/20hp Parvalux electrical motor (type SD13GWS Parvalux Electric Motors, Bournemouth, Hants, UK) to exert rotational force (torque, Nm) to the drum and induce movement of the gravel and the 'entrained' artifact replicas. The motor rotates the barrel at a constant speed of 15rpm.

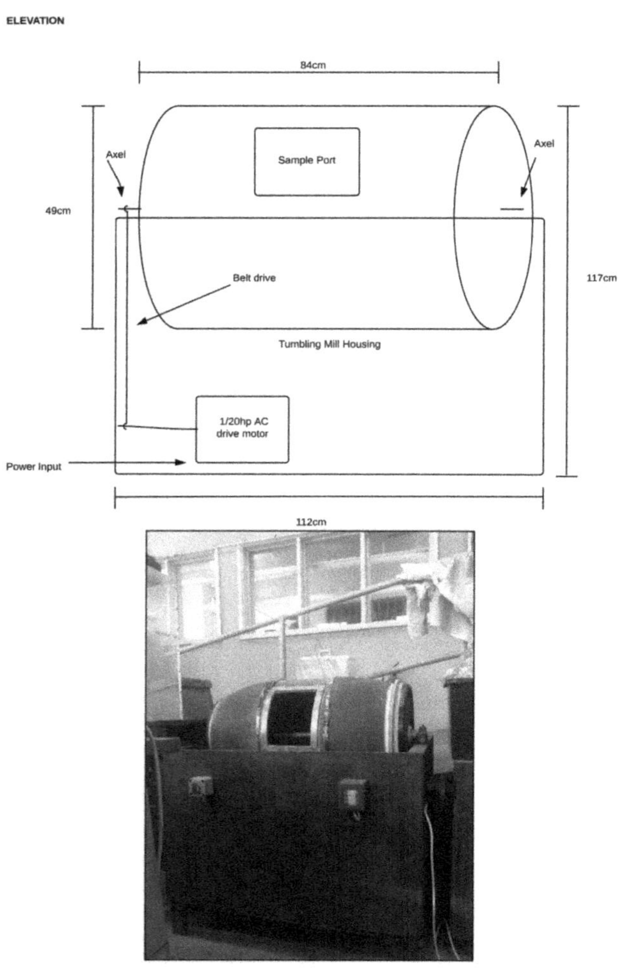

Figure 40: The tumbling barrel: schematic showing equipment dimensions (*not to scale*) and a photograph of the tumbling barrel setup (below) during experimental procedures.

3.3.3. Experimental techniques

Tumbling mills are the standard equipment used to induce fluvial abrasion, as they are capable of abrading sediments of a wide variety of sizes and lithologies cheaply. In recent years, their use has been criticized as poor simulators of fluvial entrainment, suggesting that they did not produce analogous results when compared to real world experiments (Kuenen 1956; Mikoš and Jaeggi 1995; Šolc *et al.* 2012). Thompson *et al.* (2011) have also suggested that tumbling barrels cannot accurately model aquatic sediment transport processes.

Comparing tumbling barrels, flumes, and real world experiments, Lewin and Brewer (2002) found that setups in tumbling barrels indeed simulate fluvial particle-to-particle abrasion processes, but 'of a complex kind as may occur during avalanching, rolling, working through the clast pack, 'vibrating' and overpassing' (Lewin and Brewer 2002). However, they found that these processes affected estimates of travel distances, impact velocities and individual abrasion event histories, suggesting that in terms of morphological abrasion, tumbling barrels *do* simulate clast-clast processes, albeit through an unnatural sediment motion (Kuenen 1956).

3.3.4. The sediments

Experimental artifacts were abraded with three different sets of sediments in the tumbling barrel (Figure 42): one sample was recovered from three test-pits at the site of Ham Yard in Central London, *ca.* 150m north of Piccadilly Circus (Ham Yard). The British Geological Survey (BGS; 1:50,000 Sheet 256 north London 1994) indicates these were part of the Middle Pleistocene Lynch Hill Gravel formation of the Middle/Lower Thames (Green 2012). The sediment is poorly sorted, comprised of yellowish-brown sandy gravel of predominantly sub-angular flint (up to 100mm a-axis) and well-rounded chatter-marked flint pebbles. Many well-rounded vein quartzite pebbles are present. The Ham Yard size distributions are reported in Figure 41.

The sediment was dried in a 70°C oven and sieved for artifacts. Though no artifacts were found within the sample, the Lynch Hill formation is regarded as the richest source of artifacts within the Thames terrace deposits (Wymer 1988) and bifaces were recovered from excavations of the same level at the nearby Regent Palace Hotel (Wymer 1968).

A second sample of Pleistocene gravels was acquired from the Upper Trent Valley formations at New Bold Quarry in Needwood, Staffordshire (Trent Valley; Figure 42). The sediments are characteristic of the Trent's lithology, as they are rich in high sphericity, well-rounded pebbles of predominantly silicified limestone, Lower Carboniferous chert, 'bunter' quartz/quartzite, flint and sandstone

Figure 41: Size distribution of Ham Yard sediments.

Aperture		
φ	Microns	Percentage
-6.0 to -5.5	63000	1%
-5.5 to -5.0	45000	4%
-5.0 to -4.5	31500	10%
-4.5 to -4.0	22400	16%
-4.0 to -3.5	16000	17%
-3.5 to -3.0	11200	11%
-3.0 to -2.5	8000	9%
-2.5 to -2.0	5600	7%
-2.0 and below	3	26%
TOTAL		**100%**

(SEARCH British Geological Survey Rock Name Database, n.d.). This sample was also archaeologically sterile, though the Trent terraces have long been a source of Palaeolithic finds and are of significant interest because recent scholarship has suggested that taphonomy has played a role in shaping its Palaeolithic archaeological narrative (Knight and Howard 2004; White *et al.* 2009; see Section 7.5.5 for discussion).

The third sample was a commercially available Pleistocene gravel of unknown origins (Kennet; Figure 42) though its origins are likely of the Kennet Valley (J. Allen 2013 *pers. comm*). The lithology was comprised entirely of brown and gray (7YR 3/5.4 to 5.5GY 4.2/04; Munsell 1905), low sphericity sub-angular and well rounded flint pebbles.

To ascertain the influence of particle size (diameter) on abrasion, sediment sets were sorted with test sieves into Krumbein *phi* (φ) scale sizes (Krumbein and Sloss 1963) as listed in Figure 43.

3.3.5. Tumbling barrel experiment protocol

Experimental artifacts were placed inside the tumbling barrel with a variable amount of sediment and 4l of water. Abrasion was induced for a total of three hours and interrupted at eight time-steps (1, 5, 10, 15, 30, 60, 120, 180 minutes). Three hours was selected as an adequate stopping point because macroscopic abrasion of the artifact appeared to be negligible beyond that point in the largest sediment size classes when tested up to five hours (Figure 44).

Primary metrics for each artifact were initially measured, and re-recorded at each time-step to the nearest 0.01mm using a digital vernier caliper. These measurements are described in Figure 27. The axes of these measurements were marked with paint so that they could be identified in subsequent time-steps. Samples were weighed using a digital scale to the nearest 0.1g.

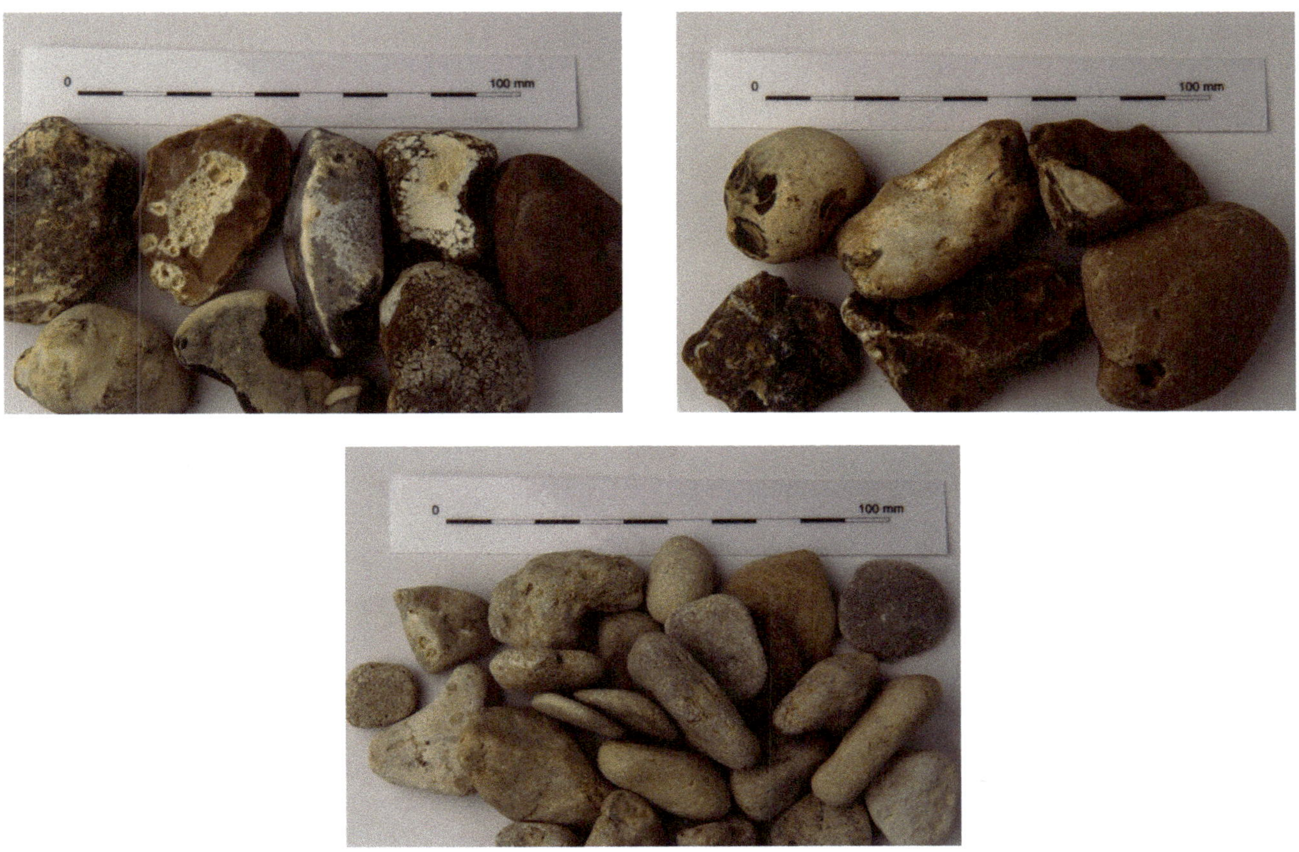

Figure 42: A sample of the sediment lithologies used in the tumbling experiments. Top left photo illustrates the Kennet sediments (φ=-6.0 to -5.5). Top right photo illustrates the Ham Yard sediments (φ=-5.5 to -5.0). Bottom left illustrates the Trent Valley sediments (φ=-4.5 to -4.0).

Figure 43: Tumbling barrel experimental variables.

Dependent variables		Independent variables						Controlled variables	
Primary Metrics (Figure 27)	*Artifact condition*	*Time (minutes)*	*Artifact class*	*Artifact lithology*	*Sediment size (Udden-Wentworth scale)*	*Sediment lithology*	*Sediment volume (kg)*	*Barrel speed*	*Water volume*
Measured 0.01mm	Fresh	0	Flakes	Flint	None (control)	Ham Yard	None (control)	15rpm	4l
	Slightly rolled	1	Handaxes	Quartzite	Aggregate (-6 to 0φ)	Trent Valley	1.0		
	Rolled	5		Patinated Flint	Cobble (-7.0 to -6.5φ)	Kennet	2.0		
	Very rolled	15			Very coarse gravel (-6 to -5.5φ)				
		30			Very coarse gravel (-5.5 to -5.0φ)				
		60			Very coarse gravel/coarse gravel (-5.5 to -4.5φ)				
		120			Coarse gravel (-5.0 to -4.5φ)				
		180			Coarse gravel (-4.5 to -4.0φ)				
					Medium gravel (4.0 to 3.5φ)				
					Medium Gravel (-3.5 to -3.0φ)				

For artifacts in this study, maximum length, maximum width, maximum thickness, and weight were measured.

The dorsal, ventral, and lateral faces of each artifact were photographed before, and during each time-step of its trial to record artifact condition and for *post-hoc* measuring. This method was modeled on the protocol set forth by McPherron and Dibble (1999). Specimens were fixed to a plastic gridded photograph board with a digital FUJIFILM 12 Megapixel Finepix S1800 affixed to a camera stand 37cm from the photography surface and was leveled with a spirit level. All photographs were taken at a focal distance of 50mm and saved in JPEG format, with an image size of 2664 x 4000 pixels.

For the handaxes in this study, primary metrics were recalculated from the photographs with the aid of NIH ImageJ software (http://rsbweb.nih.gov/ij/). In addition, L_1, Breadth at L_1, BA, and BB were recorded to calculate Roe's edge shape indices (Figure 45; Roe 1968).

Additionally, time-step artifact photographs were compared to the pre-trial photographs to determine the extent and nature, if any, of damage incurred by tumbling. Artifacts' edges and ridges were visually classified into four broad classes of condition (Fresh, slightly rolled, rolled, very rolled) according to the system described by Ashton (1998; Figure 46). A control sample of 10 flake artifacts and 3 handaxes were tumbled in 6l of water for 180 minutes. This was to see if the tumbling barrel motion had an effect on abrasion.

3.3.6. Data post-processing procedures

Following their collection, artifact measurements and other data were joined in a spreadsheet using Microsoft Excel

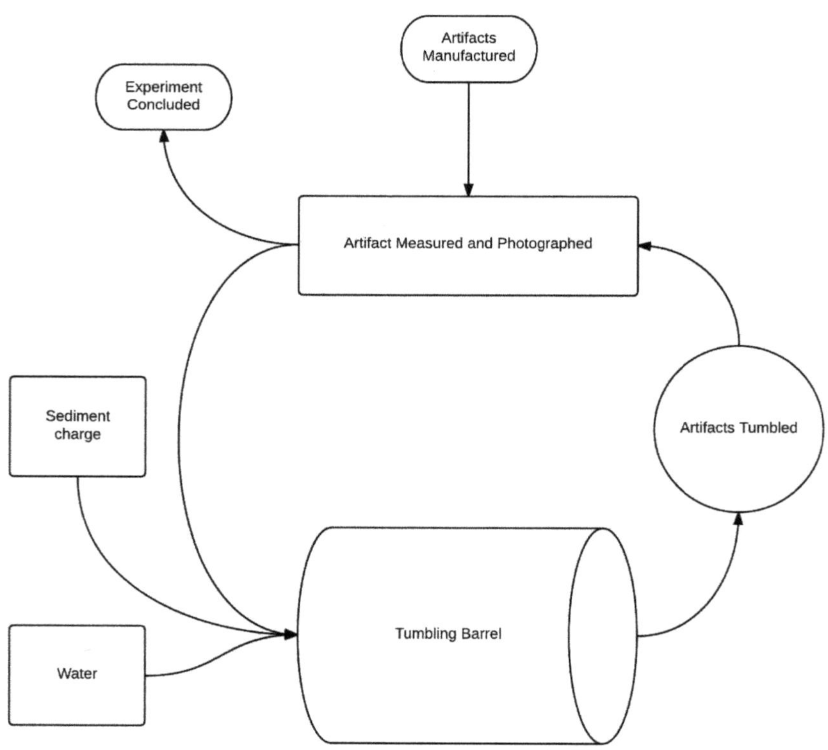

Figure 44: Tumbling barrel experimental protocol.

Figure 45: Additional handaxe measurements for the tumbling barrel experiments.

L1 (mm)	Length from the butt to the widest point of the biface as measured from edge to edge (Breadth at L1).
Breadth at L1 (mm)	Breadth from edge to edge at L1. Note that this is different from Breadth, which is that of the fitted rectangle.
BA (mm)	Breadth at 20% down from the tip or bit.
BB (mm)	Breadth at 20% up from the butt.

Figure 46: Description of artifact abrasion conditions (Ashton 1998).

(1) Fresh	As freshly knapped with edges razor sharp and tiny scars crisply defined.
(2) Slightly rolled	Fresh to handle, with very slight damage/abrasion to sharp edges, ridges and scars.
(3) Rolled	Some notching and battering of sharper edges with ridges and scars slightly abraded.
(4) Very rolled	Intense notching and battering of all edges, plus heavy abrasion of ridges and flake scars.

2011. Data were coded as listed in Figure 43. Artifact numerical values were exported to SPSS 20.0 for statistical analysis.

3.4. Flume experiments

This section presents the materials and methods used in the flume experiment, the results of which are presented in Chapter 6. In this study, replica Palaeolithic artifacts were placed in annular flumes along with small caliber Pleistocene sediments to simulate low-energy fluvial abrasion. The goal was to test the hypothesis that sediment size, artifact transport mode, and time significantly affect the development of microscopic abrasion on replica artifacts.

3.4.1. Sample preparation

To observe the effect of flume abrasion on replica lithic artifacts, 27 freshly knapped replica handaxe thinning flakes (F1–F27) were manufactured using a reindeer (*Rangifer tarandus*) antler billet from a nodule of English chalk flint (Brandon flint; Figure 47). Thinning flakes were used because they have fine edges and obtuse ridge angles analagous to those of handaxes. English chalk flint was selected because of its common use in experimental studies as a prototypical chert, but also for its relative purity and fine-grained cryptocrystalline texture. Flakes with lengths shorter than the flume channel width were selected to minimize edge effects.

After manufacture, flakes were cleaned by soft brushing in soapy water and rinsed in acetone, following the 'mild cleaning' protocol outlined in Evans and Donahue (2005) and Shipman and Rose (1983) for micro-wear analysis. Artifacts were subsequently stored in plastic bags when not in use, and handled with rubber gloves to prevent the acquisition of dirt and/or grease on the flakes' surfaces.

Flakes were then marked with a minimum of nine unique 'microscopic observation areas' on the dorsal side for pre- and post-wear image analysis. Microscopic observation areas were 2mm across and were placed approximately 20mm apart in a 3x3 grid.

3.4.2. The flumes

Wear experiments were undertaken in two fully calibrated laboratory annular flumes at the National Oceanographic Centre (Southampton, UK), with proven records in investigations of solid-transmitted stresses: the 'Miniflume' (Thompson *et al.* 2011; Thompson and Amos 2004; Thompson and Amos 2002; Amos *et al.* 2000; Figure 48) and the 'Lab Carousel' (Amos *et al.* 1998; Figure 49). The Miniflume, used for grain sizes up to -3.25φ, is a 0.3m diameter flume with working channel width of 0.045m and water depth of 0.2m. The Lab Carousel, used for grain sizes above -3.5φ, is a 2m diameter flume with a 0.15m wide channel and water depth of 0.4m. In both, equidistant paddles on the flume lids drive a unidirectional current, and their unique geometry insures a fully developed benthic boundary layer, essential for application of the results to the natural environment.

The Miniflume is an annular benthic mini flume constructed from acrylic, with an acrylic base. The walls and base are smooth with no internal protuberances. Four equidistant square paddles attached to a rotating lid induce a unidirectional current measured by a Nortek Vectrino velocimeter (Nortek AS, Rud, Norway). This flume has published success in the investigation of solid transmitted stresses as well as abrasion experiments of bone materials (Thompson and Amos 2002; 2004; Thompson *et al.* 2011).

The Miniflume was incapable of initiating sediment motion for the largest sediment category (τ_0=0.41 Pa) so the Lab Carousel, a larger annular flume, was used for these sediment sizes (Figure 49). Similar to the Miniflume, the Lab Carousel induces flow with eight equidistant

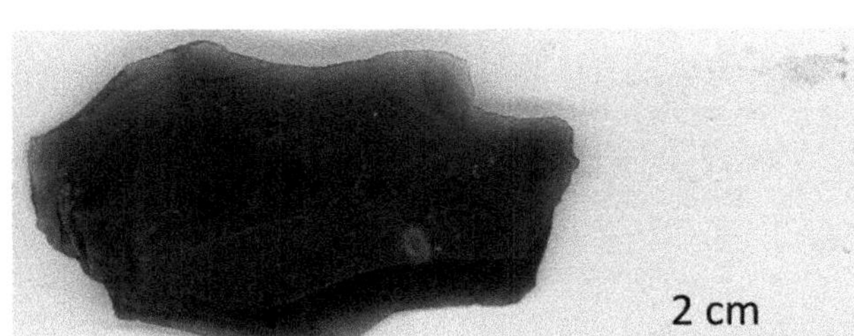

Figure 47: Example of an artifact (F10) used in the flume experiment.

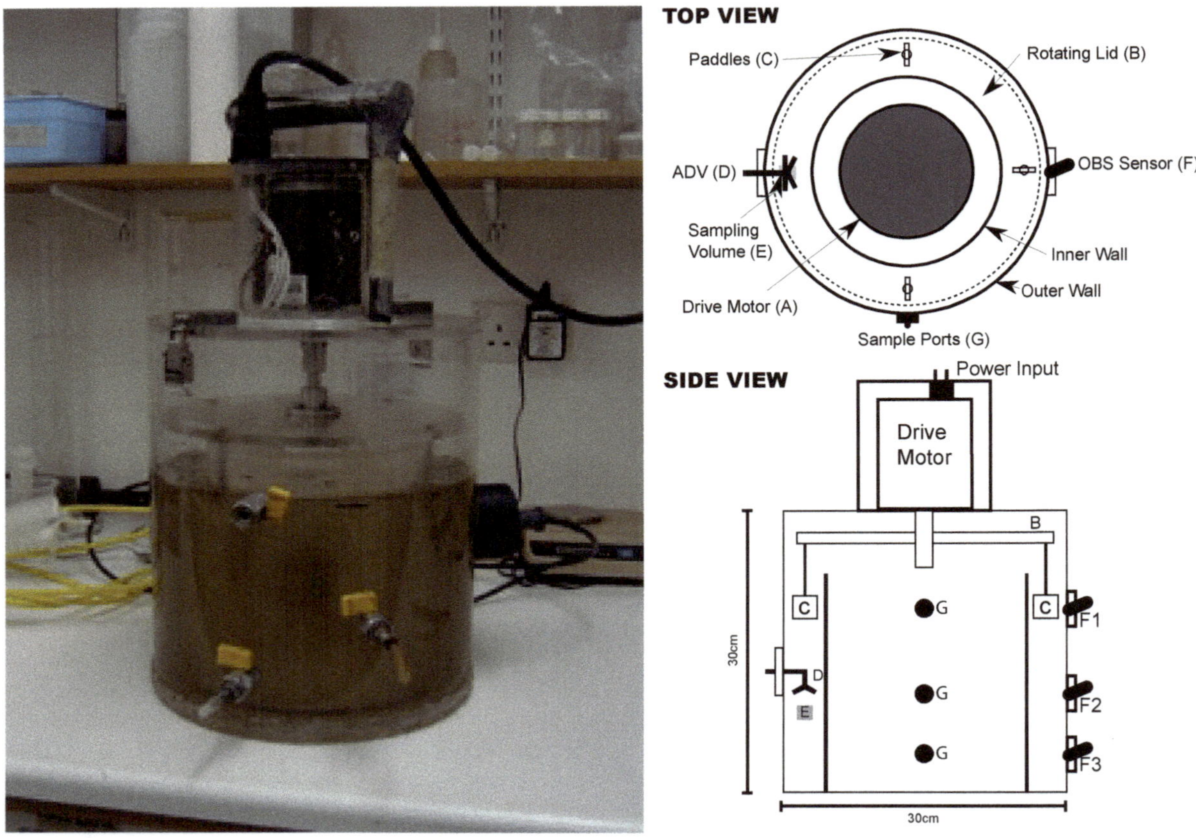

Figure 48: The Miniflume and diagram (Redrawn after Thompson *et al.* 2011, courtesy C. Thompson).

spaced square paddles fixed to a rotating lid. The speed of the lid was controlled by an E-track® AC inverter motor controller and a 1-dimensional class IIIb Helium-Neon 10 mW Laser Doppler Velocimeter (LDV) measured the tangential component of flow at various heights above the flume bed.

3.4.3. Experimental techniques

The use of an annular flume was advantageous for fine-grained experiments as it sharply controls the movement of sediments within a benthic boundary, rather than using tumblers that cannot produce a realistic situation (e.g., interaction within the tumbler). Annular flumes allow this, while also allowing for the theoretically infinite bombardment of the artifacts with a limited sediment supply (Thompson *et al.* 2011).

SEM microscopy was also used in this experiment to analyze lithic tool surfaces. SEM microscopy has been successfully applied to fluvial abrasion of fossilized bone (Bromage 1984; Fernández-Jalvo and Andrews 2003; Shipman 1981; Shipman and Rose 1983; Thompson *et al.* 2011) and to identify use-wear on flaked lithic artifacts (Evans and Donahue 2005; Faulks *et al.* 2011; Keeley 1980; Kimball *et al.* 1995), but has seldom been used for fluvial lithic taphonomy (cf., Fernandes *et al.* 2007). While in the past, SEM sample preparation has been costly, recent generations of SEM no longer require sample preparation (i.e., gold sputter coating) leading to cheaper and quicker analysis. SEM also provides a high-resolution micrograph when compared to traditional optical microscopy.

3.4.4. The sediments

The sediments used in this experiment were all quartzite sands from unknown locations along the southern coast of the United Kingdom.

Figure 49: The Lab Carousel (courtesy C. Thompson).

3.4.5. Flume experiment protocol

Pre- and post-abrasion scanning electron (SEM) images were taken after sample preparations and again following abrasion in the flume. The surface morphologies of the observation areas on the flakes were recorded with a Hitachi TM1000 Scanning Electron Microscope (SEM; Hitachi High-Technologies, Corporation, Tokyo, Japan). Scans were recorded at 24, 72, 96, and 168 hours. Image analysis was undertaken at magnifications of 100 (effective field of view (FOV) 1.6 x 1.2mm) and 1000 (FOV 160 x 120mm).

Artifacts were then equidistantly placed in the flume to allow temporal effects of abrasion to be assessed under identical flow and transport conditions, and either fixed in place or left mobile to assess *in situ* v. *ex situ* wear. Artifacts were removed at prescribed intervals. A 1cm layer of well-sorted sediment was added to the flume, and a flow induced to maintain sediment transport (Figure 50). As sediment-induced wear has been demonstrated to depend on the sediment transport mode (Thompson *et al.* 2011), flows were chosen to maintain all sediment sizes in the saltation phase, which has been shown to induce the greatest wear on both sediment beds (Thompson and Amos 2004) and modern and archaeological bone sections (Thompson *et al.* 2011), and would therefore maximize wear development while keeping experiment time to a minimum (>24h).

Elemental analysis was also performed on 18 microscopic observation areas across 4 flakes (F5, F6, F7, F8) via Energy-Dispersive X-ray spectroscopy (EDX) to understand the chemical composition of the samples and any changes they might undergo. The premise behind EDX is that it stimulates the output of X-rays specific to the elemental composition of the sample by bombarding it with a high-energy beam of charged particles. The resulting X-rays are measured by energy dispersive spectrometry or an X-ray detector (Goldstein 2003). The instrument used was a SwiftED-TM (Oxford Instruments Link INCA) coupled with the SEM. Samples were affixed to adhesive carbon discs, on an aluminum stub fixed at 22° from the EDX beam. The detector type is a Silicon drift detector, with an energy resolution of 165eV, which allowed the detection of elements from Na to U. All samples were observed under high vacuum for 60 seconds by a 15kV electron beam.

Two additional control flakes with nine microscopic observation areas each were placed in the flume for 168 hours to assess the impact of moving water only on artifact abrasion. These showed no visible damage and were not included in the analysis.

3.4.6. Data post-processing procedures

Post-trial micro-graphs were compared to pre-trial micro-graphs to assess abrasion levels. They were assigned a category of either no damage (no change), slightly abraded

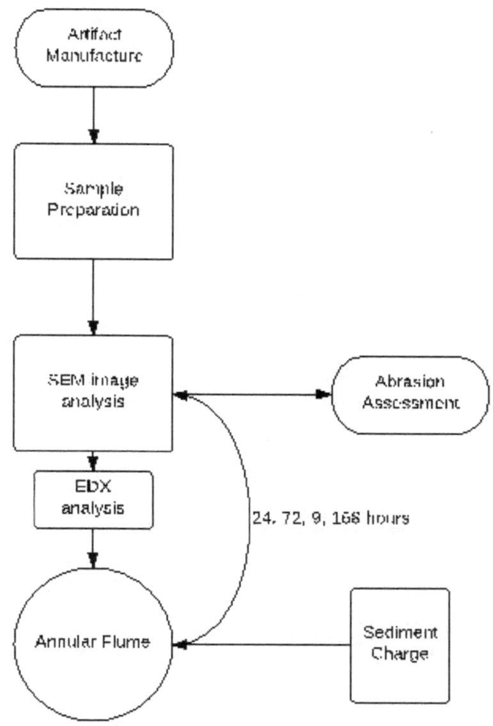

Figure 50: Flume experiment methodology flowchart.

(less than 33% of original surface removed), moderately abraded (between 33% and 66% of original surface removed), and highly abraded (66% or more of original surface removed). While computerized methods have been developed to assess lithic wear with mixed success (Bietti 1996; Durham *et al.* 1995; González-Urquijo and Ibáñez-Estévez 2003; Grace 1989; Grace *et al.* 1985), assessing them visually is more accurate. When present, ridges were measured with NIH *ImageJ* software (http://rsbweb.nih.gov/ij/). Individual artifacts were also assigned an edge damage category of no damage, light damage (micro-flaking <1mm), and heavy damage (micro-flaking 1–2mm in maximum diameter).

Values were then joined in a SQL spreadsheet and coded as in Figure 51. All statistical analyses were performed in SPSS 20.0.

3.5. Analytical methods

In this study, various statistical tests were used to analyze the data gathered in the three experiments. These statistical analyses were performed using R and SPSS and are explained in the sections following.

3.5.1. t-test

In some analyses, means were compared with a robust (bootstrapped) independent samples t-test that assesses if the means of two groups are *statistically significantly* different from each other. This analysis is especially appropriate for two-group randomized experimental designs. To make the dependent variable suitable for parametric testing, values were resampled using the bootstrap technique based on

Figure 51: Flume experiment variables.

Dependent variables			Independent variables				Controlled variables		
Surface wear	*Ridge width*	*Edge damage*	*Time*	*Sediment size (Udden-Wentworth scale)*	*Flow velocity (m/s, 5cm above bed)*	*Artifact Transport mode*	*Sediment Transport mode*	*Artifact lithology*	*Artifact class*
No abrasion	Measured (μm)	(1) No damage	0 (control)	None (Control)		Fixed	Saltation	Flint	Unretouched thinning flake
Slight abrasion		(2) Light damage	24 h	Fine sand (2.50 to 2.00φ)	0.34	Mobile			
Moderate abrasion		(3) Heavy damage	72 h	Granules (-1.75 to -2.25φ)	1.11				
High abrasion			96 h	Pebbles (medium) (-3.00 to -3.25φ)	1.7				
			168 h	Pebbles (medium coarse) (-3.50 to -4.25φ)	1.10				

1000 samples. When variances were unequal or unknown, Welch's t-test, which does not assume equal variance, was reported. All analyses were performed with the statistical software R (v2.15).

3.5.2. Mixed model

Mixed models (MIXED) were used to investigate relationships between artifact transport distances and their sedimentary conditions. Conceptually, MIXED is a multiple regression model suitable for linear and nonlinear trends. MIXED therefore models the relationship between a scalar dependent variable (e.g., transport distance) and six explanatory variables (e.g., primary metrics, scatter location, trampling, burial, artifact trapping, and river high flow events). The output of MIXED is a *p*-value for each explanatory variable that, if below or equal to a pre-stated statistical threshold or α (α=0.05), suggests that the observed relationship reflects a *statistically significant* pattern, rather than random chance.

Analytically, MIXED is a multi-level hierarchical model that analyzes explanatory variables by *nesting* dependent variables of an individual subject's (i.e., artifact's) change over time at one level, and at the second level, individual differences (e.g., scatter insertion locations) in the starting conditions which may explain patterning in the dependent variable. There are four reasons why MIXED was used in these analyses:

1. MIXED analyzes models that use both repeated measures of a dependent variable (e.g., artifact transport distances of the same artifact which may change over time) and non-repeated predictor variables (e.g., artifact primary metrics) by isolating them at different nested levels, thereby satisfying the assumption of independence. Traditionally, such experimental designs were analyzed with *repeated measures analyses of variance* (rANOVA or split-plot ANOVA). However, rANOVAs are better suited for balanced datasets. When missing data occur during rANOVA procedures, the individual case and all the associated values are excluded from the analysis while MIXED still includes all data entries with missing values. For this study, a rANOVA would have resulted in significant data loss, reducing the predictive capability of the model.
2. MIXED allows the study of both intra- and inter-individual differences in its parameters. rANOVAs assume that regression parameters are fixed while MIXED combines both a fixed effects model and a random effects model. This means that a function is fitted to the entire sample but the exact slope (e.g., steep or flat) and intercept values (e.g., river discharge at initial river insertion time-point) are allowed to vary for individual artifacts (random coefficients). This is advantageous because it accounts for the variance in the size and shape of the individual artifacts used in this study.
3. MIXED is more powerful than traditional rANOVA variants in examining the effects associated with repeated measures as it models different covariate relationships rather than imposing a certain type of structure as commonly used in traditional univariate and multivariate approaches. This means that MIXED is capable of fitting predictive models more accurately.
4. Lastly, MIXED does not assume sphericity or compound symmetry in the model. This means that there is no need for the data to satisfy these assumptions, as there is for rANOVA.

MIXED analysis was performed in SPSS 20.0. Restricted Maximum Likelihood (REML) criteria were used to estimate the model fit and the correlation matrix was set to *diagonal* because the variances were assumed independent. Two-way interactions were reported. Where interactions were significant, Bonferoni post-hoc tests assessed individual differences.

3.5.3. Rayleigh's test

To test uniform distributions against modal distributions for artifact orientations, the rotational statistic Rayleigh's test was performed. The Rayleigh statistic tests the null hypothesis that artifacts are isotropic with an unknown mean direction and unknown mean resultant length. The alternative hypothesis is an anisotropic distribution (a directional preference). Tests were followed up with a Watson test that assesses the goodness of fit test for the von Mises or circular uniform distribution.

3.5.4. AOV circular

A rotational analysis of variance (AOVCircular) is used to compare the dip values for artifacts under three trampling conditions: no trampling, trampling by sheep, and trampling by cattle. An AOVCircular is a form of ANOVA—a parametric statistical hypothesis test used to compare multiple means of rotational measurements. AOVCircular differs from ANOVA in that it uses the *circular* normal distribution (von Mises) rather than the normal distribution, which is appropriate for rotational measurements. The output of the ANOVA is a *p*-value that, if equal to or below a threshold (α), suggests the means are significantly different and not the result of random distribution. Because circular statistics are used for unidirectional periodicity (1–360°) and this data was only from 1–90°, for statistical tests, all values were transformed 4x as is standard for bounded directional datasets (Kluskens 1995). All calculations were performed in R v2.15 using the package 'circular.'

3.5.5. Binary ordinal regression

A binary logistic regression (logit; OLR) was conducted to provide statistical support for the observations of flaked lithic surface micro-wear using sediment size, time, and mobility as predictors.

Analytically, OLR is analogous to linear regression but they differ in that OLR measures the relationship between continuous independent variables and an ordinal dependent variable by converting the dependent variable to probability scores. OLR compares the observed data to theoretical models where predictors have no effect to determine if the independent variables predict the outcome more accurately. The output for this comparison is a test statistic (α) that, if below a certain critical threshold (0.05), indicates that the result is statistically significant, i.e., the predictors make a significant difference to the outcome.

3.5.6. Chi-squared test

A chi-squared test was used in Section 4.4.7 to test the effects of trampling on edge breakage. A chi-square is a non-parametric statistical hypothesis test of independence that assesses whether paired observations on two variables are independent of each other. The output is a *p*-value that, if below a certain threshold ($\alpha<0.05$), suggests that the two groups are *statistically significantly* different from each other.

3.5.7. ANOVA

Section 6.3 uses a robust (bootstrapped) ANOVA to test the hypothesis that if time, sediment size, and transport mode affect the abrasion of artifacts over time, then the mean ridge widths between groups should be statistically different from each other.

Analytically, a robust ANOVA is the same as a traditional ANOVA and is a statistical model commonly used to test the equality of multiple means. A robust ANOVA differs by bootstrapping the outcome dataset (ridge widths). Bootstrapping is a powerful resampling technique that analyzes a dataset derived from the means of samples from the original dataset. In effect, this increases sample size and normalizes data distribution through the Central Limit Theorem without altering the original parameters of the dataset. In essence, bootstrapping makes non-normal and unequal datasets suitable for parametric testing. Bootstraps in this analysis are based on 1000 samples.

Where significant results were found, *post-hoc* tests were used to make comparisons between the datasets. Because sample sizes differed, Hochberg's GT2 correction was used to control familywise error rates (Field 2009: 341).

3.5.8. Mixed ANOVA

Mixed-design Analysis of Variance (ANOVA), or Split-plot ANOVA, is used to determine the effect of factors influencing primary metrics in Sections 5.2 to 5.6. Mixed-design ANOVAs are typically used in longitudinal studies when repeated measures (random factors, between subject factors, e.g., time) are combined with fixed independent factors (e.g., sediment mass).

Analytically, ANOVA partitions the observed variance in a particular variable (e.g., change in weight) into components attributable to different sources of variation. The output of this test is a *p*-value that, if below a significance threshold (α), justifies the rejection of the null hypothesis (i.e., that there is no statistical difference). In the following experiments, the significance level used was 0.05, indicating that this would be the expected result in 95% of occurrences. Conservatively, datasets were selectively chosen so that only sample groups where datasets were equal in size were compared.

Additionally, where multiple groups were compared and statistically significant effects were found, ANOVAs were followed up with *a priori* comparisons tests and *a posteriori post-hoc* tests. Analytically, both procedures compare group means to identify where two means differ more than the standard error would expect, allowing the identification of statistically different means and thus groups labeled (A, B, C,...). They define which groups are different from each other and establish the nature of the relationships, suggesting which specific conditions can be predicted. Planned comparisons differ from *post-hoc* tests in that the former predicts specific relationships from the outset, while the latter compares *all* means to locate unpredicted differences.

In the following experiments, *a priori* tests were used to compare time-steps because they were expected to have an effect, while Tukey's range tests were used for all other independent variables because the relationships were unclear. Tukey's range test was selected because it is highly conservative and controls well for Type I errors (where true null hypotheses are incorrectly rejected).

It is worth highlighting the statistical challenges of the mixed ANOVA in Chapter 5, particularly the repeated measures aspect of the analysis that cannot be analyzed with traditional statistical methods. This experiment was devised to be analyzed with a mixed model (Section 3.5.2). However, the model could not be made to converge, producing inaccurate results. Therefore mixed ANOVAs were used to analyze the data.

Test assumptions

Textbook analyses with mixed ANOVAs (i.e., Field 2011) are based on three assumptions:

1. The first is that univariate data is normally distributed. The Kolmogorov-Smirnov test was used to test data normality for parametric tests. These analyses compare values to a standard normal distribution and reject their similarity at the $p<0.05$ levels.
2. A second criterion is the homoscedasticity or homogeneity of variance. ANOVA assumes that data are similarly distributed around the mean at different levels of measurement through the variable. This was tested with Levene's test, which examines the null hypothesis that population variances are not significantly different ($p<0.05$). Therefore, when the test is insignificant, the levels are homogeneous and suitable for testing.
3. Finally, mixed-design ANOVAs need to satisfy the assumption of sphericity that refers to the condition where the variances of the difference between the levels of repeated measures are equal (Field 2009: 459). Sphericity was measured using Mauchly's test of Sphericity. When Mauchly's test is insignificant ($p>0.05$), the variances are statistically equal, which leads to the conclusion that the assumption has not been violated. When Mauchly's test is significant, modified Greenhouse-Geisser, and Huynh-Feldt procedures are used that do not assume sphericity and modify the degrees of freedom to produce an F-ratio that corrects Type I error rates.

For the tests performed in sections 5.2 to 5.6, many, but not all of these assumptions were met. In consultation with the University of Reading Statistical Services department, the results are presented nevertheless because mixed ANOVAs are widely acknowledged to be robust against violations of non-normally distributed data and equal variance especially when comparing equal sample sizes (Glass *et al.* 1972; Harwell *et al.* 1992; Lix *et al.* 1996).

3.5.9. Ordinal logistic regression

An ordinal logistic regression (logit; OLR) was conducted in sections 6.2 and 6.3 to provide statistical support for the observations of flaked lithic surface micro-wear using sediment size, time, and mobility as predictors.

Analytically, OLR is analogous to linear regression but it differs in that OLR measures the relationship between multiple continuous independent variables and an ordinal dependent variable by converting the dependent variable to probability scores. OLR compares the observed data to theoretical models, in which predictors have no effect to determine if the independent variables predict the outcome more accurately. The output for this comparison is a test statistic (α) that, if below a certain critical threshold (0.05), indicates that the result is statistically significant.

3.5.10. Kruskall-Wallis test

A Kruskall-Wallis test is used in Section 6.4 to test the effect of sediment size on edge damage. A Kruskall-Wallis is a non-parametric variant of ANOVA that tests if the observed frequency distribution differs significantly from a theoretical distribution where predictors have no effect. Where significant differences were found, non-parametric *post-hoc* tests were followed up with Bonferoni corrected Mann-Whitney U tests.

3.6. Chapter summary

This chapter outlined the materials, experimental methods, and analytical methods for the fluvial (Chapter 4), tumbling barrel (Chapter 5) and flume experiments (Chapter 6). In the next three chapters, the results of these experiments will be given.

Chapter 4

Field-based experiments

4.1. Introduction

This chapter presents the results of an 18-month experiment examining the effects of fluvial processes on Palaeolithic assemblages (refer to Section 3.2 for a detailed discussion of methods). Painted replica flint artifacts were inserted into the Glenderamackin river system (Thelkeld, Cumbria, UK, national grid reference NY3124) at nine different locations (Figure 52) and were monitored from February 2011 to July 2012 for changes to location (vertical and horizontal), orientation, dip, and morphology. Midway through the experiment, four artifact scatters were subjected to animal trampling. The descriptive statistics of the artifact scatters are detailed in Figure 53.

This chapter is divided into two sections. The first section gives descriptive taphonomic histories of each of the nine experimental scatters. The second section uses data from these scatters to test hypotheses pertaining to artifact movement and morphological change outlined in Section 2.3.

4.2. River flow summary

River discharges were gauged by the Environment Agency below the site at the Threlkeld gauging station. The discharge values of the Glenderamackin were highest in December 2011 with an average monthly discharge of $7.86m^3s^{-1}$ (Figure 54). The lowest average monthly discharge was $0.88m^3s^{-1}$ observed in June 2011 (Figure 54).

4.3. Scatter histories

This section describes the taphonomic histories of Scatters 1–9 from their insertion in February 2011 to their recovery in July 2012. In the following subsections, scatters are described first at insertion and followed up with observations at each subsequent monitoring period. All 9 sites were exposed to river flow between each monitoring period as evidenced by a combination of personal observation, regular deposition of fluvial sandy silts, and gauging station readouts indicating that river levels exceeded artifact heights. During the course of this experiment, artifact recovery was often hampered by burial and thick vegetation. Consequently, artifacts were frequently not recovered in one monitoring period, only to be recovered in later phases or upon excavation.

In this first section, bootstrapped paired t-tests were used to compare mean artifact horizontal and vertical movements between monitoring periods. In addition, bootstrapped independent sample t-tests were used to compare the primary metrics of recovered and unrecovered artifact at the end of the project. Welch's corrections are reported where the assumption of equal variance was not met. *P*-values are reported significant at 0.01 to reduce inflation error of the

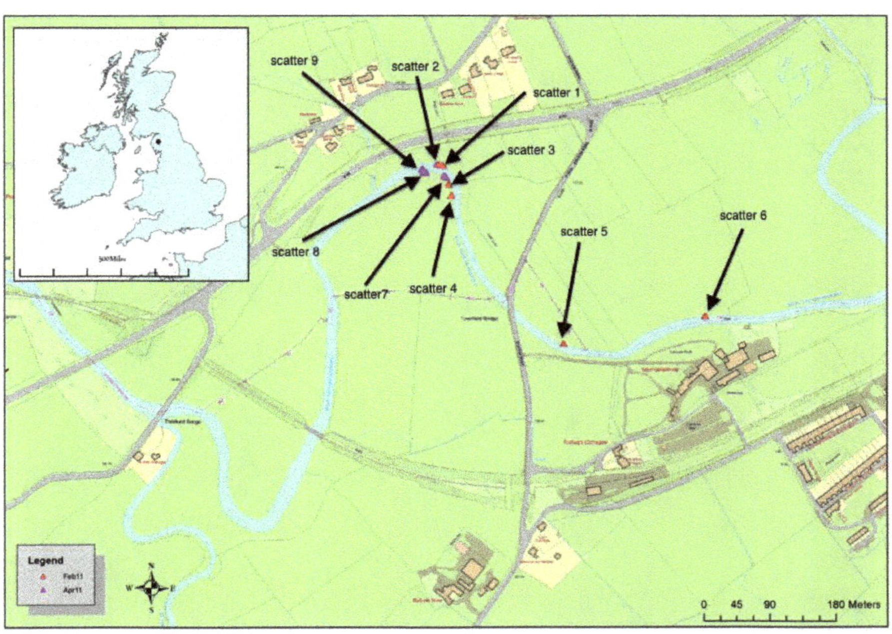

Figure 52: The project area of the fluvial experiments showing the initial insertion points of artifacts in February and April 2011. Note: the river flows from east to west (Figure is the same as Figure 29 and repeated for the reader's convenience).

Figure 53: Descriptive statistics including primary metrics for the nine experimental scatters. Note: Maximum width, and thickness are for flakes and handaxes only.

Scatter	*N*	Handaxe (*N*)	Core (*N*)	Flake (*N*)	Length (mm)	Width (mm)	Thickness (mm)	Weight (g)
1	50	1	1	48	62.99 ±19.32	40.17 ±12.17	11.83 ±5.28	34.29 ±59.47
2	50	2	0	48	58.60 ±19.93	40.42 ±13.19	9.82 ±7.08	27.49 ±55.56
3	50	0	0	50	77.45 ±23.94	54.35 ±19.52	17.57 ±12.73	75.05 ±96.46
4	50	0	0	50	78.14 ±29.14	53.40 ±26.50	19.01 ±15.86	133.02 ±246.18
5	50	0	5	45	87.95 ±19.66	62.43 ±17.29	23.17 ±11.01	136.30 ±113.98
6	50	0	1	49	81.69 ±19.57	55.04 ±16.87	24.33 ±23.59	84.01 ±93.97
7	50	0	0	50	65.08 ±15.95	44.33 ±12.80	14.58 ±11.30	30.13 ±23.05
8	50	0	0	50	46.53 ±10.92	33.06 ±7.35	8.38 ±3.54	11.56 ±12.56
9	56	0	0	56	47.14 ±11.38	33.30 ±8.05	9.07 ±4.71	20.33 ±28.77

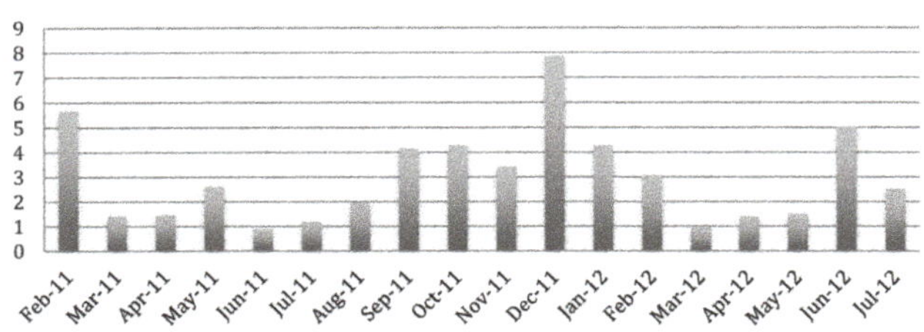

Figure 54: Average monthly discharges (m^3s^{-1}) discharge of the River Glenderamackin at the Threlkeld gauging station from Feburary 2011 to July 2012.

multiple t-tests used (Field 2009: 348). It should be noted however that the standard *p*-value of 0.05 would have resulted in nearly identical results.

4.3.1. Scatter 1

Scatter 1 (Artifacts 1–50) was located at the edge of a northern erosion bank of the channel at the beginning of an outside bend of the river, approximately 50cm above the channel bed (Figure 55 and Figure 56). The scatter was composed entirely of flakes (Figure 53) except for a single ovate handaxe (ID=1) and a single Mode 1 style core (ID=3). Mean height at insertion was 135.55±0.44MAOD (Figure 84).

April 2011

After insertion, the site was exposed to river flow and a fine layer of sandy silts (<1cm) was deposited on top of the artifacts. 40 artifacts were recovered (80% of the original

Figure 55: Scatter 1 at the first recovery period in April 2011; western viewpoint. ***Note the artifacts in the lower right hand corner behind the small gravel/earth island.***

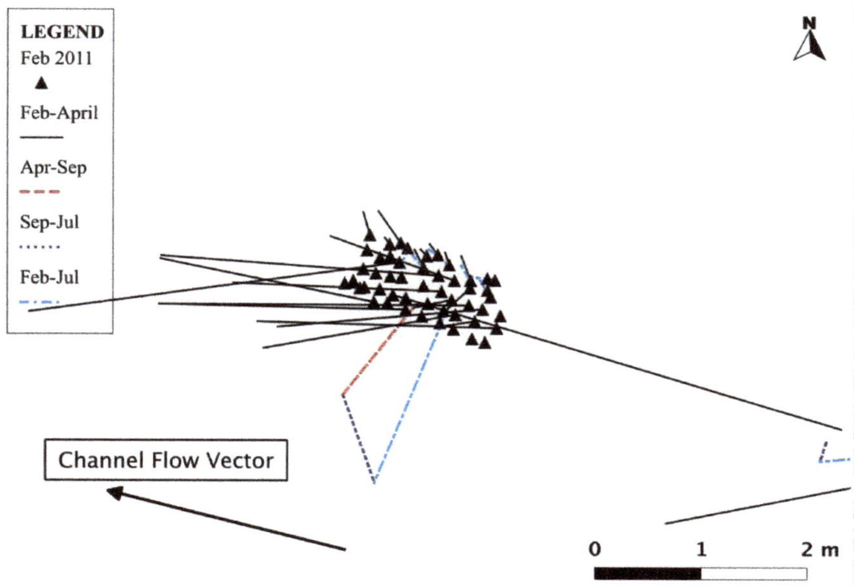

Figure 56: Artifact horizontal movement in Scatter 1 from insertion in February 2011 to excavation in July 2012.

scatter), at an average of 0.79±0.76m from their insertion locations. At least 17 artifacts had been transported off the grassy erosion bank into the gravel bed of the stream several meters away from their original context resulting in an average artifact height loss of 48cm (Figure 84; $t(38)=6.096$, $p<0.001$). These artifacts were found behind a small 'island' (2.6m x 1.3m) of clasts (approximately 20cm in b-axis) and grassy earth (Figure 55). This feature may have protected artifacts from further transport in the channel as the stream flow was observed to be lower in this area. Four of these artifacts traveled opposite to the direction of river flow suggesting that slope processes and runoff had influenced their redeposition. One other artifact (ID-39) was located in a small rivulet 1.6m downstream from its original location. Artifact long-axis orientations had a mean vector of 135°±52° (with 0° being the upstream direction of river flow), and demonstrated statistical isotropy (Rayleigh's, Watson's and Kuiper's tests; $p>0.05$; Appendix 3).

September 2011

Sometime before the second monitoring period, the river further inundated the site and more artifacts were removed from the scatter. The site was overgrown with tall (<15cm) grass. 18 artifacts were recovered (36% of the original scatter), transported an average of 0.47±0.72m from their previous locations. This was an average of 32cm less than the previous monitoring period, though the difference was statistically insignificant ($t(14)=0.332$, $p=0.75$). At least 10 artifacts had been completely or partially buried in the northeastern part of the site by sandy silts (>1cm). Mean recovered artifact height increased by 9cm (Figure 84). This was a statistically insignificant difference from the previous monitoring period ($t(16)=2.22$, $p<0.041$). Artifact long axes orientations had a mean vector of 96°±54°, and demonstrated isotropy (Rayleigh's, Watson's and Kuiper's tests; $p>0.05$; Appendix 3).

July 2012

At the final monitoring, the insertion site was buried in 8 to 12cm of fluvial sandy silts and artifacts were excavated in an approximate 4 square meter area over the original site. 12 artifacts were recovered (24% of the original scatter), transported an average of 0.10±0.10m from their previous location. This was a statistically similar distance compared to the previous monitoring period ($t(5)=3.12$, $p=0.026$). Mean recovered artifact height increased by 19cm (Figure 84), because the lower artifacts in the gravel bed were not recovered, however this result was statistically insignificantly different from the previous monitoring period ($t(6)=0.306$, $p=0.770$).

Long axes artifact orientation showed a mean orientation vector of 74°±67° (with 0° being the upstream direction of river flow) and a mean dip vector of 8°±10°, and demonstrated statistical isotropy (Rayleigh's, Watson's and Kuiper's tests; $p>0.05$; Appendix 3).

Unrecovered artifacts' sizes were statistically indistinguishable from recovered artifacts suggesting that recovery was not size dependent ($p>0.01$; Figure 57). Recovered artifacts showed no damage, as was also evidenced by the fact that there was no change to their weight.

Figure 57: Comparison of recovered and unrecovered artifacts from Scatter 1 in July 2012.

	Recovered *N*=12	Unrecovered *N*=38
Length (mm)	64.89±18.47	60.32±15.38
Width (mm)	40.70±11.40	39.05±11.27
Thickness (mm)	14.18±4.64	11.07±5.30
Weight (g)	33.38±26.86	25.59±37.96

*Rows marked *indicated a significant bootstrapped t-test, p<0.01*

Figure 58: Scatter 2 on February 6, 2013 at insertion; southern viewpoint across the river.

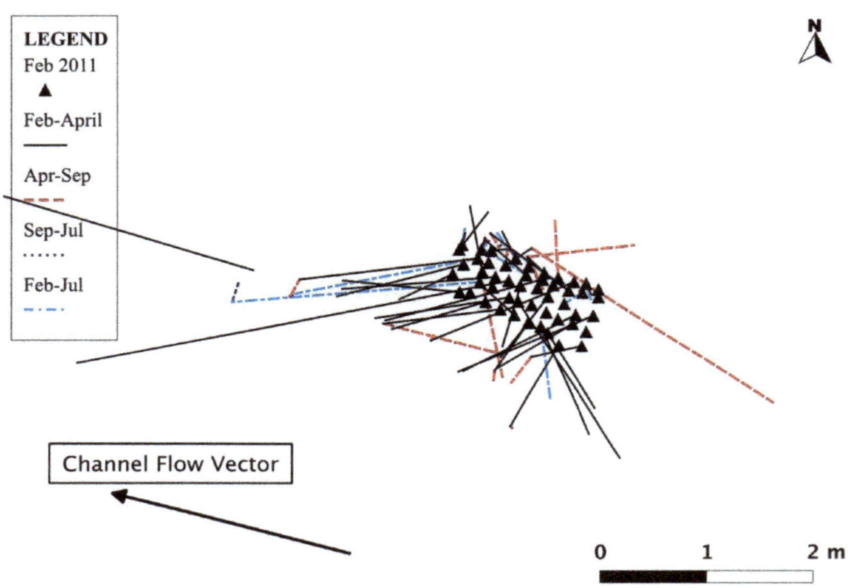

Figure 59: Artifact horizontal movement in Scatter 2 from insertion in February 2011 to July 2012.

4.3.2. Scatter 2

Scatter 2 (Artifacts 51–100) was located at the edge of a northern erosion bank several meters upstream from Scatter 1 (Figure 58 and Figure 59). The scatter was composed of 48 flakes and two handaxes (ID=74, 99; Figure 53). Mean height at insertion was 135.54±0.05mAOD (Figure 84).

April 2011

Similar to Scatter 1, Scatter 2 was fluvially disturbed as evidenced by the accumulation of fluvial sandy silts on the site. 24 artifacts were recovered (48% of the original scatter), transported downstream an average of 1.24±1.28m. Many artifacts were redistributed several meters downstream. Nine of these, including an ovate handaxe (ID=74), were recovered within the active channel. Three other artifacts were recovered in a small rivulet (10cm across). Mean artifact height decreased 33cm (Figure 84) since insertion, resulting from the movement of nine artifacts off the erosion bank and into the channel bed ($t(23)=3.54$, $p<0.002$) that connected the bank and the channel. Orientation of artifact long axes showed a mean vector of 75°±55°, and demonstrated statistical isotropy (Rayleigh's, Watson's and Kuiper's tests; $p>0.05$; Appendix 3).

September 2011

At the second monitoring period, the entire site was covered with a fine layer of sandy silts (>1cm) and artifact recovery was low ($n=2$). None of the artifacts recovered in the channel bed in April 2011 were still present, and they had

Figure 60: Comparison of recovered and unrecovered artifacts from Scatter 2 at recovery in July 2012.

	Recovered n=9	Unrecovered n=41
Length (mm)	64.00±18.69	57.41±20.21
Width (mm)	43.74±14.55	39.69±12.96
Thickness (mm)	11.98±6.80	9.34±7.13
Weight (g)	65.69	26.34±57.48

*Rows marked *indicated a significant bootstrapped t-test, $p<0.01$*

possibly been transported downstream out of the project area. Only two artifacts were recovered (4% of the original scatter). One of them was the handaxe (ID=74), which had been transported 1.07m from its previous location (Figure 84).

July 2012

At the final monitoring, the entire insertion site was buried under approximately 10cm of sandy silts. 4 square meters were excavated over the site. Nine artifacts from the original scatter were recovered (18% of the original scatter). Eight of these artifacts had not been monitored since their insertion as they had been buried before the first monitoring period. Handaxe 74 was recovered behind the same grassy island described in Scatter 1, having traveled 0.90m since September 2011. Mean recovered artifact height was 135.32±37cm. Orientation of artifact long axes showed a mean vector of 134°±42°, while artifact dips showed a mean vector of 21°±8°, neither demonstrating a directional bias (Rayleigh's, Watson's and Kuiper's tests; $p>0.05$; Appendix 3). Comparisons of primary metrics showed that while unrecovered artifacts were consistently smaller in all dimensions, they were not statistically significantly different (Figure 60), suggesting that artifact recovery from the site was not related to size.

Figure 61: Artifact 74 (ovate handaxe) after a minimum transport of 2.36m along the gravel bed.

No recovered artifacts were damaged as was also evidenced by a lack of change to their weight. Examination of handaxe 74 did not show any appreciable damage, also evidenced by the largely intact orange paint (Figure 61).

4.3.3. Scatter 3

Scatter 3 (Artifacts 101–150) was located at the edge of a low-lying southern erosion bank (Figure 62 and Figure 63). The scatter was composed entirely of flakes (Figure 53). Mean height at insertion was 135.98±0.03mAOD (Figure 84).

April 2011

At the first monitoring period, 49 artifacts were recovered (98% of the original scatter), at an average of 0.15±0.20m from their insertion locations. Many of the artifacts had

Figure 62: Scatter 3 on April 7, 2013; western viewpoint.

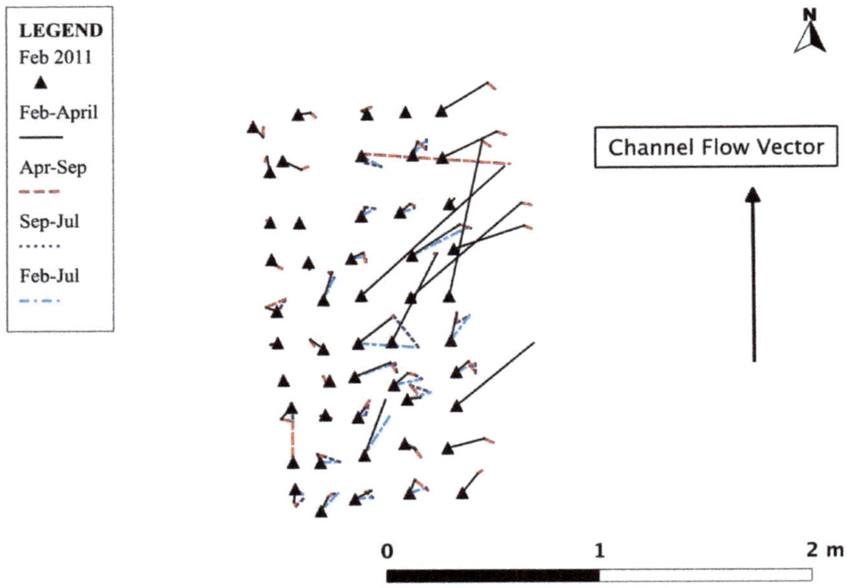

Figure 63: Artifact horizontal movement in Scatter 3 from insertion in February 2011 to July 2012.

moved northeast, towards the channel in the general direction of the stream flow. Mean artifact height increased significantly by 3cm (Figure 84; $t(47)=0.49$, $p=0.002$). This was possibly attributed to their movement to higher local topographies as they moved downstream. Orientation of artifact long axes showed a mean vector of 75°±55° with no directional bias (Rayleigh's, Watson's and Kuiper's tests; $p>0.05$; Appendix 3).

September 2011

At the second monitoring period, artifacts were overgrown by tall grasses (<10cm) and were surrounded by sandy silts (<1cm). Cattle were witnessed trampling the site. 45 artifacts were recovered (90% of the original scatter), transported an average of 0.06±0.10m. This was a statistically significant average decrease of 9cm from the previous monitoring period ($t(44)=3.53$, $p=0.001$). Artifact height decreased 2cm from the previous monitoring period (Figure 84; $t(44)=4.54$, $p<0.001$). This decrease in movement from April may reflect trampling; the pushing of the artifacts into the ground may have inhibited their lateral movement. Orientation of artifact long axes showed a mean vector of 99°±47°, and demonstrated an isotropic orientation (Rayleigh's, Watson's and Kuiper's tests; $p>0.05$; Appendix 3).

July 2012

At the final recovery, artifacts were covered in 8 to 12cm of sandy silts and the site was overgrown with tall grass (<10cm). 4 square meters were excavated over the site. 24 artifacts were recovered (48% of the original scatter), transported an average of 0.05±0.04m. This was a statistically insignificant difference from the previous monitoring period ($t(22)=0.861$, $p=0.43$). Primary metrics comparisons showed that unrecovered artifacts were not statistically significantly different from recovered artifacts (Figure 64), suggesting that recovery was not related to size. Mean height of the artifacts increased by 4cm (Figure 84), a statistically significant difference from the previous monitoring period ($t(22)=3.55$, $p=0.001$). It is not entirely clear what the increase in height may reflect, however it could be the result of a post-trampling soil rebound or reflect vegetation regrowth (Belshaw 1960). Artifact long axes orientations showed a mean vector of 93°±56° while dips showed a mean vector of 8°±13°, both demonstrated an isotropic orientation (Rayleigh's, Watson's and Kuiper's tests; $p>0.05$; Appendix 3).

Examination of the recovered artifacts showed damage to four different artifacts (17% of the original scatter). Two artifacts (8% of the original scatter), numbers 135 and 138 had *snap fractures* or bending fractures that resulted in a clean breaks across the side of the artifacts (Fischer *et al.* 1984) resulting in losses of 0.7 and 0.8 grams respectively. Artifacts 106 and 118 both had *impact notches,* single smooth semi-circular, unretouched notches (Lombard and Pargeter 2008; Yaroshevich *et al.* 2010; approximately 10mm along the edge) removed from their edges (Figure 65).

4.3.4. Scatter 4

Scatter 4 (Artifacts 151–200) was a sister site to Scatter 3, located 13m away at the edge of the same low-lying southern bank of the river (Figure 66 and Figure 67). The scatter was composed entirely of flakes (Figure 53). Mean height at insertion was 135.58±0.05mAOD (Figure 84).

April 2011

Most artifacts (n=49; 98%) were recovered having been slightly transported (an average of 0.08±0.03m), generally in a downstream direction towards the river. Mean height of the artifacts increased by 1cm (Figure 84), a statistically

Figure 64: Comparison of recovered and unrecovered artifacts from Scatter 3 at recovery in July 2012.

	Recovered n=24	Unrecovered n=26
Length (mm)	83.28±26.7	72.28±20.30
Width (mm)	58.30±18.05	48.50±16.13
Thickness (mm)	18.26±10.10	16.07±14.29
Weight (g)	87.38±85.61	50.65±78.13

*Rows marked *indicated a significant bootstrapped t-test, p<0.01*

insignificant ($t(47)=1.461$, $p=0.159$) difference from their insertion.

Orientation of artifact long axes showed a mean vector of 86°±46°, and demonstrated isotropic orientation (Rayleigh's, Watson's and Kuiper's tests; $p>0.05$; Appendix 3).

September 2011

At the second monitoring period, artifacts were overgrown by tall grasses (>10cm) and a layer of sandy silts was found on site. Cattle had also been witnessed trampling the site. 38 artifacts were recovered (76% of the original scatter), transported an average of 0.09±0.22m from their previous locations. This was a statistically insignificant difference from the previous monitoring period ($t(37)=0.726$, $p=0.719$).

Mean artifact height increased by 1cm (Figure 84). This was a statistically insignificant difference from the previous monitoring period ($t(37)=1.65$, $p<0.107$). Orientation of artifact long axes showed a mean vector of 103°±49°, and demonstrated isotropy (Rayleigh's, Watson's and Kuiper's tests; $p>0.05$; Appendix 3).

Figure 65: An example of an impact notch removed by trampling (Artifact 118).

July 2012

At the final recovery, artifacts were covered with 8 to 12cm of sandy silts and tall grass (10cm). 36 artifacts were recovered (72% of the original scatter), transported

Figure 66: Scatter 4 at insertion in February 2011; southern viewpoint.

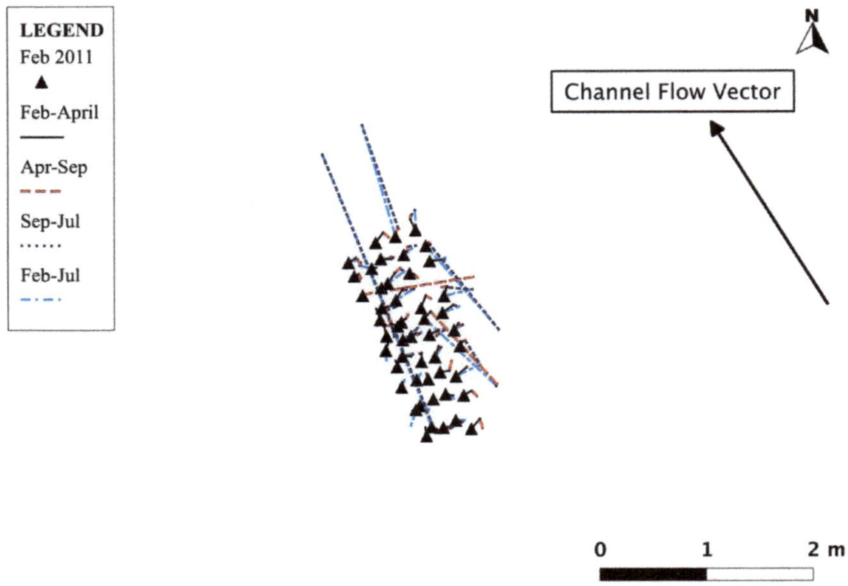

Figure 67: Artifact horizontal movement in Scatter 4 from insertion in February 2011 to July 2012.

an average of 0.28±0.61m. This was a statistically insignificant difference from the previous monitoring period ($t(24)=1.62$, $p=0.119$).

Unrecovered and recovered artifact primary metrics were not statistically significantly different ($p>0.01$; Figure 68), suggesting that artifact recovery from the site was not related to size. Mean artifact height dropped by 3cm (Figure 84). This was a statistically insignificant ($t(24)=1.74$, $p=0.095$) difference from the previous monitoring period.

Orientation of artifact long axes showed a mean vector of 94°±54°, and demonstrated an isotropic orientation for Kuiper's test ($p>0.05$; Appendix 3) but Rayleigh's and Watson's tests demonstrated anisotropy. It is unclear what the reason for this was. Artifact orientations are notoriously difficult to detect, especially for lower sample sizes (Fisher 1995). The issue of orientation will be further discussed in Section 4.4.4.

Figure 68: Comparison of recovered and unrecovered artifacts from Scatter 4 at recovery in July 2012.

	Recovered n=36	Unrecovered n=14
Length (mm)	74.87±27.40	87.97±33.45
Width (mm)	51.07±26.89	60.10±26.30
Thickness (mm)	17.07±16.20	24.34±14.83
Weight (g)	108.84±244.36	198.131±248.76

** Rows marked *indicated a significant bootstrapped t-test, p<0.01*

Figure 69: Comparison of the pre-insertion primary metrics between damaged and all artifacts in Scatter 4.

	Scatter 4 n=50	Damaged n=5
Length (mm)	78.14±29.14	57.00±7.01
Width (mm)	53.40±26.50	40.20±10.85
Thickness (mm)	19.01±15.86	5.79±2.03
Weight (g)	133.02±246.18	9.5±2.2

Artifact dips showed a mean vector of 20°±13°, with the distribution demonstrating no clear directional bias (Rayleigh's, Watson's and Kuiper's tests; $p>0.05$; Appendix 3). Examination of the recovered artifacts showed damage to eight artifacts. Five artifacts (ID=178, 184, 185, 188 and 190) had at least one transverse snap, resulting in an average weight loss of 23%. Three of the biproducts of these breakages were recovered *in situ*. All of them were less than 2cm in their maximum axes (Figure

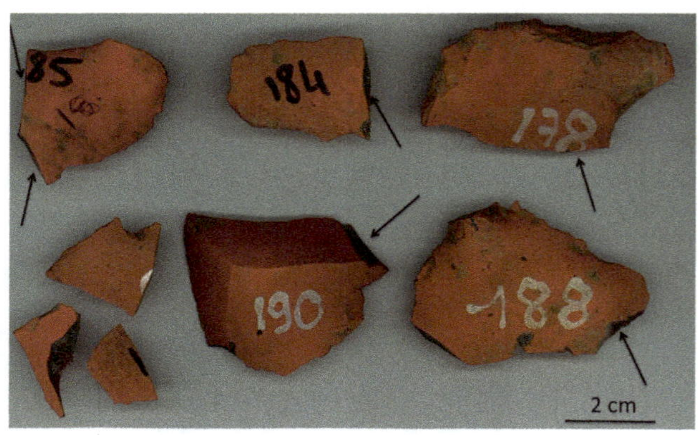

Figure 70: Artifacts 178, 184, 185, 188, 190 and their recovered broken pieces demonstrating the results of cattle trampling.

Figure 71: Artifacts 172, 173, and 187 showing damage attributed to trampling.

Figure 72: Scatter 5 at insertion in February 2011; northern viewpoint.

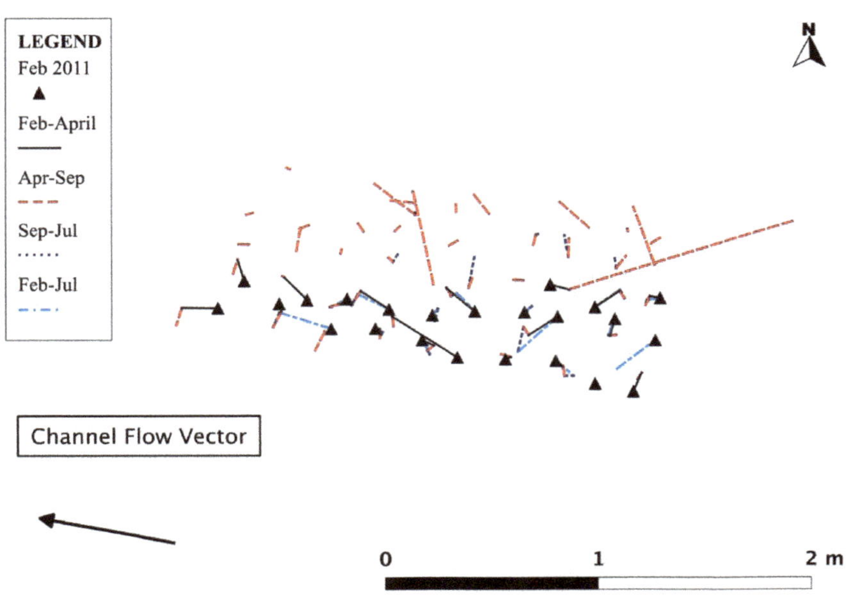

Figure 73: Artifact horizontal movement in Scatter 5 from insertion in February 2011 to July 2012.

70). Compared to the entire scatter, broken artifacts were smaller in all primary metrics, but particularly in original weight and thickness (Figure 69). Three other artifacts demonstrated other forms of damage (Figure 71). Artifact 187 had a small impact notch removed from the side of the artifact. Artifact 172 had a small chip taken out of its tip, while artifact 173 had a large impact notch (21mm in width; Figure 71).

4.3.5. Scatter 5

Scatter 5 (Artifacts 201–250) was located at the edge of a low-lying northern bank on a straight part of the channel (Figure 72 and Figure 73). The site was composed of 45 flakes except for five Mode 1 cores (ID=214, 216, 221, 223, 247; Figure 53). Mean height at insertion was 136.59±0.06mAOD (Figure 84). Due to technical difficulties, artifact locations were not recorded at the initial visit for artifacts 224–250, however they were recorded at subsequent monitoring periods.

April 2011

At the first monitoring period, the site was covered in dense tall grass (>15cm). 46 artifacts were recovered (92% of the original scatter), and recorded artifacts had been transported an average of 0.10±0.08m in no clear direction.

Mean height of the artifacts was 136.59±0.04cm (Figure 84). This was a statistically insignificant difference from their insertion ($t(20)=-2.03$, $p=0.055$). Orientation of artifact long axes showed a mean vector of 104°±60°, and

Figure 74: Comparison of recovered and unrecovered artifacts from Scatter 5 at recovery in July 2012.

	Recovered n=21	Unrecovered n=29
Length (mm)	84.32±15.57	90.57±22.04
Width (mm)	58.73±16.42	65.03±17.70
Thickness (mm)	22.74±8.56	23.48±12.60
Weight (g)	139.98±118.00	123.30±139.35

*Rows marked *indicated a significant bootstrapped t-test, p<0.01*

demonstrated isotropic orientation (Rayleigh's, Watson's and Kuiper's tests; p>0.05; Appendix 3).

September 2011

At the second monitoring period, artifacts were covered in tall grass and had been trampled by sheep as evidenced by feces on the site. 46 artifacts were recovered (92% of the original scatter), transported an average of 0.09±0.17m. This was a statistically insignificant difference from the previous monitoring period (t(19)=0.945, p=0.945).

Mean artifact height increased by 1cm (Figure 84). This was a statistically insignificant difference from the previous monitoring period (t(44)=-5.28, p=0.600). Orientation of artifact long axes showed a mean vector of 108°±51°, and demonstrated an isotropic orientation (Rayleigh's, Watson's and Kuiper's tests; p>0.05; Appendix 3).

July 2012

At the final recovery, artifacts were covered with 8 to 12cm of sandy silts and tall grass. 21 artifacts were recovered (42% of the original scatter), and were transported an average of 0.06±0.04m. This was a statistically insignificant difference from the previous monitoring period (t(18)=0.619, p=0.602). Primary metrics comparisons showed that, while unrecovered artifacts were consistently larger in most dimensions, they were statistically similar (Figure 74), suggesting that artifact recovery at the site was not related to size. Mean artifact height increased by 3cm (Figure 84), a statistically insignificant difference from the previous monitoring period (t(18)=1.23, p=0.234).

Figure 75: Artifact 207 showing damage attributed to trampling.

Artifact long axes orientations showed a mean vector of 111°±40°, and demonstrated anisotropy when measured by Rayleigh's and Kuiper's tests, however Watson's test was insignificant (p>0.05; Appendix 3). It is unclear what the reasons for these findings are, though they should be treated with caution as the sample size is small. As indicated earlier, scatter isotropy will be discussed further in Section

Figure 76: Scatter 6 in April 2011; northern viewpoint.

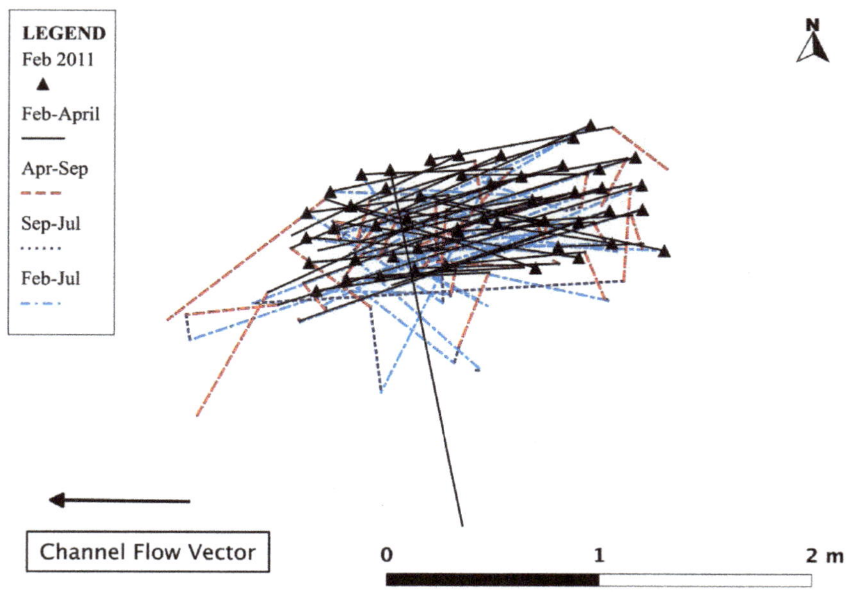

Figure 77: Artifact horizontal movement in Scatter 6 from insertion in February 2011 to July 2012.

4.4.4. Artifact dips showed a mean vector of 10°±16°, and demonstrated no statistical bias (Rayleigh's, Watson's and Kuiper's tests; $p>0.05$; Appendix 3).

Recovered artifacts showed damage to only one artifact (ID=207), which had a small impact notch removed from the ventral side of the flake. The notch was approximately 10mm in size (Figure 75).

4.3.6. Scatter 6

Scatter 6 (Artifacts 251–300) was located at the edge of a steep northern bank of the river on a straight part of the channel on a steep slope, approximately 20° (Figure 76 and Figure 77). The site was composed entirely of flakes except for a single core (ID=251; Figure 53). Mean height at insertion was 137.61±0.06mAOD (Figure 84).

April 2011

Artifacts were found stochastically transported several meters downstream on the riverbank. One artifact was found lodged directly below the site in the gravel bed (ID=292). 42 artifacts were recovered (84% of the original scatter), transported an average of 0.89±0.29m. Mean heights of the artifacts were 137.60±0.11cm (Figure 84), a statistically insignificant ($t(45)=0.382$, $p=0.722$) difference from their insertion. Artifact long axes orientations showed a mean vector of 97°±55°, and demonstrated isotropy (Rayleigh's, Watson's and Kuiper's tests; $p>0.05$; Appendix 3).

September 2011

At the second monitoring period, artifacts were covered in tall grass and were trampled by sheep, as evidenced by feces on the site. 36 artifacts were recovered (72% of the original scatter), transported an average of 0.26±0.22m. This was a statistically significant difference from the previous monitoring period ($t(34)=11.202$, $p<0.001$). Mean height of the artifacts increased by 1cm (Figure 84), a statistically insignificant ($t(34)=0.956$, $p=0.346$) difference from the previous monitoring period. Orientation of artifact long axes showed a mean vector of 85°±47°, and the distribution was statistically isotropic (Rayleigh's, Watson's and Kuiper's tests; $p>0.05$; Appendix 3).

July 2012

At the final recovery, the site had been affected by river flow as it was covered with 8 to 12cm of sandy silts and tall grass. 38 artifacts were recovered (76% of the original scatter), transported an average of 0.07±0.08m. This was a statistically significant ($t(27)=4.88$, $p=0.002$) difference from the previous monitoring period. Primary metrics of unrecovered artifacts were not statistically significantly different from recovered artifacts (Figure 78). This suggested that artifact transport from the site was not related to size. Mean height of the artifacts dropped by 3cm (Figure 84). This was a statistically significant ($t(29)=4.39$, $p<0.001$) difference from the previous monitoring period. This may have been due to slight movement downhill.

Orientation of artifact long axes showed a mean vector of 88°±51° (with 0° being the upstream direction of river flow) and a dip of 19°±16°, both statistically isotropic (Rayleigh's, Watson's and Kuiper's tests; $p>0.05$; Appendix

Figure 78: Comparison of recovered and unrecovered artifacts from Scatter 6 at recovery in July 2012.

	Recovered n=38	Unrecovered n=12
Length (mm)	81.65±20.61	81.84±15.93
Width (mm)	54.71±17.30	56.28±15.90
Thickness (mm)	25.56±26.17	19.69±7.83
Weight (g)	74.57±71.29	119.90±153.02

*Rows marked *indicated a significant bootstrapped t-test, $p<0.01$*

Figure 79: Artifact 283, a notched artifact from Scatter 6.

3). Examination of the recovered artifacts showed no damage, evidenced by no change to their primary metrics except for artifact 283, which had an impact notch removed from the side of the artifact, probably a result of trampling (Figure 79).

4.3.7. Scatter 7

Scatter 7 (Artifacts 301–350) was located at the inside bend of the river on a depositional sand and gravel bar (Figure 80, Figure 81, Figure 82). The scatter was composed entirely of flakes (Figure 53). Mean height at insertion was 134.60±0.11mAOD (Figure 84).

September 2011

Water had coursed down the channel and winnowed artifacts on the southern gravel bar between 8cm and 61m downstream. 25 artifacts were recovered (50% of the original scatter), transported an average of 7.37±14.02m downstream. Mean height of the artifacts decreased by 6cm (Figure 84) though this difference was statistically significant ($t(24)=3.05$, $p=0.005$) from their insertion height, due to artifacts traveling downstream. Orientation of artifact long axes showed a mean vector of 93°±50°, and the distribution was statistically isotropic (Rayleigh's, Watson's and Kuiper's tests; $p>0.05$; Appendix 3). Examination of the recovered artifacts in the field showed nearly continuous small (approximately 1mm in all dimensions) bifacial *feather terminating* (Crabtree 1972; Ho Ho Committee 1979) fracturing across the lateral margins (*micro-flaking;* i.e., Hosfield and Chambers 2005c) of two artifacts (ID=321, 325). One artifact (ID=334) had been transversely broken (Figure 83).

July 2012

At the final recovery period (July 2012), no artifacts were recovered though the gravel bar on which they were inserted appeared to have been largely unaltered otherwise.

Figure 80: The insertion location of Scatters 7–9 in April 2011; Northern viewpoint. The gravel patch in the center right part of the photograph is where Scatter 7 was located (circled). Scatters 8 and 9 insertion locations were submerged on this monitoring visit.

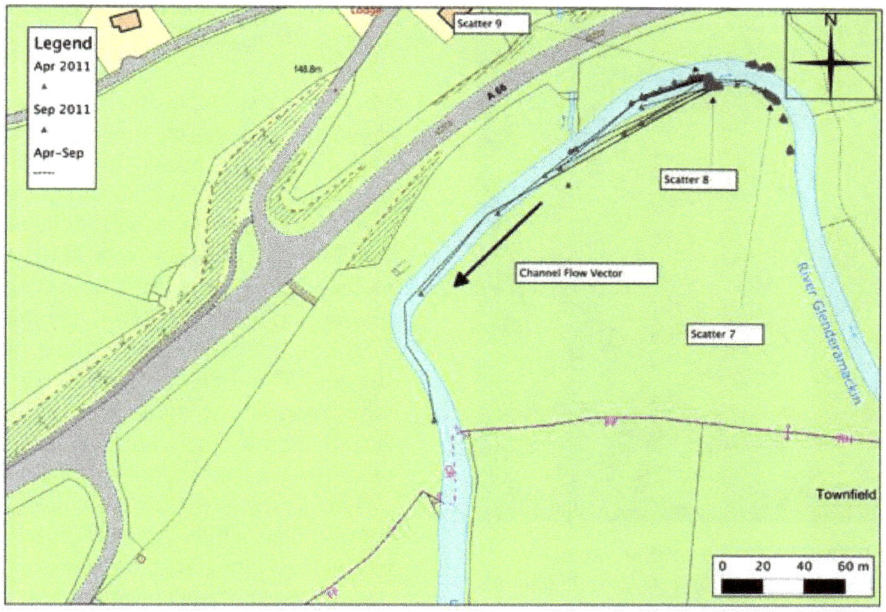

Figure 81: Artifact horizontal movement in Scatters 7–9 from insertion in April 2011 to September 2011.

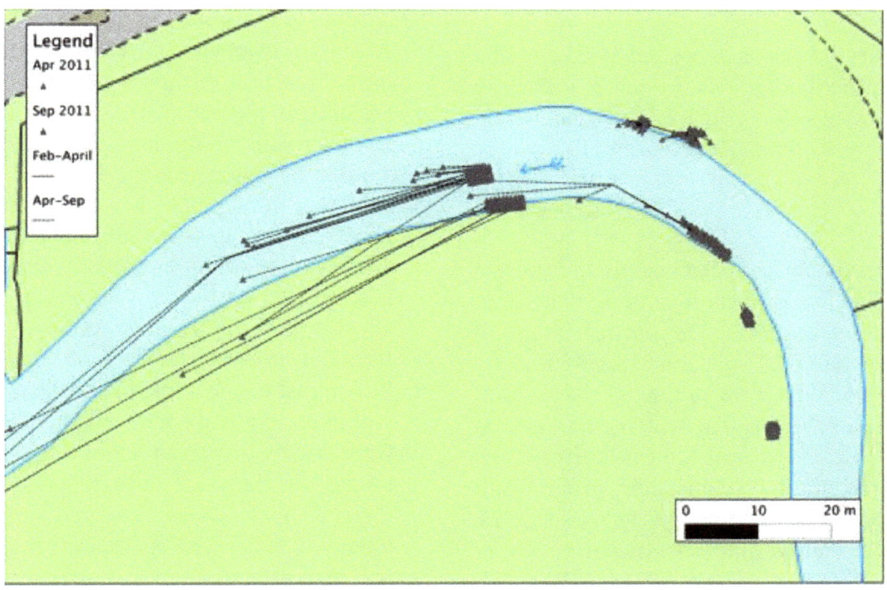

Figure 82: Artifact horizontal movement in Scatters 7–9 from insertion in April 2011 to September 2011 (at closer scale).

4.3.8. Scatter 8

Scatter 8 (Artifacts 351–400) was deposited in the middle of a gravel bar in April 2011 (Figure 80, Figure 81, Figure 82). The site was composed entirely of flakes (Figure 53). Mean height at insertion was 134.60±0.11mAOD (Figure 84).

September 2011

Between April and September 2011, artifacts were heavily transported downstream. 10 artifacts were recovered (20% of the original scatter), transported an average of 55.28±66.49m downstream. Two artifacts (364b, 373) were found 179 and 250m respectively downstream in small scour pits. Mean height of the artifacts was 134.26mAOD (Figure 84), a statistically insignificant difference from their insertion ($t(9)=1.94, p=0.083$). This was because most of the artifacts were recovered on the same bar downstream as their discard location. Orientation of artifact long axes showed a mean vector of 86°±70°, and demonstrated an isotropic orientation (Rayleigh's, Watson's and Kuiper's tests; $p>0.05$; Appendix 3). Artifacts' micro-flaking was noted in the field for five of the artifacts.

July 2012

At the final recovery period, no artifacts were recovered, though the site appeared to have been largely unaltered otherwise (Figure 81 and Figure 82).

Figure 83: Artifact 334 in September 2011 that had been snapped and exhibited micro-flaking after at least 35.63m of transport.

4.3.9. Scatter 9

Scatter 9 (Artifacts 401–456) was deposited at the end of a point gravel bar (Figure 80, Figure 81, Figure 82). The site was composed entirely of flakes (Figure 53). Mean height at insertion was 134.17±0.45mAOD (Figure 84).

September 2011

By September 2011, artifacts had been heavily transported downstream with all of the artifacts resting on the higher southern gravel bank of the river. One artifact that had been transported 123m (ID=417) was found approximately 25cm behind a large boulder (*c.* 50cm a-axis). 14 artifacts were recovered (25% of the original scatter), transported an average of 32.8±35.21m. Mean height of artifacts was 134.17±0.45mAOD (Figure 84), a statistically insignificant difference from their insertion ($t(13)=0.885$, $p=0.392$). Artifact long axes orientation showed a mean vector of 86°±70°, and demonstrated isotropy (Rayleigh's, Watson's and Kuiper's tests; $p>0.05$; Appendix 3). Examination of the recovered artifacts in the field showed micro-flaking to all of the artifacts except for one (Artifact 455) that had only been transported approximately 7m.

July 2012

At the final recovery period (July 2012), no artifacts were recovered though the site appeared to have been largely unaltered otherwise (Figure 81 and Figure 82).

4.3.10. Summary

The nine scatters placed in the River Glenderamackin showed different patterns of spatial and morphological change, dependent on their insertion location and taphonomic history. These results and taphonomic effects discussed above are summarized in Figure 84. A synthesis of these observations at the scatters demonstrates five key recurrent patterns:

1. Most artifacts deposited on the banks of the river (Scatters 1–6) stayed relatively *in situ*, as recovered artifacts did not experience major horizontal (<5m) or vertical (<50cm) movement. In general, vertical and horizontal movements did not differ statistically significantly between monitorings, suggesting that variations in mean or high flow events did not influence transport (Figure 84). This, combined with the often stochastic directions of artifact movement strongly suggests that transport may have been the result of other taphonomic processes such as slope processes, surface runoff, and trampling in some cases (Bertran *et al.* 2012; Lenoble 2005a; Lenoble *et al.* 2008).
   Two cases were an exception: Scatters 1 and 2 showed that, while slope processes may have played a part in their movement, there was greater initial average movement and re-deposition into the active channel zone (0.79 and 1.24m respectively; Figure 84), probably as a result of their location on the faster flowing outer bend of the river. Pieces were then likely lost in the channel and transported longer distances downstream. Artifact movements in Scatter 6 also showed greater initial transport distances (average=0.89m; Figure 84), possibly as a result of the steep inclination of the site, however values still did not generally exceed 1m.
2. Among bankside scatters, recovery levels consistently declined throughout the course of the experiment. However, by the end of the experiment recovered artifact sizes were statistically indistinguishable from unrecovered artifacts, suggesting that whatever the agent responsible for their removal, artifact winnowing from the site was not size dependant.
3. In contrast to the bankside artifacts, scatters deposited within the gravel-bedded stream channel (Scatters 7–9) were widely dispersed and transported as far as 250m. Except for the three farthest transported pieces, all of these artifacts were recovered on an inside channel point gravel bar. Transport along the gravel bed resulted in a high proportion (37%) of edge micro-flaking along the acute lateral edges of the artifacts.
4. Long-axis orientations and the dips of all artifacts were not clearly statistically clustered.
5. Some smaller artifacts (n=14) sustained damage while remaining relatively *in situ*. These were observed as snap fractures and impact notches to flakes and were likely the result of animal trampling as they were only observed among trampled scatters.

4.4. Analytical case studies

So far, this chapter has described nine different artifact scatters and observed how their spatial and morphological composition changed during the course of 18 months in the River Glenderamackin. The behaviors of these scatters were different, because of their different contexts and artifact types. Inter-scatter comparisons allow us to answer specific questions about lithic artifact scatter behavior in

Figure 84: Descriptive summary of the scatter taphonomy. Values are mean±SD. *indicates a significant change from the previous monitoring with a bootstrapped paired samples t-test of available pairs p<0.01. 'a' indicates either insufficient values or no valid pairs. † indicates estimated values. ‡ indicates a significant result of Rayleigh's, Kuiper's and/or Watson's tests (p<0.05): Values and Rose and Schmitt diagrams for these are presented in Appendix 3. In addition to symbols, all significant values (p<0.05) are bolded. Mean daily flows, high flows, and max river heights relate to periods between the last monitoring.

	Insertion				**First monitoring**							
	Feb 3–5, 2011				April 15–16, 2011							
	N/A				Mean daily flow=3.16m³/s							
	N/A				High flow=28.1m³/s							
	N/A				Max river height=137.20mAOD							
Scatter	*n (recovered)*	*Height (mAOD)*	*Orientation (0–180°)*	*Dip (0–90°)*	*n (recovered)*	*Transport distance (m)*	*Min transport distance (m)*	*Max transport distance (m)*	*Height (mAOD)*	*Orientation (0–180°)*	*Dip (0–90°)*	*Trampled*
1	50	135.53 ±0.06	55 ±44	0†	40	.79 ±.76[a]	.01	4.13	**135.04 ±0.54***	84 ±52	N/A	No
2	50	135.54 ±0.05	54 ±45	0†	24	1.24 ±1.28[a]	.03	4.42	**135.21 ±0.47***	75 ±55	N/A	No
3	50	135.98 ±0.03	37 ±45	0†	49	.15 ±.20[a]	.01	.92	**135.98 ±0.06**	76 ±50	N/A	No
4	50	135.58 ±0.05	56 ±44	0†	49	.08 ±.03[a]	.01	.13	135.59 ±0.04	86 ±46	N/A	No
5	50	136.59 ±0.06	51 ±46	0†	48	.10 ±.08[a]	.01	.37	136.59 ±0.04	104 ±60	N/A	No
6	50	137.61 ±0.06	53 ±45	0†	46	.89 ±.29[a]	.04	1.75	137.60 ±0.11	97 ±55	N/A	No
7	-	-	-	-	50	N/A			134.66 ±0.06	48 ±48	N/A	N/A
8	-	-	-	-	50	N/A			134.82 ±0.11	45 ±45	N/A	N/A
9	-	-	-	-	56	N/A			134.45 ±0.07	42 ±45	N/A	N/A
Total	300				412							0

	Second monitoring							
	September 1–2 2011							
	Mean daily flow=1.50m³/s							
	High flow=15.6m³/s							
	Max river height=137.41mAOD							
Scatter	*n (recovered)*	*Transport distance (m)*	*Min transport distance (m)*	*Max transport distance (m)*	*Height (mAOD)*	*Orientation (0–180°)*	*Dip (0–90°)*	*Trampled*
1	18	0.47 ±0.72	.02	2.73	135.13 ±0.57	96 ±54	N/A	No
2	2	N/A	1.07	1.07	N/A	N/A	N/A	No
3	45	**0.06 ±0.10**	0	.71	**135.96 ±0.05**	99 ±47	N/A	Cattle
4	38	0.09 ±0.22	0	1.06	135.60 ±0.07	103 ±49	N/A	Cattle
5	46	0.09 ±0.17	.01	1.01	136.60 ±0.05	108 ±51	N/A	Sheep
6	36	**0.26 ±0.22**	.01	.94	137.61 ±0.10	85 ±47	N/A	Sheep
7	25	7.37 ±14.02	.08	61.32	**134.60 ±0.11**	93 ±50	N/A	No
8	10	55.28 ±66.49	.21	250.10	134.26 ±0.78	86 ±70	N/A	No
9	14	32.80 ±35.21	4.99	123.55	134.17 ±0.45	90 ±53	N/A	No
Total	234							165

	Third monitoring								
	July 13–15, 2012								
	Mean daily flow=3.54m³/s								
	High flow=30m³/s								
	Max river height=137.76mAOD								
Scatter	*n (recovered)*	*Transport distance (m)*	*Min transport distance (m)*	*Max transport distance (m)*	*Height (mAOD)*	*Orientation (0–180°)*	*Dip (0–90°)*	*Trampled*	*Damage*
1	12	0.10 ±0.10	.02	.28	135.32 ±0.46	74 ±67	8 ±10	No	0
2	9	.90[a]	.90	.90	135.32 ±.37[a]	134 ±42	21 ±8	No	0
3	24	0.05 ±0.04	.01	.19	**136.00 ±0.03**	93 ±56	8 ±13	No	4
4	36	0.28 ±0.61	.02	2.90	135.57 ±0.06	94 ±54‡	20 ±13	No	8
5	21	0.06 ±0.04	.01	.15	136.63 ±0.04	111 ±40‡	10 ±16	No	1
6	38	0.07 ±0.08	0	.42	**137.58 ±0.08**	88 ±51	19 ±16	No	1
7	-	-	-	-	-	-	-	-	-
8	-	-	-	-	-	-	-	-	-
9	-	-	-	-	-	-	-	-	-
Total	140							0	14

fluvial settings. The goal of the following section is to statistically test outstanding hypotheses set forth in Section 2.6.1, as well as examine new ones, such as trampling, that were presented because of the unpredictable nature of the fluvial experiments.

Specifically, the questions addressed in the following analyses are:

- How do the metrics of artifacts, their insertion location, burial conditions, trapping within the gravel bed, and river discharge levels affect the spatial rearrangement of flaked stone assemblages over time (Sections 4.4.1 and 4.4.2)?
- How does artifact loss within the fluvial system affect our interpretation of lithic transport experiments (Section 4.4.3)?
- How does fluvial transport in a meandering gravel-bedded river affect the imbrication (orientation) and dip of lithic artifact long-axes (Section 4.4.4)?
- How does post-depositional trampling of artifacts affect the dip of artifacts in bankside fluvial settings (Section 4.4.5)?

- How does transport distance affect the characteristic micro-flaking of unretouched fluvially re-worked artifacts (Section 4.4.6)?
- Does artifact trampling by variously sized animals in bankside fluvial settings affect morphology of recovered artifacts (Section 4.4.7)?

4.4.1. Artifact transport distance and independent variables

The aim of this section is to determine the effect of artifact primary metrics (maximum length, maximum width, maximum thickness, and weight), scatter location, trampling, burial, artifact trapping, and river high flow and average flow events on artifact horizontal displacement. If artifact morphology and/or the aforementioned taphonomic variables (Bridgland and Harding 1993; Bunn *et al.* 1980; Schick 1986; 1987) affect artifact transport, then statistically significant differences should exist between the transport distance data for artifacts with variable primary metric conditions and/or starting conditions at the beginning of the experiment.

Formally stated, the hypotheses for this section were as follows:

> H_0=Artifact primary metrics, scatter location, trampling, burial, artifact trapping, and river high flow and average flow levels do not significantly affect artifact transport distances.
>
> H_1=Artifact primary metrics, scatter location, trampling, burial, artifact trapping, and river high flow and average flow levels significantly affect artifact transport distances.

Artifact transport distances under different conditions were compared using a Mixed Model (MIXED), a type of regression model that combines both repeated and non-repeated measures (refer to Section 3.5.2). Where interactions were significant, Bonferoni *post-hoc* tests assessed individual differences.

In this analysis, transport distance was recorded for 456 flakes between 0 and 4 times depending on recovery (Figure 85). Repeated measures of transport distances were nested in a different level so time was treated as the Level 1 predictor. Maximum length, maximum width, maximum thickness, weight, maximum/average river discharge, trampling, trapping, burial, and scatter locations were all treated as Level 2 predictors. All primary metrics and river discharge were treated as covariates (because they were continuous values), while the rest were considered fixed factors (because they were only fixed conditions). Artifact lithology (flint) was kept constant to reduce the effect of additional variables.

Mixed model assumptions are similar to those of a linear regression (Field 2005). The data collected in this analysis are appropriate for testing because all of the dependent and independent variables in this examination are unbounded, quantitative, or categorical.

One assumption of a mixed model is that there is no strong linear relationship between two predictor variables (multicollinearity). To evaluate this, a Pearson's correlation matrix was tabulated between predictor variables in this test (Appendix 2). The results showed that there were no highly correlated independent variables ($r>0.80$), indicating that the assumption of non-multicollinearity was met. Because MIXED is a parametric test, it is assumed that there are normally distributed errors. These errors are best assessed with a visual verification of data and when plotted out, they were approximately normally distributed (Figure 86).

Figure 87 reports the results of the mixed model. Figure 88 to Figure 91 show graphs of the statistically significant

Figure 85: List of experimental conditions for the transport distance analysis.

Dependent variable	Independent variables											Controlled variables
	Between groups factor (fixed factor)										Within groups factor (Repeated measure)	
Transport distance	*Maximum length*	*Maximum width*	*Maximum thickness*	*Weight*	*Max river discharge*	*Average river discharge*	*Trampling*	*Trapping*	*Burial*	*Scatter*	*Time*	*Artifact lithology*
Measured (1.0cm)	Measured (0.01mm)	Measured (0.01mm)	Measured (0.01mm)	Measured (0.1g)	Measure (0.001m^3/s)	Measure (0.001m^3/s)	None	No	No	1	Apr 2011	Flint
							Large animal (size 4)	By clast	Partially	2	Sep 2011	
							Medium animal (size 3)	By vegetation	Yes	3	Jul 2012	
										4		
										5		
										6		
										7		
										8		
										9		

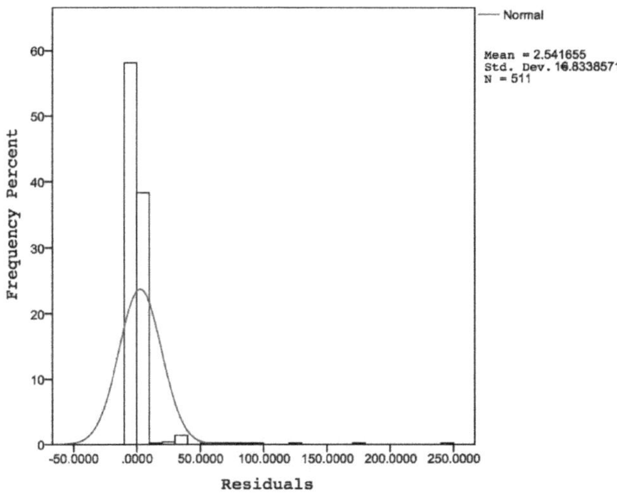

Figure 86: Plot of residuals for the MIXED procedure.

variables affecting artifact transport. All statistically significant effects are reported where $p<0.05$.

The results in Figure 87 show a statistically significant effect of scatter location, burial condition, maximum length, and maximum width on artifact transport distance ($p<0.05$). Thickness, weight, maximum/average river discharge, and trapping were statistically insignificant (Figure 87; $p>0.05$).

Figure 87: Estimates of the effects of independent variables in the transport distance analysis (see Figure 85 for definitions).

Variable	Numerator df	Denominator df	F	Sig.
Scatter location	**8**	**190.153**	**3.538**	**.001**
Maximum length	**1**	**182.671**	**9.210**	**.003**
Maximum width	**1**	**182.618**	**8.321**	**.004**
Maximum thickness	1	182.806	.057	.811
Weight	1	182.607	1.007	.317
Burial	**2**	**185.403**	**16.701**	**.000**
Trapping (vegetation, clasts)	2	196.955	.955	.387
Trampled	2	219.291	.058	.944
Average river flow	1	288.832	.003	.957
River high flow	1	303.693	.086	.770

Scatter location had a statistically significant effect on transport distances (Figure 87 and Figure 88; $p=0.001$). This was principally apparent in the contrast between bankside scatters (1–6) and channel scatters (7–9). A bootstrapped independent samples Welch's t-test comparing the mean transport distances of recovered artifacts in bankside scatters (Scatters 1–6; $M=0.159$m, $SD=0.30$) to artifacts deposited within the channel (Scatters 7–9; $M=27.88$m, $SD=48.86$) for the only comparable period (April to September 2011) showed they were statistically significantly different ($t(47)=-3.932$, $p<0.001$), confirming the results of the mixed model. It is likely that transport distances for the final monitoring period would have confirmed these findings (September 2011 to July 2011), however there

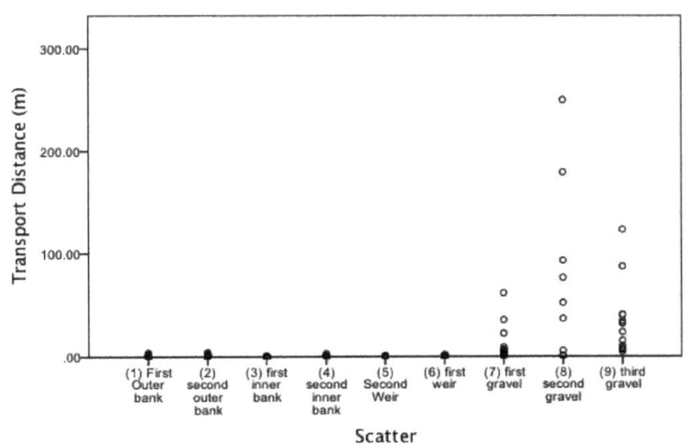

Figure 88: Transport distance v. scatter location; $p=0.001$. Refer to Figure 52 for locations.

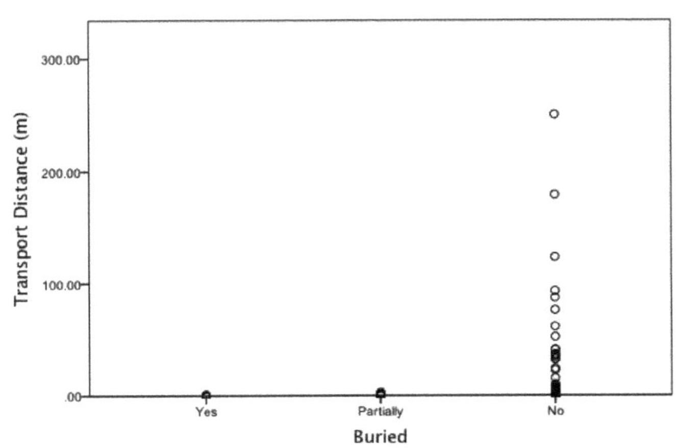

Figure 89: Transport distance v. burial $p<0.001$.

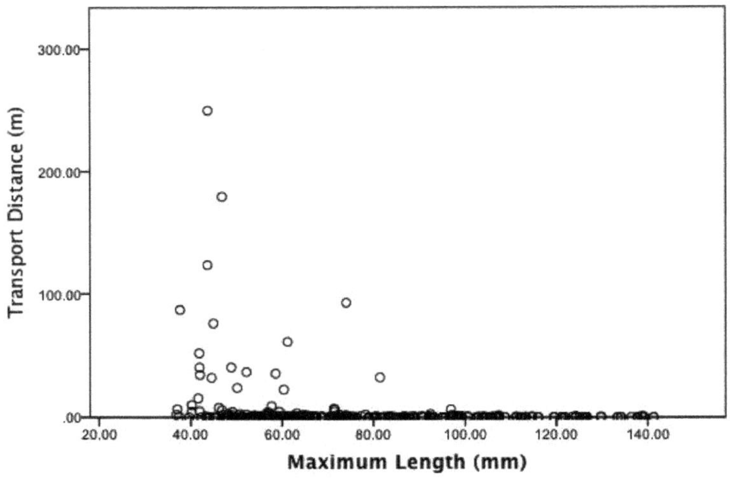

Figure 90: Transport distance v. maximum length; *p*=0.003.

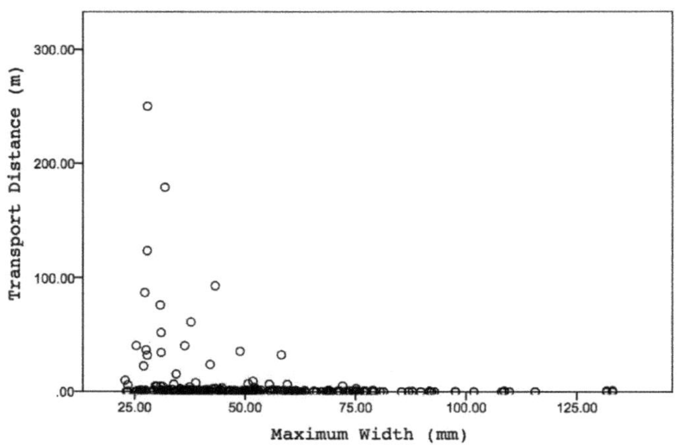

Figure 91: Transport distance v. maximum width; *p*=0.004.

were no values for transport distances among channel artifacts, presumably because they were transported out of the project area or buried within the channel.

Results also show a statistically significant effect of burial condition on transport distances (Figure 87 and Figure 89; $p<0.001$). Artifacts that were buried or partially buried exhibited low transport distances (<3m) while those that were unburied showed widely varying transport distances (between 0 and 250m; Figure 89). There were no artifacts buried at initial emplacement. The results report that maximum flake length had a statistically significant effect on transport distance, with shorter flakes being transported longer distances than longer ones (Figure 87 and Figure 90; $p=0.003$). Maximum flake width also showed a statistically significant relationship with transportation distance, with narrower flakes being transported longer distances (Figure 87 and Figure 91; $p=0.004$).

Four statistically insignificant predictors were also observed:

1. Thickness and weight did not statistically significantly affect artifact transport distances (Figure 87; $p>0.05$).
2. Neither cattle nor sheep trampling statistically significantly affected artifact transport distances (Figure 87; $p>0.05$).
3. Maximum river discharge or average river discharge had no statistically significant effect on artifact transport distances (Figure 87; $p>0.05$).
4. Artifact transport distances were not statistically significantly influenced by artifact trapping by other clasts (Figure 87; $p>0.05$).

The hypothesis made in this analysis was that horizontal artifact displacement in a fluvial system would be impacted by artifact primary metrics and average and maximum river discharge. It was also expected that scatter location, trampling, and artifact trapping (*sensu* Hosfield and Chambers 2005c) might influence artifact transport distances. This hypothesis was partially verified by statistical analysis and the following conclusions were drawn:

1. Scatter location had a statistically significant impact on the transport distance of artifacts with channel artifacts moving a statistically significant mean distance of 27.72m farther than bankside artifacts.

Figure 92: List of experimental conditions for the transport distance analysis.

Dependent variable	Independent variables						Controlled variable
Transport distance	*Maximum length*	*Maximum width*	*Maximum thickness*	*Weight*	*Elongation*	*Refinement*	*Artifact lithology*
Measured (0.01m)	Measured (0.01mm)	Measured (0.01mm)	Measured (0.01mm)	Measured (0.1g)	Measured (0.01mm)	Measured (0.01mm)	Flint

2. Artifact burial was a significant predictor of artifact transport. Artifacts that were buried or partially buried in floodplain sediments were limited from moving substantial distances.
3. Increasing length, width, thickness, and weight were related to shorter artifact transport distances but only length and width showed significant correlations, suggesting that they are better predictors of artifact transport than weight or thickness.
4. Trampling by cattle and sheep did not significantly affect artifact transport distances.
5. Neither average or maximum river discharge significantly affected the transport of artifacts though this result is only applicable to bankside artifacts. It is suggested that the previously described factors such as burial or the presence of 'binding' vegetation may prevent artifacts from becoming entrained in spite of higher discharge values.

4.4.2. Artifact transport distance and artifact primary metrics (bar artifacts)

Relating artifact transport distances to artifact dimensions was a main goal of this study. As outlined in Section 2.3.3, current literature reports conflicting results as to the effects of artifact size on fluvial transport distances. Because the role of fluvial transport was often unclear among Scatters 1–6, this analysis targets Scatters 7–9 that were undoubtedly fluvially transported.

The benefits of redoing this analysis were threefold. First, they allowed us to recheck the results of the mixed model with a targeted dataset. Second, they provided the opportunity to examine the effects of artifact shapes (elongation, refinement) on heavily transported scatters. Lastly, they produce a coefficient of determination (R^2) to assess the strength of the relationship. In short, this analysis aims to determine if there is a relationship between artifact primary metrics (maximum length, maximum width, maximum thickness and weight, elongation, refinement) and artifact transport, over large transport distances (up to 250m). If artifact morphology affects artifact transport distance, then statistically significant relationships should exist between the transport distance data for artifacts with variable primary metric conditions (Bridgland and Harding 1993; Bunn *et al.* 1980; Schick 1986; 1987).

Formally stated, the hypotheses for this section were as follows:

H_0=Artifact primary metrics do not significantly affect artifact transport distances.

Figure 93: Durbin-Watson statistics for the linear regression analyses of artifact primary metrics.

	Durbin-Watson statistic
Maximum length	1.279
Maximum width	1.245
Maximum thickness	1.162
Weight	0.199
Elongation	1.122
Refinement	1.022

H_1=Artifact primary metrics significantly affect artifact transport distances.

The individual relationships between primary metrics, elongation and refinement, and transport distances were tested with bootstrapped (robust) single linear regressions. Regression analyses were followed up with one-way ANOVAs to test the statistical significance of these relationships.

The data used in these analyses were artifact primary metrics and the transport distances from Scatters 7–9 between April 2011 and September 2011 (n=49; Figure 92). Artifacts from the gravel bar (Scatters 7–9) presented a unique opportunity within this study to explore the effects of artifact primary metrics on transport distances since they were fluvially transported over long distances (up to 250m), were exposed to the same hydrological flows, and occupied the same insertion context as they were located on the same gravel bar. Furthermore, recovery rates for these scatters were low ($\leq$50%), so combining these scatters increased the sample size (n=48) and therefore the statistical power of the test.

Durbin Watson tests were performed to evaluate the independence of errors in the single linear regression analyses (Figure 93). Values between 1 and 3 verify the independence of errors (Field 2009: 221). As weight did not satisfy these criteria, it suggests signs of autocorrelation and should therefore be treated with caution.

Figure 94 to Figure 99 are scatter plots displaying and reporting the results of bootstrapped linear regressions comparing transport distance to maximum length, maximum width, maximum thickness, weight, elongation and refinement. All significant effects are reported where p<0.05.

Between April 2011 and September 2011, artifacts from Scatters 7–9 were transported an average of 27.87±48.86m.

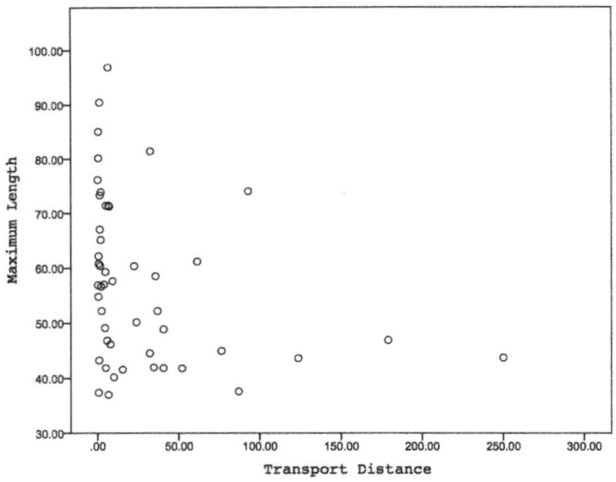

Figure 94: Scatter plot of artifact transport distance and flake maximum length (Scatters 7–9; April 2011–September 2011). Bootstrapped linear regression; R2=0.09, F(1, 46)=4.50, *p*<0.05.

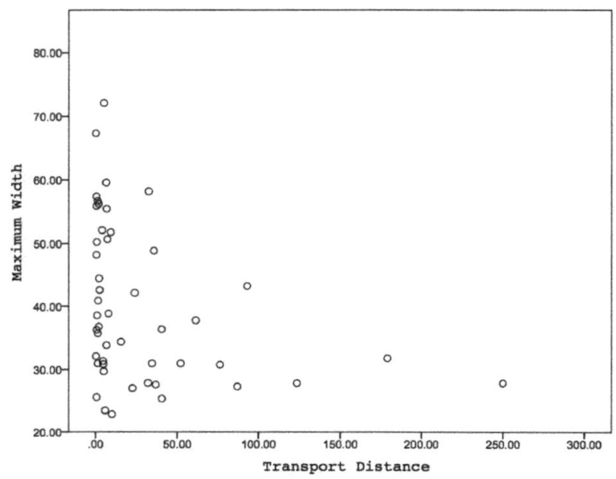

Figure 95: Scatter plot of artifact transport distance and flake maximum width (Scatters 7–9; April 2011–September 2011). Bootstrapped linear regression; R2=0.10, F(1, 46)=5.499, *p*<0.05.

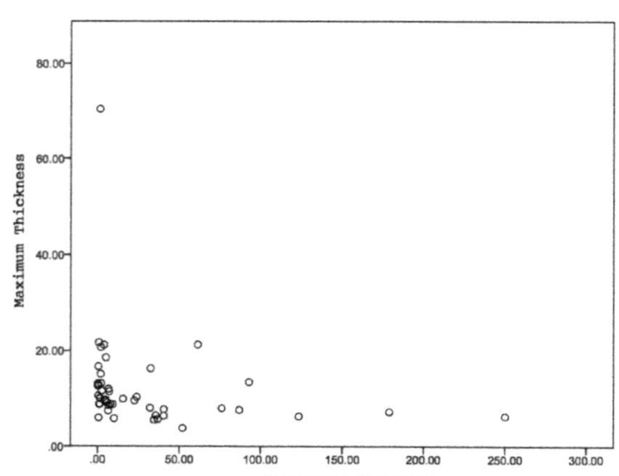

Figure 96: Scatter plot of artifact transport distance and flake maximum thickness (Scatters 7–9; April 2011–September 2011). Bootstrapped linear regression; R2=0.04, F(1, 46)=2.01, *p*=0.155, *n.s.*

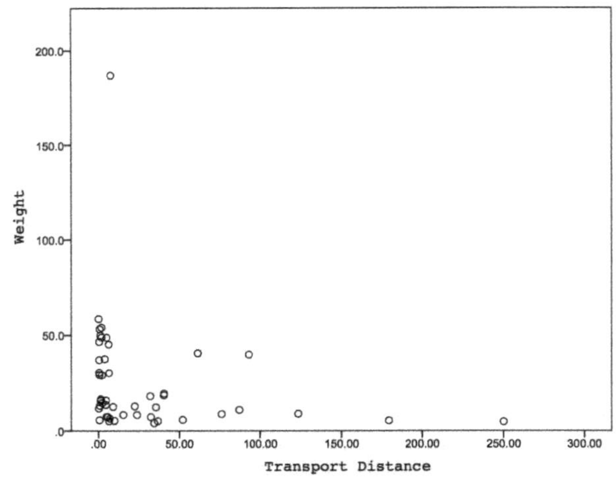

Figure 97: Scatter plot of artifact transport distance and weight (Scatters 7–9; April 2011–September 2011). Bootstrapped linear regression; R2=0.05, F(1, 46)=2.37, *p*=0.131, *n.s.*

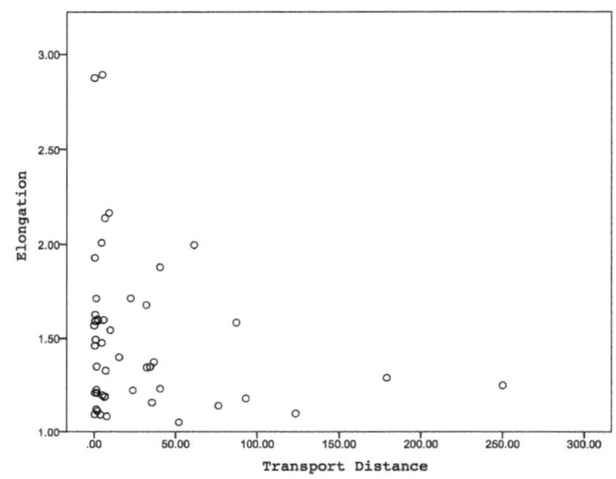

Figure 98: Scatter plot of artifact transport distance and elongation (Scatters 7–9; April 2011–September 2011). Bootstrapped linear regression; R2=0.204, F(1, 46)=1.99, *p*=0.164, *n.s.*

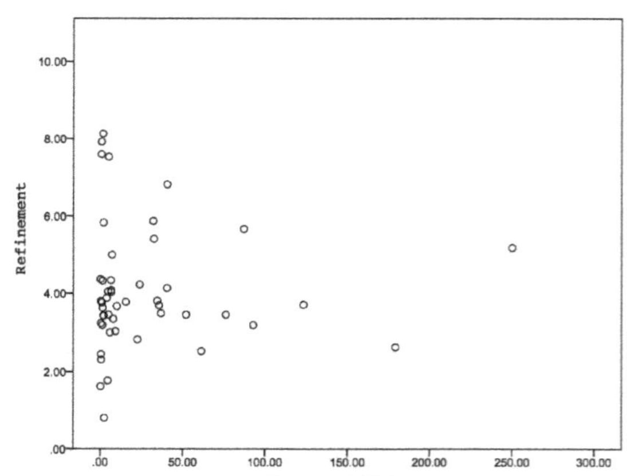

Figure 99: Scatter plot of artifact transport distance and flake refinement (Scatters 7–9; April 2011–September 2011). Bootstrapped linear regression; R2=0.00, F(1, 46)=0.000, *p*=0.994, *n.s.*

Figure 100: List of experimental conditions for the artifact recovery analysis.

Dependent variables				Independent variable	Controlled variable
Maximum length	*Maximum width*	*Maximum thickness*	*Weight*	*Recovery*	*Artifact lithology*
Measured (0.01mm)	Measured (0.01mm)	Measured (0.01mm)	Measured (0.1g)	Recovered artifacts	Flint
				Unrecovered artifacts	

The results of the bootstrapped linear regressions showed that transport distance was weakly, but significantly, predicted by maximum length (R^2=0.09; Figure 94; $p<0.05$) and maximum width (R^2=0.10; Figure 95; $p<0.05$). Weight, thickness, elongation, and refinement were statistically insignificant predictors of artifact transport (Figure 96 to Figure 99; $p>0.05$).

The results of the bootstrapped linear regressions suggest that transport distances of flint artifacts in gravel-bedded rivers are partially dependent, albeit weakly, on overall length and width of artifacts. Longer and wider artifacts were not transported as far as shorter and narrower ones.

The results also indicate that weight and thickness are not significant predictors of artifact transport, suggesting that artifact transport is better predicted by overall dimensions than weight (cf., Hosfield and Chambers 2005c). These results confirm the results of the previous mixed model.

Elongation and thickness were also insignificant predictors of artifact transport distances indicating that 'dimensionless' artifact shapes (e.g., short squat flakes or long thin flakes) played no statistically significant role in their transport.

4.4.3. A recovery bias in the transport distance analysis?

The analyses in the previous sections (4.4.2 and 4.4.3) indicate that artifact transport is influenced by the length and width of artifacts. These values were calculated on the transport distances of recovered artifacts. However, for the time period examined (April to Sept 2011) 107 (69%) of the artifacts were unrecovered after their insertion at the experimental site. The aim of this analysis was to determine if artifact recovery systematically biased the results of the previous section.

This is important because missing data might reduce the representativeness of the sample and distort subsequent inferences about the artifact population. However, statistical bias might also provide meaningful information about transport processes by indicating that recovery was influenced by an external factor. To this end, primary metrics of recovered and unrecovered artifacts from Scatters 7–9 were compared. If recovery was biased due to artifact metrics (e.g., if smaller artifacts were consistently transported out of the search area) then their means should differ statistically significantly. If the two groups did not differ statistically significantly, it can be assumed that the recovery of these artifacts was random with reference to primary metrics.

Formally stated, the hypotheses for this section were as follows:

H_0=Recovered artifact's primary metrics do not differ statistically significantly from unrecovered artifacts.

H_1=Recovered artifact's primary metrics differ statistically significantly from unrecovered artifacts.

Artifact primary metric means were compared with a robust (bootstrapped) independent samples t-test assessing if recovered and unrecovered artifacts were *statistically significantly* different from each other.

In this analysis, the primary metrics (maximum length, maximum width, maximum thickness, and weight; Figure 100) of the 456 artifacts inserted in the river were analyzed as the dependent variables. Their recovery at monitoring events was the independent variable being tested. Artifact lithology (flint) was kept constant to reduce the effect of additional variables.

The data collected was appropriate for t-tests because recovered artifacts and unrecovered artifacts were independent and mutually exclusive of each other. Bootstrapping eliminated the need for dependent variables to be tested for normal distribution. Levene's test showed significant variance ($p<0.05$) between the two groups being compared, so Welch's t-test estimates are reported where the equal variances of datasets are not assumed.

Figure 101 shows the primary metrics of recovered and unrecovered artifacts from Scatters 7–9.

Mean measurements of length, width, thickness, and weight were slightly smaller but this difference was only statistically significant for length and width ($p<0.05$;

Figure 101: Descriptive statistics of recovered and unrecovered artifacts for Scatters 7–9 from April 2011 to September 2011.

	Recovered artifacts (mean±SD)	Unrecovered artifacts (mean±SD)
n	48	108
Length (mm)	57.46 ± 15.07	50.57±15.07*
Width (mm)	40.10±12.45	35.27±9.85*
Thickness (mm)	12.08±9.69	9.96±6.62
Weight (g)	24.86±29.03	18.79±20.97
Elongation	1.49±.42	1.45±.32
Refinement	4.06±1.58	4.30±1.98

**p<0.05 paired bootstrapped t-test.*

Figure 102: List of experimental conditions for the transport distance analysis. ***Note: Rayleigh's test is a one-sample statistic and therefore only evaluates the dependent variable.***

Dependent variable	Independent variable (*for t-test only*)	Controlled variables
Long artifact (l/w≥1.5) orientation	*Scatter ID*	*Artifact lithology*
Measured (0-179°)	Scatters in riverbank (Scatters 1-6)	Flint
	Scatters in channel (Scatters 7-9)	

Figure 101). Length and width were on average between 5 and 7mm smaller, respectively, in the case of unrecovered flakes.

Differences in primary measurements of transported artifacts were statistically insignificant across the recovered and unrecovered groups except for length and width, which were significantly smaller for unrecovered artifacts. These results suggest that the recovery of flakes in Scatters 7–9 were biased towards longer and wider flakes. This suggests that shorter and narrower flakes were transported out of the project area, confirming the results of the mixed model in Section 4.4.1 and the linear regression in 4.4.2.

4.4.4. Artifact transport and orientations

Previous research has suggested that biases in the long-axis orientation of archaeological artifacts (imbrication) can serve as indicators of both the vector and intensity of fluvial reworking at an archaeological site (e.g., Benito-Calvo and de la Torre 2011; Bertran and Texier 1995; Domínguez-Rodrigo *et al.* 2012). Yet, the number of experimental datasets to have directly tested this remains limited (cf., Benito-Calvo *et al.* 2011; Bertran *et al.* 2010; Lenoble *et al.* 2008; Schick 1986: 111). The aim of this analysis was to determine if fluvial disturbance affected orientations of distinctly elongated flakes (i.e., where length/width≥1.5) in scatters located in the channel (Scatters 7–9) and those located on the channel banks (Scatters 1–6). If so, it should be possible to observe *statistically* significant patterns in artifact orientations in the transformed assemblages.

Formally stated, the hypotheses for this section were as follows:

H_0=Artifact orientation distributions in transformed assemblages are random.

H_1=Artifact orientation distributions in transformed assemblages are not random.

Kuiper's test was performed to test for any trend towards non-uniform distribution. To test uniform against modal distributions, the rotational statistic Rayleigh's test was performed (see Section 3.5.3), followed up with a Watson test that assessed the goodness of fit test for the von Mises or circular uniform distribution (see Section 3.5.3).

Paired t-tests were also used in this analysis to compare the means of artifact orientations at the different monitoring periods. This is a standard practice for directional data comparison in the archaeological literature to verify whether a mean scatter orientation changes significantly from one monitoring period to the next (e.g., Eren *et al.* 2010).

The dependent variable used was the bidirectional measurement of orientation for all artifacts where 1.5≤l/w. Circular statistical tests are stronger with larger sample sizes. Therefore, channel deposited artifacts (Scatters 7–9) and the bankside artifacts (Scatters 1–6) were combined in two respective groups, to test the different contexts. This was not problematic since the contexts were largely similar (scatters were only 5m apart). Splitting each scatter would have produced small samples (<50), which would have been less statistically reliable.

To standardize data, artifact orientations (0–179°) were calculated in relation to the direction of river flow at their local position (Figure 102). The orientation 0°, therefore, represents an artifact whose long axis is parallel to the flow of the river rather than to the North-South azimuth. It is worth restating that the original artifact orientations were either perpendicular or parallel to the stream flow to standardize the experimental orientation (see details in Section 3.2.3), thereby eliminating the need to perform initial tests of randomness.

Because Rayleigh's test is designed for unidirectional periodicity (1–360°), for the purposes of this test, bidirectional data was transformed (2x) to render it suitable for testing (C. Agostinelli 2012, *pers. comm.*, 7 Dec; S. McPherron 2012, *pers. comm.*, Dec 10). All statistical operations were performed in R v2.15, using the package 'circstats'. Rose and Schmitt diagrams were created using Stereonet 8.0 and OSXStereonet 2.8.

A prerequisite for Rayleigh's test is that it follows the Von Mises distribution. This is evaluated with Kuiper's test, which tests the null hypothesis that the datasets are sampled from a von Mises distribution. For all of the scatters, the null hypothesis could not be rejected indicating that they were drawn from von Mises distributions and were suitable for Rayleigh's test (Figure 103).

Paired t-tests were used to compare changes to orientations though time because observations were on the same artifacts. Bootstrapping was also used to normalize the distribution of values.

Figure 103 reports the Rose diagrams of artifact orientations for bankside and channel deposited artifacts

Figure 103: Rose and Schmidt diagrams, and statistical tests of isotropy applied to scatters initially emplaced on the riverbanks (Scatters 1–6) and scatters initially emplaced within the channel (Scatters 7–9) and their significance (*p*). *p* values <0.05 indicate anisotropy.

		February	**April**	**September**	**July**
Scat-ters 1-6		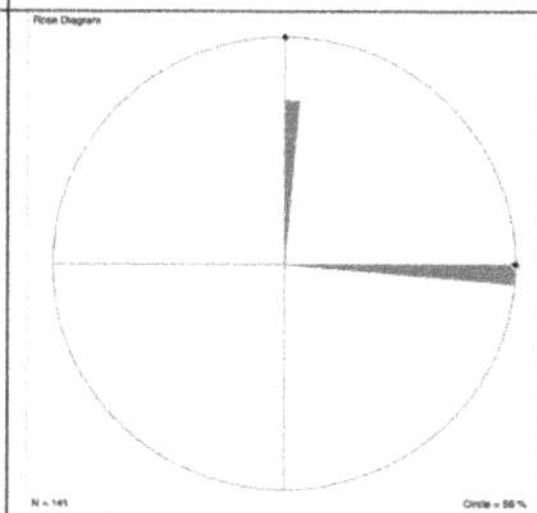	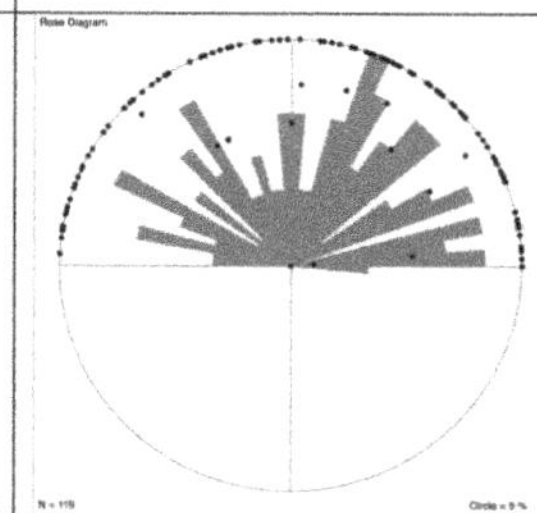	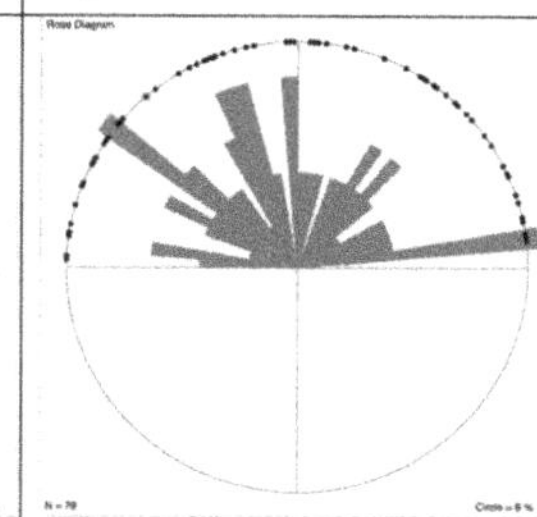	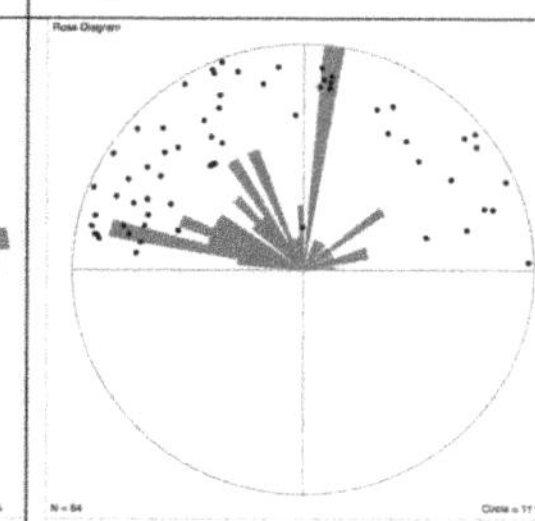
	Orientation trend, Length, Concentration factor k	54.3 0.7165 k=3.5017	8.5 0.6147 k=2.5733	355.1 0.6484 k=2.8078	7.0 0.6908 k=3.0802
	t-test	**t(117)=-4.863, *p*<0.001**			
			t(73)=-1.586, *p*=0.117		
				t(44)=-0.66, *p*=0.948	
	Rayleigh	N/A	R=0.0577, *p*>0.05	R=0.0795, *p*>0.05	R=0.2437, *p*>0.05
	Kuiper	N/A	V=3.1928, *p*>0.05	V=0.9976, *p*>0.05	V=2.0917 *p*>0.05
	Watson	N/A	U^2=0.5262, *p*>0.05	U^2=0.0198, *p*>0.05	U^2=0.0356, *p*>0.05

Scat-ters 7-9		Not deposited	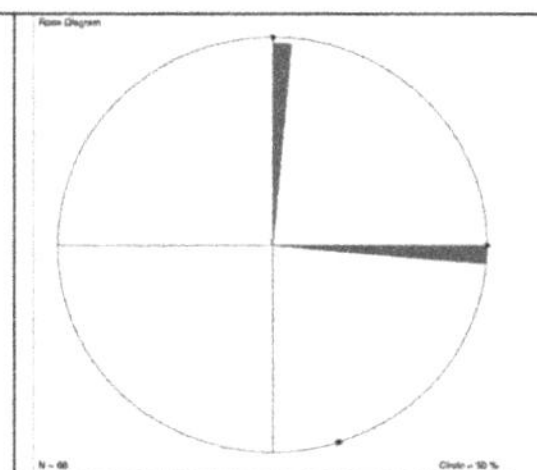	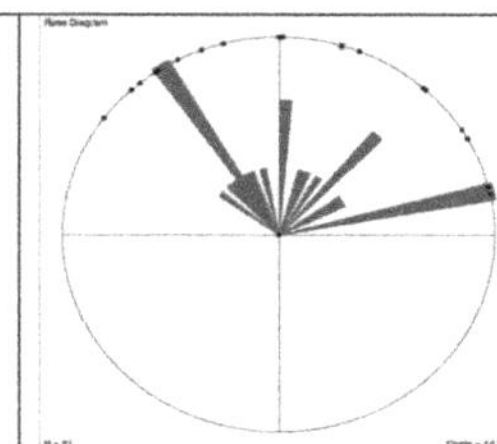	Not recovered
	Orientation trend, Length, Concentration factor k	N/A	47.0 0.6904 k=3.1828	7.0 0.6908 k=3.0802	
	t-test	N/A			
			t(21)=-2.413, *p*=0.025		
				No artifacts recovered	
	Rayleigh	N/A	-	R=0.02099, *p*=0.38	-
	Kuiper	N/A	-	V=1.3824, *p*>0.15	-
	Watson	N/A	-	U^2=0.0418, *p*>0.10	-

at each monitoring stage of the experiment. It also displays the t-tests comparing mean orientations by scatter between monitoring periods and the Rayleigh, Kuiper, and Watson tests, which test the null hypothesis of isotropy (uniformity in all directions).

The paired t-tests in Figure 103 showed that artifacts on the riverbanks significantly changed their orientation between their initial insertion point in February 2011 and the first monitoring period in April 2011 (*p*<0.001). The overall data between April 2011 and September 2011 and September 2011 and July 2012 showed that there was no significant change to the mean orientation data (*p*>0.05). The paired t-test for the long artifacts in the channel revealed the same relationship between their initial insertion in April 2011 and their only recovery period in September 2012.

Rayleigh, Kuiper, and Watson tests in Figure 103 showed that, at all post-insertion monitoring periods, orientations were statistically isotropic (*p*>0.05). The results suggest that long-axis artifacts changed orientations between their insertion and the first monitoring period (*p*<0.05). However, following the initial change, they remained stable (*p*>0.05).

Rayleigh, Kuiper. and Watson tests all indicate that, in spite of this initial change, artifact orientations were isotropic suggesting that their change was random. This may have been due to the non-random orientation of the artifacts at their initial insertion. However, given that they remained isotropic in subsequent monitoring periods, it is likely that the initial insertion orientation did not affect subsequent orientation patterns.

Taken together, these results suggest that artifacts in bankside contexts (Scatters 1–6) that are minimally transported (<2m) may change their orientations, but that these orientations may not result in anisotropic patterns. The same may be said for artifacts transported within the gravel-bedded river: long distance transport (>7m) may not result in the anisotropic orientation of artifacts.

4.4.5. Trampling and artifact dip

Previous work has shown that trampling impacts artifact dip in saturated sediments (Eren *et al.* 2010). The animal trampling episodes allowed the opportunity to evaluate the effect of trampling by large and small classes of animals on artifact dip. This is detailed in the methodology chapter (Section 3.2.2). Artifacts were trampled as evidenced by either visual confirmation or the presence of animal feces on or near the artifacts. The aim of this section is to identify if the artifacts in this experiment demonstrated significantly different dip values when trampled in the experimental site. If trampling does affect the dip of artifacts, one would expect to find statistically significant differences between the means of trampled and non-trampled scatters.

Formally stated, the hypotheses for this section were as follows:

> H_0=Animal trampling does not significantly affect artifact dip.
>
> H_1=Animal trampling significantly affects artifact dip.

A rotational ANOVA (AOVCircular) is used here to compare the dip values for artifacts under three trampling conditions: no trampling, trampling by sheep, and trampling by cattle (see Section 3.5.4).

The dependent variable in this analysis is the dip (i.e., *plunge* or *maximum inclination*) of the artifact plane defined by angle in comparison to the horizon of the axis perpendicular to the long axis (Figure 104). Dip is therefore a measure from 0° to 90° with the former indicating a horizontal position and the latter indicating a vertical position. Dip was recorded at all monitoring visits, but in April and September 2011 it was only visually assessed as 'flat' when less than 45° and 'vertical' when greater than 45°. Therefore, the data used in this analysis are dip values from July 2012, which were quantified. Artifact dip was recorded to the nearest degree. Values were tested against the known record of animal trampling by both cattle and sheep.

AOVCircular samples are assumed to be drawn from von Mises populations with equal concentration parameters (k). Both of these assumptions can be tested with a Kuiper's test, a rotational variant of the Kolmogorov-Smirnov test (see Section 3.5.8). The results show that for all three groups, $p>0.15$ indicating they were suitable for parametric testing. Additionally, k values were all close, suggesting equal concentration parameters (Figure 105).

Figure 104: List of experimental conditions for the dip and trampling analysis.

Dependent variable	Independent variables	Controlled variables
Dip	Trampling	Artifact lithology
Measured (1.0°)	*No trampling*	*Flint*
	Sheep trampling	
	Cattle trampling	

Figure 105: Results of Kuiper's Test of Uniformity for the dip analysis.

	K	*p*
No trampling	1.45	>0.15
Sheep trampling	1.17	>0.15
Cattle trampling	1.25	>0.15

Figure 106 reports the descriptive statistics of artifact dips while Figure 107 shows comparative box plots between the three conditions.

Artifact dip in all trampling conditions showed a mean dip between 18 and 20° (Figure 106). Standard deviations of artifact dips also showed similar variations (SD between 8.6 and 13.6 degrees; Figure 106), increasing from no trampling to sheep and then cattle trampling. There was only one exception: one artifact (ID=265) with a nearly vertical dip of 75° that had been trampled by sheep. Statistical analysis of the trampling conditions confirmed that there was no significant change to dip values because of trampling; $F(2, 130)=1.07$, $p=0.34$ (Figure 106).

The results of this analysis report little variation between the average dips of artifacts under different trampling conditions. This indicates that trampling by different sizes of animals had no impact on the dip of artifacts in this experiment.

4.4.6. Transport distance and artifact micro-flaking

A total of 18 artifacts (37%) of the heavily transported Scatters 7–9 were micro-flaked (*sensu* Hosfield and Chambers 2005c), as characterized by the continuous bifacial flaking of the acute margins of the flake. Flake scars were no larger than 1mm in all dimensions (Figure 108). Because micro-flaking was observed only on highly transported flakes, this section explores whether primary metrics or transport distance affected the appearance of micro-flaking on artifacts.

Figure 106: Descriptive statistics for the artifact dips (degrees) in trampling conditions.

	Mean	SD
None	20.48	8.6
Sheep	17.65	11.3
Cow	18.46	13.6

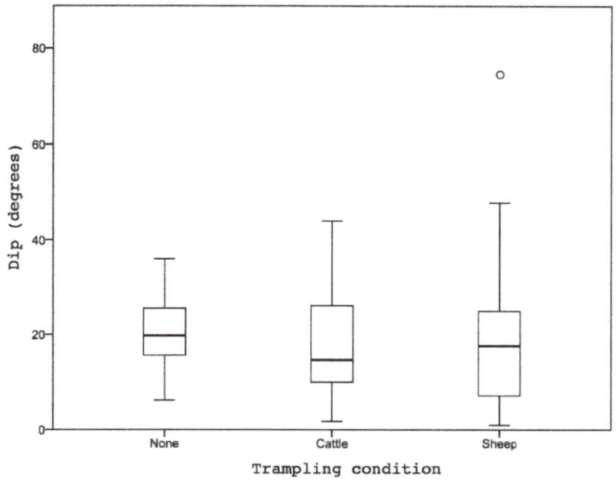

Figure 107: Dips of artifacts at the final recovery period (July 2012) sorted by the trampling conditions they underwent.

Formally stated, the hypotheses for this section were as follows:

H_0=Artifact primary metrics and transport distances do not significantly affect artifact micro-flaking.

H_1=Artifact primary metrics and transport distances significantly affect artifact micro-flaking.

The method used here was a binary ordinal regression that statistically analyzes the relationship between one or more continuous variables and a binary ordinal outcome variable (Figure 109; see Section 3.5.5).

The dependent variable in this analysis is the presence or absence of micro-flaking. Micro-flaking was only found on Scatters 7–9 that were recovered between April and September 2011. Therefore, that is the only period used in this analysis. The independent variables were their primary metrics (maximum length, maximum width, maximum thickness, and weight) and transport distance. The artifact lithology was kept constant to reduce extra variables.

As with all logistic regressions, the data must satisfy the Hosmer-Lemeshow test, which is a statistical test for goodness of fit. The null hypothesis is that the observed event rates match expected event rates in subgroups of the model population. This model could not reject the null hypothesis (χ^2=(7)=11.377, p=0.123), indicating that the model was suitable for testing.

Figure 108: Photographic examples of micro-flaking on fluvially transported artifacts; Artifacts 417, 373, and 364.

Figure 110 reports on the descriptive statistics of micro-flaked artifacts while Figure 111 presents the results of the binary logistic regression.

The results of the binary logistic regression are found in Figure 111. A Pseudo-R^2 value of 0.713 indicates that the model explains *c.* 70% of the data patterning and that distance is a highly significant predictor of edge damage (p=0.003). Flake dimensions were not significant predictors of micro-flaking (p>0.05; Figure 60).

The micro-flaked artifacts from Scatters 7–9 showed an average transport distance of 63.99 ± 15.23m, with a minimum of value of 7.11m and a maximum of 250.10m (Figure 110). None of the artifacts transported less than 7m showed edge damage, while of those that were transported further than 7m, 18 out of 22 (82%) exhibited edge micro-flaking.

The results of this experiment found that micro-flaking was often a result of fluvial transport. This analysis showed that the presence/absence of micro-flaking was highly correlated with transport distance, with artifacts being micro-flaked in as little as 7m. This analysis also showed that the actual dimensions of the flakes did not effect whether artifacts were micro-flaked.

Though it has been suggested that abrasion can significantly affect the morphology of artifacts (Harding *et al.* 1987: 125; Hosfield and Chambers 2005c; Schick 1986: 112), it was unclear to what extent micro-flaking affected artifact morphology in this experiment. Three heavily transported artifacts (>123m) were recovered from Scatters 7–9 in September 2011 that provide some limited insight. They were all similar in starting dimensions (Figure 112). Weight loss appeared to increase with transport distance, up to more than 8% (Figure 112). However, no clear patterns emerged from other dimensions.

Figure 109: List of experimental conditions for the micro-flaking analysis.

Dependent variable	Independent variables					Controlled variables	
Micro-flaking	*Maximum length*	*Maximum width*	*Maximum thickness*	*Weight*	*Transport distance*	*Time*	*Artifact lithology*
Yes	Measured (0.01mm)	Measured (0.01mm)	Measured (0.01mm)	Measured (0.1g)	Measured (0.01m)	Apr to Sep 2011	Flint
No							

Figure 110: Descriptive statistics for micro-flaked artifacts from Scatters 7-9 after transport from April 2011 to September 2011.

	Micro-flaked	Undamaged
n	19	30
Transport distance* (m)	62.50±63.12	5.22±10.22
Length (mm)	52.19±13.07	61.67±15.33
Width (mm)	22.64±40.29	43.02±13.05
Thickness (mm)	9.30±4.04	14.09±11.62
Weight (g)	22.64±40.30	27.17±17.43
		**p<0.05 binary logistic regression*

This suggests that changes to artifact primary metrics because of local transport follow no clear patterns and may depend on other variables such as their mode of transport and unexplored aspects of their morphology such as ridges and edge angles.

4.4.7. Flake damage and trampling

14 damaged flakes were recovered from trampled assemblages (Figure 113). However, 12 out of 14 of flakes had been subject to trampling by cattle, while only 2 out of 14 had been subject to trampling by sheep. Of the entire 119 recovered trampled artifacts, sheep trampled 59 artifacts while cattle trampled 60; none were significantly fluvially transported (<1m). The aim of this analysis was to see if breakage patterns differed according to animal trampling size.

Formally stated, the hypotheses for this section were as follows:

> H_0=Animal size does not significantly affect trampled artifact-breakage patterns.
>
> H_1=Animal size significantly affects trampled artifact-breakage patterns.

The test used here is a 2-sample Chi-Squared test (see Section 3.5.6). The data observations are independent of each other so Fisher's exact test is appropriate for this test.

Figure 114 reports the variables used in the flake damage analysis.

Figure 111: Binary logistic regression for micro-flaking.

	B	S.E.	Wald	df	Sig.	Exp(B)
Max Length	.030	.051	.351	1	.554	1.031
Max Thickness	-.002	.080	.001	1	.982	.998
Weight	-.015	.014	1.192	1	.275	.985
Width	.015	.056	.074	1	.786	1.015
Transport distance	**-.099**	**.034**	**8.677**	**1**	**.003**	**.906**
						Pseudo R^2 (Nagelkerke)=0.713

Figure 115 compares the number of damaged flakes under different trampling conditions.

Trampling resulted in the transverse snapping of flakes. Those that were trampled by cattle were broken 25% of the time, while those trampled by sheep were broken only 4% of the time (see Section 3.2.2 for tramping details). A Fisher's exact test showed that the two conditions differed significantly (Figure 115; p<0.01).

The results of this test show that there were different breakage frequencies for artifacts that were trampled in different conditions. Those that were trampled by cattle had statistically significantly (Figure 115; p<0.05) more broken artifacts, even though there were more sheep trampling a roughly equivalent plot of land (approximately 2000m^2). An intuitive conclusion is that the weight of cattle compared to that of the sheep is an important factor in the breakage of artifacts. The significance of these results will be discussed in more detail in Section 7.2.8.

4.5. Chapter summary

The goal of the analytical section was to statistically test some observations made on the fluvial scatters in the Glenderamackin. In these analyses, the effects of artifact primary metrics (maximum length, maximum width, maximum thickness and weight), river discharge levels, trampling, artifact trapping, burial, insertion location, and monitoring events on artifact recovery, transport distances, orientations, dip, micro-flaking, and breakages were studied.

Figure 112: Changes to the primary metrics of artifacts 364(b), 373, and 417 between April 2011 and September 2011.

Artifact ID	417		364(b)		373	
Transport distance (m)	123.55		179.20		250.10	
	Original	*Recovered*	*Original*	*Recovered*	*Original*	*Recovered*
Length (mm)	43.66	42.95 (98.4%)	46.91	42.28 (90.1%)	43.77	42.64 (97.5%)
Width (mm)	37.83	36.51 (96.0%)	35.19	34.47 (98.0%)	27.87	27.60 (99.0%)
Thickness (mm)	6.36	6.00 (95.2%)	10.20	10.04 (98.4%)	6.33	6.33 (100%)
Weight (g)	9.0	8.8 (97.8%)	9.0	8.3 (92.2%)	4.8	4.4 (91.7%)
Elongation (l/w)	1.15	1.18	1.33	1.22 (109.0%)	1.57	1.54 (102.0%)
Refinement (w/th)	5.95	6.85 (86.8%)	3.45	3.43 (100.5%)	4.40	4.36 (101.0%)

Figure 113: Photographic example of an artifact damaged by trampling.

Of these, the significant results were:

1. Scatter location had an influence on artifact transport distances, with artifacts deposited in the channel being transported farther than those deposited on the riverbanks.
2. Burial had a significant effect on inhibiting artifact transport.
3. Length and width of heavily transported artifacts were weakly, but significantly, inversely proportional to transport distances—shorter, narrower flakes were more likely to be transported longer distances downstream.
4. Animal trampling resulted in notches and transverse breakages to smaller artifacts. Increasing animal size affected the frequency of these breakages.
5. The lateral edges of artifacts in the gravel-bedded channel were micro-flaked when transport exceeded 7m, resulting in observed mass losses as large as 8%.

Observations and statistical analyses of the site data demonstrated the following scenario:

Artifact scatters discarded on river margins can remain largely intact for extended periods of time, as a result of fine-grained fluvial deposition and vegetation growth. During this time, artifacts may be affected by taphonomic processes such as trampling, resulting in pseudo-anthropogenic edge- or tool-types. Over time, artifacts, independent of size, may be removed from the site through a probable combination of high-level flooding and slope processes, and surface runoff. If artifacts enter the stream, they are widely distributed over long distances (partially dependent on their overall length and width) and redeposited on low flow zones such as point bars. These artifacts also sustain characteristic edge micro-flaking damage.

The significance, synthesis, and limitations of the fluvial field results will be discussed in more detail in Section 7.2.

Figure 114: List of experimental conditions for the flake breakage experiment.

Dependent variable	Independent variables	Controlled variables
Flake damage	*Trampling*	*Artifact lithology*
Yes	Large animal (size 4)	Flint
No	Medium animal (size 3)	

Figure 115: Descriptive statistics comparing trampling conditions. Fisher exact test, p=0.008.

		Edge damage		Total
		None	*Yes*	
Trampled	Cattle	48	12(25%)	60
	Sheep	57	2(4%)	59
Total		105	14	119

Chapter 5

Tumbling barrel experiments

5.1. Introduction

While the previous chapter examined the influences of fluvial processes on the spatial displacement of lithic artifacts, this chapter aims to understand the effects of natural flaking by fluvial processes on flaked lithic artifacts through a series of targeted laboratory experiments (refer to Section 3.3).

Though it has long been acknowledged that fluvial disturbance can alter the appearance of flaked lithic artifacts (Harding et al. 1987; Hosfield and Chambers 2005a; Prestwich 1860), the effects are still only summarily understood. Researchers have recently re-emphasized the significant effects powerful fluvial processes may have on the morphology of flaked lithic artifacts, demonstrating that aspects of artifact morphology, size, and transport mode can affect the severity and type of damage sustained by lithic artifacts. These arguments have underscored the awareness that fluvial processes are dynamic and complex, with multiple factors playing significant roles (e.g., Grosman et al. 2011; Harding et al. 1987; Hosfield 1999; Hosfield and Chambers 2003; 2004; 2005).

An understudied aspect of fluvial abrasion that these studies point to is the role of local depositional environments. This chapter presents results from five longitudinal studies exploring the influence of time and impact frequency, time and artifact lithology, time and sediment size, time and sediment lithology, and time and artifact class on the percentage of original weight (%originalweight), percentage of original length (%originallength), percentage of original width (%originalwidth), elongation (length/width), refinement (width/thickness; Bordes 1961; 1988)[1], and artifact condition (fresh, lightly abraded, abraded, heavily abraded). Thickness was also measured but it was clear from early experiments that changes were minimal and less than the measurement error of the vernier calipers. Thickness values were therefore only initially recorded and used as part of the refinement ratio.

This chapter presents the results of the analyses only. Discussion of these results is detailed in Section 7.3.

5.2. Sediment mass, time, and artifact abrasion

The aim of this section is to determine the effect of sedimentary impact and time on the abrasion of lithic artifacts. If the frequency of sediment impacting plays a role in artifact abrasion as some authors have suggested (Thompson et al. 2011), then greater sediment mass inside the tumbling barrel should increase the rate of abrasion. Using the two flake datasets *ƒ* and ¢ (Appendix 4), a test was devised to assess the impact of time and sediment mass on the development of primary metrics. This was done by comparing (1) 10 flint flakes in 1kg of fluvial sediment and (2) 10 flint flakes in 2kg of fluvial sediments that were tumbled and repeatedly measured 8 times over the course of 2 hours.

Formally stated, the hypotheses for this section were as follows:

> H_0=There is no significant change to primary metrics over time between artifacts tumbled in 1kg of sediment and 2kg of sediment.
>
> H_1=There is a significant change to primary metrics over time between flakes tumbled in 1kg of sediment and 2kg of sediment.

If sediment mass affects flake abrasion over time, then the mean rates of changes to primary metrics between groups should be statistically different from each other. This experiment tests each primary metric through a two-way mixed design (mixed ANOVA), with time as the repeated factor and sediment mass as the fixed factor. Repeated planned contrasts were determined, comparing each time-step to the previous one for all primary metrics (see Section 3.5.8). Since artifact conditions did not vary at time-steps, they were not statistically treated.

Six dependent variables (%originalweight, %originallength, %originalwidth, elongation, refinement, and artifact condition; Figure 116) were measured on two groups (*ƒ*, ¢; see Appendix 4) of 10 flint flakes over eight time-steps (t=0, 1, 5, 15, 30, 60, 120[2]). A bootstrapped Welch's t-test showed that each group of 10 flakes was statistically the same in terms of their initial flake mass, length, width, thickness, elongation, and refinement (Appendix 5).

Time was analyzed over the same artifacts, as a within-subjects factor (a repeated or random factor). Sediment mass varied between groups of flake artifacts (1kg and 2kg) as a between-subjects factor (fixed factor). To reduce

[1] Bordes (1961) definitions were used for elongation and refinement but comparisons with Roe's (1964) classification scheme are valid since the only difference is the inversion of width/thickness to thickness/width (Pettitt and White 2012: 504).

[2] All of the other experiments in this chapter were conducted until they reached 180 minutes, however the tumbling barrel broke an axle after 120 minutes during the tumbling of group *ƒ* (1kg sediments). Therefore it was decided to not finish the experiment since it was unknown how long beyond 120 minutes they had been tumbled.

Figure 116: List of experimental conditions for the artifact sediment mass experiment.

Dependent variables	Independent variables		Controlled variables
	Fixed	*Random*	
%originalweight	Sediment mass	Time	Tumbling barrel speed (15rpm)
%originallength	(1 and 2kg)		Water volume (5l)
%originalwidth			Artifact class (unretouched flakes)
Elongation			*Phi* group (φ=-5.5 to 5.0)
Refinement			Sediment lithology (Kennet)
Artifact condition			Artifact lithology (flint)

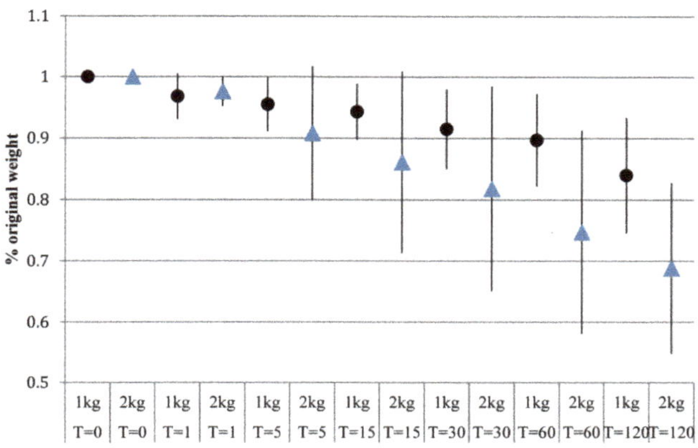

Figure 117: Impact of time on weight of flakes tumbled in 1kg of sediment (*n*=10) and 2kg of sediment (*n*=10). Error bars represent 1 SD in either direction.

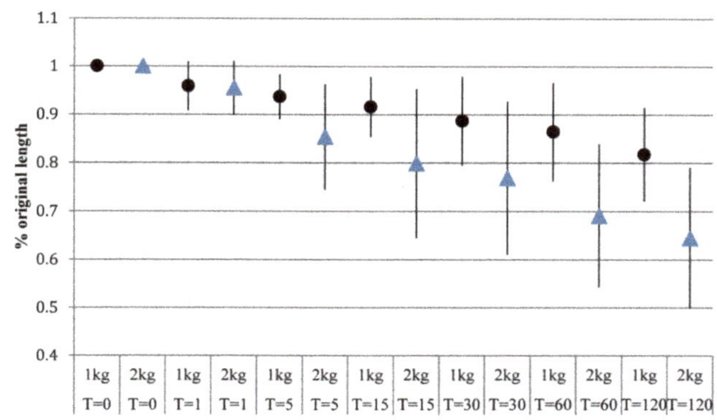

Figure 118: Impact of time on length of flakes tumbled in 1kg of sediment (*n*=10) and 2kg of sediment (*n*=10). Error bars represent 1 SD in either direction.

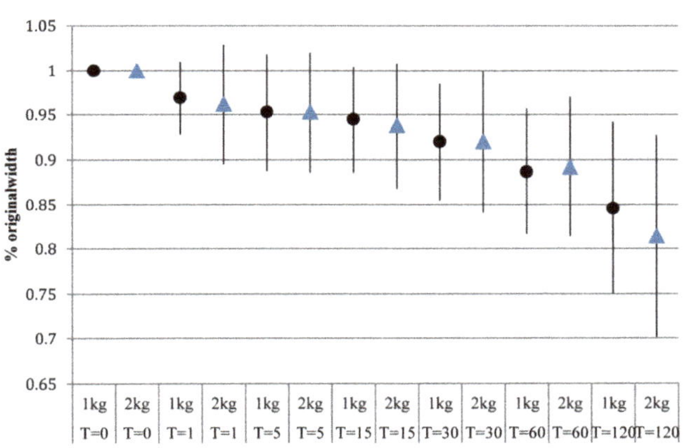

Figure 119: Impact of time on width of flakes tumbled in 1kg (*n*=10) of sediment and 2kg (*n*=10) of sediment. Error bars represent 1 SD in either direction.

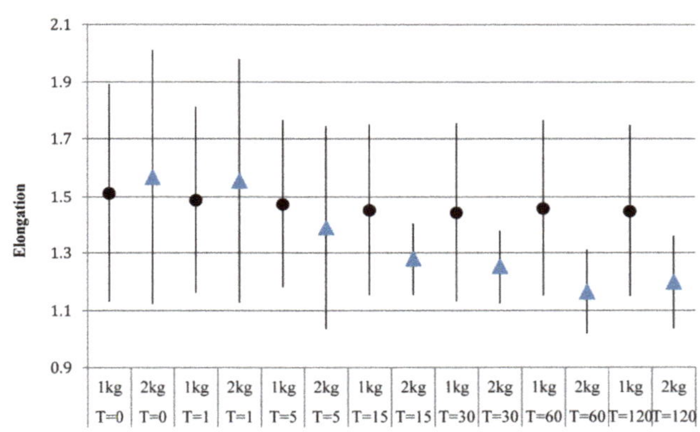

Figure 120: Impact of time on elongation of flakes tumbled in 1kg of sediment (*n*=10) and 2kg of sediment (*n*=10). Error bars represent 1 SD in either direction.

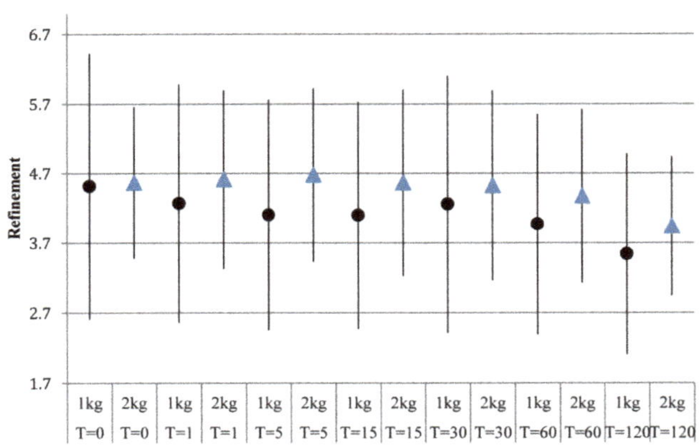

Figure 121: Impact of time on refinement of flakes tumbled in 1kg of sediment (*n*=10) and 2kg of sediment (*n*=10). Error bars represent 1 SD in either direction.

the effect of additional variables, tumbling barrel speed (15rpm), water volume (5l), artifact class (unretouched flakes), sediment size (φ=-5.5 to 5.0), and sediment lithology (Kennet; see Section 3.3.4 for details) were treated as controlled variables and kept constant (Figure 116).

Figure 117 to Figure 123 illustrate the changes to artifact primary metrics and condition. Figure 124 reports the summary statistics of changes to artifact primary metrics and conditions. The results of the mixed-design ANOVAs are reported in Figure 125 to Figure 127. All effects are reported as significant where $p<0.05$.

The results show a significant effect of time on %originalweight, %originallength, %originalwidth,

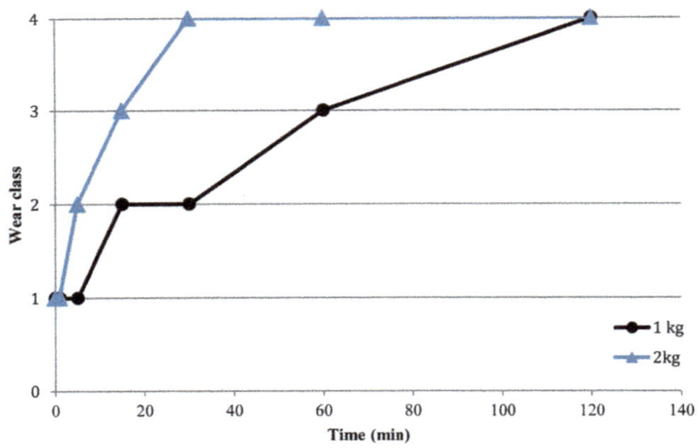

Figure 122: Impact of time on artifact condition of flakes tumbled in 1kg of sediment (*n*=10) and 2kg of sediment (*n*=10).

Figure 123: Final conditions of artifacts used in the sediment mass analysis. *Left artifact was tumbled in 1kg of sediments while the right artifact in 2kg.*

elongation, and refinement (Figure 125, Figure 117, Figure 124, Figure 119, Figure 120, Figure 121, Figure 122). The contrasts obtained for %originalweight, %originallength, and %originalwidth revealed that the effect of time was significant across all time-steps except for %originallength between 5 and 15 minutes and for width until up to 30 minutes. Elongation was significantly affected by time between 1 and 5 minutes while refinement was significant between 60 and 120 minutes (Appendix 6-time).

Results also report a significant effect of sediment mass on artifact abrasion as measured by %originalweight, %originallength, and %original width ($p<0.05$; Figure 126, Figure 117 to Figure 119), while no significant effect was found for the elongation and refinement of artifacts ($p>0.05$; Figure 127, Figure 120, Figure 121).

Sediment mass also affected the rate of change of artifact condition, with artifacts in 2kg of sediment becoming heavily abraded (artifact condition 4) after 30 minutes, while the same level of artifact condition was not reached until 120 minutes in 1kg (Figure 122; Figure 123). Statistical analyses showed a significant interaction effect between time and sediment mass on %originalweight, %originallength and %originalwidth ($p<0.05$; Figure 127).

The results of this analysis indicate that time had a significant effect on artifact abrasion, in the case of all metrics. For this experimental set, abrasion was found to impact, first, length, and then, after 30 minutes, width. This is evidenced by the significant effect of time on %originallength in the first few minutes of the experiment, leading to elongation

Figure 124: Descriptive statistics of changes to primary metrics and condition over time between artifacts tumbled in 1kg (*n*=10) and 2kg (*n*=10) sediments.

		T=0	T=1	T=5	T=15	T=30	T=60	T=120
%originalweight	1kg	1±0	0.968 ±0.037	0.955 ±0.044	0.943 ±0.046	0.915 ±0.065	0.897 ±0.075	0.84 ±0.094
	2kg	1±0	0.976 ±0.024	0.908 ±0.109	0.861 ±0.148	0.818 ±0.167	0.747 ±0.166	0.688 ±0.14
%originallength	1kg	1±0	0.959 ±0.051	0.937 ±0.047	0.916 ±0.062	0.887 ±0.092	0.865 ±0.102	0.818 ±0.097
	2kg	1±0	0.955 ±0.056	0.854 ±0.109	0.799 ±0.154	0.769 ±0.158	0.691 ±0.148	0.645 ±0.146
%originalwidth	1kg	1±0	0.969 ±0.041	0.953 ±0.065	0.945 ±0.059	0.92 ±0.065	0.887 ±0.07	0.846 ±0.096
	2kg	1±0	0.962 ±0.067	0.953 ±0.067	0.938 ±0.07	0.92 ±0.079	0.892 ±0.078	0.814 ±0.113
Elongation	1kg	1.512 ±0.38	1.487 ±0.325	1.473 ±0.292	1.452 ±0.298	1.443 ±0.311	1.458 ±0.307	1.448 ±0.299
	2kg	1.568 ±0.444	1.555 ±0.427	1.389 ±0.354	1.279 ±0.125	1.252 ±0.127	1.165 ±0.146	1.198 ±0.162
Refinement	1kg	4.519 ±1.908	4.273 ±1.709	4.108 ±1.656	4.101 ±1.628	4.26 ±1.843	3.973 ±1.578	3.545 ±1.437
	2kg	4.57 ±1.091	4.615 ±1.28	4.678 ±1.247	4.568 ±1.34	4.527 ±1.365	4.377 ±1.247	3.942 ±0.996
Artifact condition	1kg	1	1	1	2	2	3	4
	2kg	1	1	2	3	4	4	4

Figure 125: Statistical analysis of the effect of time on primary metrics (*n*=20; Greenhouse-Geisser correction).

	Type III Sum of Squares	df	df error	Mean square	F	Sig.
%originalweight	.571	2.206	39.710	.259	29.937	**.000**
%originallength	.657	2.355	42.392	.279	35.202	**.000**
%originalwidth	.251	2.081	37.462	.121	19.171	**.000**
Elongation	1.051	1.635	29.432	.643	5.4558	**.014**
Refinement	8.038	2.832	48.152	2.838	6.261	**.001**
Artifact Condition	N/A	N/A	N/A	N/A	N/A	N/A

Figure 126: Statistical analysis of the effect of sediment mass on primary metrics of 20 artifacts.

	Type III Sum of Squares	df	df error	Mean square	F	Sig.
%originalweight	.277	1	18	.227	4.746	**.043**
%originallength	.372	1	18	.372	6.991	**.016**
%originalwidth	.001	1	18	.001	.064	.804
Elongation	.539	1	18	.539	1.193	.289
Refinement	4.214	1	17	4.214	.308	.586
Artifact Condition	N/A	N/A	N/A	N/A	N/A	N/A

Figure 127: Statistical analysis of the effect of the interaction of sediment mass and time on primary metrics (Greenhouse-Geisser correction).

	Type III Sum of Squares	df	df error	Mean square	F	Sig.
%originalweight	.571	2.206	39.710	.259	29.937	**.000**
%originallength	.100	2.355	42.392	.042	5.339	**.006**
%originalwidth	.004	2.081	37.462	.121	19.171	**.000**
Elongation	.612	1.635	29.432	.374	3.176	.065
Refinement	.772	2.832	48.152	.273	.602	.608
Artifact Condition	N/A	N/A	N/A	N/A	N/A	N/A

being significantly affected between 1 and 5 minutes. This sudden length loss is probably due to breakage of the distal end of single or multiple artifacts. This early breakage was not common in the other experiments in this chapter suggesting that there was something different about the morphology of the artifacts not described by primary metrics, such as ridge scar patterns or internal stress fractures.

After the first few minutes, the non-significant effects of time on elongation suggests that with time, flakes get significantly smaller, but that these changes are scaled (as length decreases, width decreases proportionally). This suggests that, following the initial length loss (t<5 minutes), abrasion was less dramatic (length loss is not significant between 5 and 15 minutes), possibly due to a change in breakage mechanisms towards the lateral margins once the fragile termination was removed, leading to width being significantly affected after 30 minutes.

The results also demonstrated that, as expected, increased sediment mass leads to an increase in the reduction of most primary metrics, implying that more frequent impacting in a mobile sediment alters abrasion rates for at least the first 3 hours of active transport. This increased abrasion rate also affected the overall condition of the artifact, with more sediment producing more abraded conditions more rapidly.

Furthermore, results showed a significant interaction between the effects of sediment mass and time. This indicates that combined effects of sediment mass and time leads to different abrasion rates of the artifact at a given time-steps ($p<0.05$; Figure 127). This may be due to increasing impact frequency as time in the tumbling barrel increases. This interaction effect does not depend on initial shape as randomly selected artifacts (i.e., different shapes) showed similar combined effects of time and sediment size.

5.3. Artifact lithology, time, and artifact abrasion

The purpose of the following analyses is to determine if, with time, there are differences in the way different artifact lithologies are affected. Following Bridgland et al. (2006), who suggest that softer raw materials are more susceptible to fluvial abrasion, a test was devised to assess the impact of artifact lithology on primary metrics using datasets zł, ₫, and ^ (Appendix 4). Lithological variations were measured through 10 flint flakes, 10 artificially patinated flakes (see Section 3.3.1), and 10 quartzite flakes.

Figure 128: List of experimental conditions for the artifact lithology experiment.

Dependent variables	Independent variables		Controlled variables
	Fixed	*Random*	
%originalweight	Artifact lithology	Time	Tumbling barrel speed (15rpm)
%originallength	(Flint, patinated flint,		Water volume (5l)
%originalwidth	quartzite)		Artifact class (unretouched flakes)
Elongation			*Phi* Group (φ=0 to -6)
Refinement			Sediment lithology (Ham Yard)
Artifact condition			Sediment mass (2kg)

Formally stated, the hypotheses for this section were as follows:

H_0=There is no significant difference in changes to primary metrics over time between flint, patinated flint, and quartzite flakes.

H_1=There is a significant difference in changes to primary metrics over time between flint, patinated flint and quartzite artifacts.

The null hypothesis is tested on each primary metric using a 2-way mixed-design ANOVA, with time as the random factor and artifact lithology as the fixed factor. Repeated planned contrasts were also performed between each time-step (see Section 3.5.8). Since artifact condition categories did not vary across time-steps, they were not statistically treated.

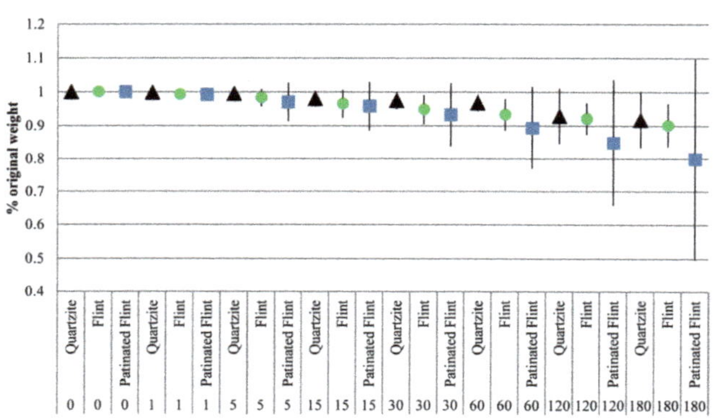

Figure 129: Impact of time on weight loss of flint flakes (*n*=10), patinated flint flakes (*n*=10), and quartzite flakes (*n*=10). Error bars represent 1 SD in either direction.

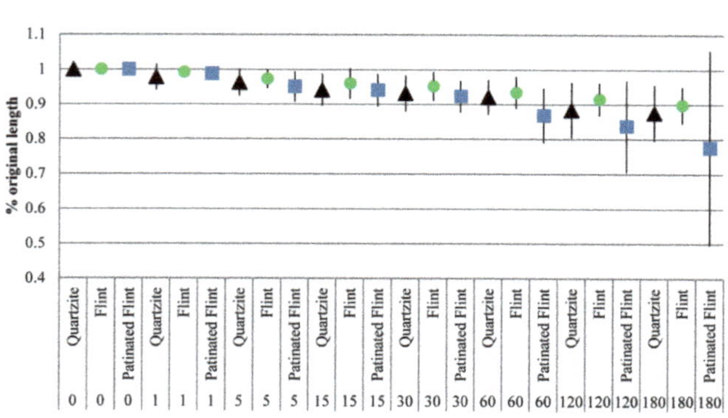

Figure 130: Impact of time on length of flint flakes (*n*=10), patinated flint flakes (*n*=10), and quartzite flakes (*n*=10). Error bars represent 1 SD in either direction.

In this experiment, six dependent variables (%originalweight, %originallength, %originalwidth, elongation, refinement, and artifact condition; Figure 128) were measured on 10 flint, 10 artificially patinated flint, and 10 quartzite flakes (sets zł, ₫, ^; see Appendix 4) over eight time-steps (0, 1, 5, 15, 30, 60, 120, and 180 minutes). Flakes were all similar in initial metrics. A bootstrapped one-way between subjects ANOVA showed that mean differences of the initial weight, length, width, thickness, elongation, and refinement of the flakes were statistically similar (Appendix 7).

Time was measured for the same artifacts as a within-subjects factor (random factor), while artifact lithology varied between the three groups of artifacts and was treated as a between-subjects factor (fixed factor). The effect of additional variables was reduced, using controlled and constant tumbling barrel speed (15rpm), water volume (5l), artifact class (unretouched flakes), sediment size (φ=-6 to 0), sediment lithology (Ham Yard; see Section 3.3.4 for details), and sediment mass (2kg; Figure 128).

Figure 129 to Figure 135 show changes to artifact primary metrics and the condition of artifacts produced in different lithologies. Figure 136 reports the summary statistics while Figure 137 to Figure 139 show the results of the mixed ANOVA.

The results showed a significant effect of time on the %originalweight, %originallength, %originalwidth, and refinement ($p<0.001$; Figure 137, Figure 129, Figure 130, Figure 131, Figure 133). Contrasts revealed that %originalweight, %originallength, and %originalwidth were consistently affected until 60 minutes to 180 minutes,

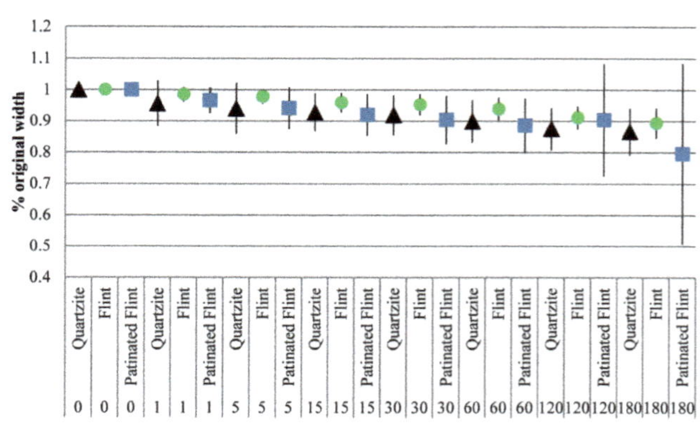

Figure 131: Impact of time on width of flint flakes (*n*=10), patinated flint flakes (*n*=10), and quartzite flakes (*n*=10). Error bars represent 1 SD in either direction.

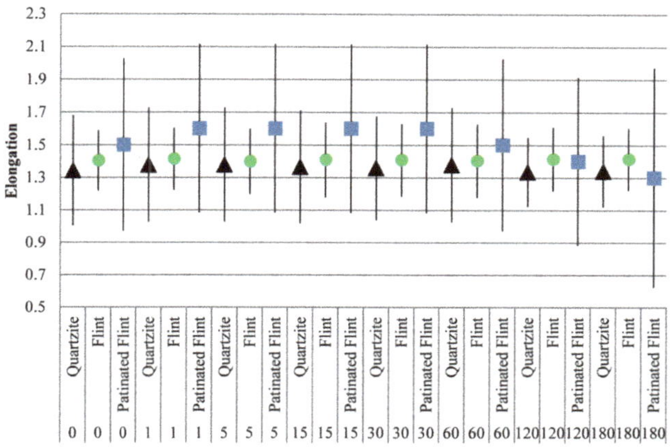

Figure 132: Impact of time on elongation of flint flakes (n=10), patinated flint flakes (n=10), and quartzite flakes (n=10). Error bars represent 1 SD in either direction.

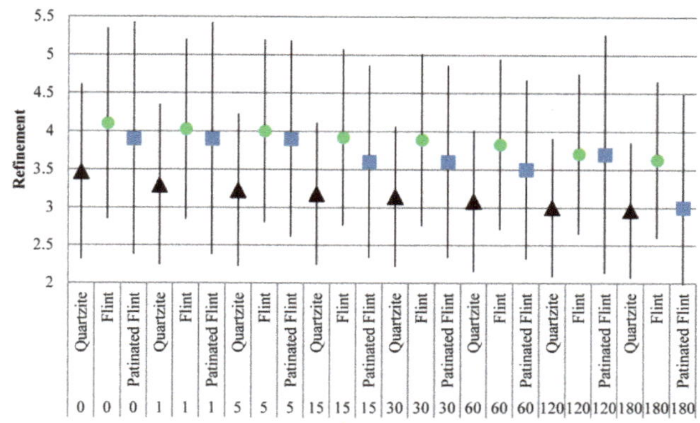

Figure 133: Impact of time on the refinement of flint flakes (n=10), patinated flint flakes (n=10), and quartzite flakes (n=10). Error bars represent 1 SD in either direction.

while time did not significantly affect elongation and refinement, in a consistent manner, over the different time-steps (Appendix 8).

Statistical analysis reported no significant effect of artifact lithology on %originalweight, %originallength, %originalwidth, and subsequently elongation and refinement (p>0.05; Figure 138, Figure 129, Figure 130, Figure 131). The results in Figure 134 however showed that the condition of artifacts significantly differed. Patinated flint artifacts became heavily abraded after 5 minutes, while quartzite flakes did not become heavily abraded until they reached 180 minutes. Patinated flint artifacts' condition deteriorated more rapidly than quartzite due to the chalky nature of the patinated neocortex (Figure 135). Flint artifacts performed between the two other groups, but closer to patinated flint, becoming heavily abraded after 30 minutes.

There was no statistically significant interaction of artifact lithology and time (Figure 139).

The results of this experiment led to three main conclusions:

1. Time had a significant effect on all dependent variables at all time-steps except elongation but contrasts revealed that these effects began to slow down after 120 minutes. This indicates that, after 2 hours, the abrasion process requires longer time-steps to alter lithic artifact morphology in a statistically significant manner, suggesting that time has a non-linear effect on abrasion. Since this experiment only measured 1 additional time-step after 120 minutes, this conclusion is tentative. It is possible that over longer time periods, this relationship may change.
2. The prediction for artifact lithology was not met for changes to morphology. The artifact lithologies explored here had no effect on the changes to primary metrics over time.
3. Artifact lithology did affect the change of artifact condition, which agrees with Bridgland et al. (2006), who suggest that lithological hardness plays a role in artifact condition. This was verified when comparing patinated flint to flint, because the abrasions of the softer neocortices of the patinated flint made it abrade rapidly. Bridgland et al.'s prediction however was not met when comparing quartzite to flint, which abraded more slowly in spite of being a softer material.

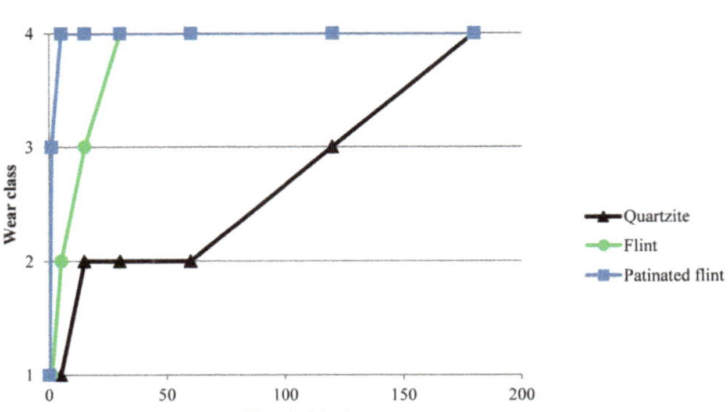

Figure 134: Impact of time on the artifact condition of flint flakes (n=10), patinated flint flakes (n=10) and quartzite flakes (n=10).

Figure 135: Examples of flint artifacts of different lithologies after 180 minutes of tumbling. *From top left to right: flint, quartzite, and patinated flint flakes.*

Figure 136: Descriptive statistics of changes to primary metrics and condition over time between flint, patinated flint, and quartzite artifacts.

	Artifact Lithology	T=0	T=1	T=5	T=15	T=30	T=60	T=120	T=180
%originalweight	Flint	1±0	0.998±0.005	0.995±0.006	0.98±0.023	0.975±0.025	0.968±0.024	0.929±0.083	0.918±0.084
	Quartzite	1±0	0.993±0.01	0.983±0.025	0.965±0.041	0.948±0.043	0.933±0.047	0.921±0.047	0.901±0.065
	Patinated Flint	1±0	0.992±0.018	0.97±0.057	0.958±0.072	0.932±0.094	0.893±0.123	0.848±0.189	0.798±0.3
%originallength	Flint	1±0	0.978±0.037	0.963±0.039	0.942±0.046	0.932±0.052	0.921±0.05	0.884±0.08	0.876±0.081
	Quartzite	1±0	0.992±0.009	0.973±0.026	0.96±0.044	0.952±0.041	0.934±0.046	0.915±0.047	0.898±0.052
	Patinated Flint	1±0	0.988±0.017	0.951±0.044	0.941±0.047	0.923±0.046	0.868±0.079	0.837±0.132	0.776±0.28
%originalwidth	Flint	1±0	0.956±0.073	0.94±0.082	0.928±0.061	0.919±0.064	0.899±0.068	0.875±0.067	0.866±0.075
	Quartzite	1±0	0.985±0.023	0.978±0.023	0.959±0.031	0.953±0.034	0.939±0.037	0.911±0.037	0.893±0.048
	Patinated Flint	1±0	0.966±0.041	0.941±0.067	0.92±0.066	0.904±0.077	0.886±0.087	0.904±0.179	0.796±0.289
Elongation	Flint	1.342±0.338	1.377±0.351	1.379±0.35	1.365±0.346	1.359±0.319	1.378±0.351	1.336±0.212	1.34±0.219
	Quartzite	1.404±0.184	1.414±0.19	1.399±0.201	1.409±0.228	1.408±0.223	1.403±0.225	1.413±0.195	1.414±0.19
	Patinated Flint	1.5±0.527	1.6±0.516	1.6±0.516	1.6±0.516	1.6±0.516	1.5±0.527	1.4±0.516	1.3±0.675
Refinement	Flint	3.465±1.152	3.295±1.057	3.226±1.003	3.179±0.935	3.144±0.924	3.085±0.933	2.999±0.911	2.964±0.897
	Quartzite	4.098±1.252	4.023±1.182	3.998±1.2	3.918±1.158	3.886±1.129	3.827±1.119	3.705±1.053	3.627±1.032
	Patinated Flint	3.9±1.524	3.9±1.524	3.9±1.287	3.6±1.265	3.6±1.265	3.5±1.179	3.7±1.567	3±1.491
Abrasion Class	Quartzite	1	1	1	2	2	2	3	4
	Flint	1	1	2	3	4	4	4	4
	Patinated Flint	1	3	4	4	4	4	4	4

Figure 137: Statistical analysis of the effect of time on primary metrics (Greenhouse-Geisser correction).

	Type III Sum of Squares	df	df error	Mean square	F	Sig.
%originalweight	.357	1.305	35.239	2.73	13.719	**.000**
%originallength	.413	1.418	38.296	.291	15.227	**.000**
%originalwidth	.275	1.954	52.745	.141	11.027	**.000**
Elongation	.360	1.889	50.999	.190	1.768	.198
Refinement	8.114	2.426	65.502	3.344	8.378	**.000**
Artifact condition	N/A	N/A	N/A	N/A	N/A	N/A

Figure 138: Statistical analysis of the effect of artifact lithology on primary metrics (Greenhouse-Geisser correction).

	Type III Sum of Squares	df	df error	Mean square	F	Sig.
%originalweight	.103	2	27	.051	1.412	.264
%originallength	.084	2	27	.042	2.039	.150
%originalwidth	.071	2	27	.035	.983	.160
Elongation	.979	2	27	.489	.518	.602
Refinement	21.118	2	27	10.559	1.029	.371
Artifact condition	N/A	N/A	N/A	N/A	N/A	N/A

Figure 139: Statistical analysis of the effect of the interaction of artifact lithology and time on primary metrics (Greenhouse-Geisser correction).

	Type III Sum of Squares	df	df error	Mean square	F	Sig.
%originalweight	.065	2.610	35.239	.025	1.246	.306
%originallength	.065	2.837	38.296	.023	1.203	.321
%originalwidth	.036	3.907	52.745	.009	.718	.580
Elongation	.553	3.778	50.999	.146	1.291	.286
Refinement	1.931	4.852	65.502	.398	.997	.425
Artifact condition	N/A	N/A	N/A	N/A	N/A	N/A

5.4. Sediment size, time, and artifact abrasion

The aim of the following analyses is to determine if sediment size affects the development of abrasion and the reduction of primary metrics over time. If sediment size plays a role in the abrasion of artifacts, then one would expect that larger sediment size classes should abrade artifacts more rapidly than smaller sediment sizes. A test was devised to assess the impact of sediment size on the development of primary metrics comparing groups of 10 flint flakes (subsets listed in heading, Appendix 4) in different phi groups consisting of a no-sediment control group (phi group=0), the natural Ham Yard sediment, and Ham Yard sediments sifted according to half-phi units from -6.0 to -3.0.

Formally stated, the hypotheses for this section were as follows:

> H_0=There is no significant difference in changes to primary metrics over time between different sediment sizes.
>
> H_1=There is a significant difference in changes to primary metrics over time between different sediment sizes.

A mixed-design ANOVA was used to test the effects of the different sediment sizes and time on all dependent variables, using time as the random factor and sediment size as the fixed factor. The influence of time was assessed, using planned comparisons that compared the means of dependent variables between two successive time-steps. Post-hoc Tukey tests were conducted to contrast individual relationships between the various phi-groups under study (see Section 3.5.8). Since artifact condition categories did not vary across time-steps, they were not statistically treated.

This experiment examined six dependent variables (%originalweight, %originallength, %originalwidth, elongation, refinement and artifact condition; Figure 140) on eight groups (see Appendix 4; sets ₣, zł, ~, #, ₲, ฿, ₴, %) of 10 flint flakes, measured at eight different time-steps (T=0, 1, 5, 15, 30, 60, 120, 180 minutes). The two independent variables tested were time and sediment size (phi-group). A bootstrapped one-way between subjects ANOVA showed that mean differences of the initial weight, length, width, thickness, elongation, and refinement of the flakes were statistically similar (Appendix 9).

Because time was measured for the same artifacts, it was considered a within-subjects factor (random factor). Sediment size varied between artifacts and was therefore considered a between-subjects factor (fixed factor). Tumbling barrel speed (15rpm), water volume (5l), artifact class (unretouched flakes), artifact lithology (flint), sediment lithology (Ham Yard), and sediment mass (2kg) were kept constant to reduce the effect of additional variables.

Figures 141 to 147 show changes to artifact primary metrics and conditions when abraded in different gravel sizes. Figure 148 reports the summary statistics and Figure 149, Figure 150, and Figure 154 report the results of the mixed ANOVA. Figure 151, 152, and 153 report the post-hoc Tukey tests showing the groupings of gravel size groups.

Results showed a significant effect of time on artifact %originalweight, %originallength, %originalwidth and refinement ($p<0.05$; Figure 149, Figure 141, Figure 142, Figure 143, Figure 145), while elongation was not significantly affected by time ($p>0.05$; Figure 149 and Figure 144). Contrasts revealed that changes in %originalweight, %originallength, %originalwidth, and refinement were consistently significant until at least 120 minutes (Appendix 10-time). The change in refinement was presumably in response to the reduction of artifact width, as thickness remained the same.

Statistical analysis showed no significant effect of sediment size on the refinement or elongation of artifacts ($p>0.05$; Figure 150, Figure 144, Figure 145). However, sediment size significantly affected other primary metrics such as %originalweight and %originalwidth and %originallength ($p<0.05$; Figure 150, Figure 141, Figure 142, Figure 143).

Post-hoc Tukey tests showed only significant differences between the effect of the smallest phi groups (up to φ=-3.5) and the largest phi group (φ=-6.0 to -5.5) on the %originalweight and %originalwidth (Figure 151 and Figure 153). Even though the ANOVA procedure revealed

Figure 140: List of experimental conditions for the sediment size experiment.

Dependent Variables	Independent Variables		Controlled Variables
	Fixed	*Random*	
%originalweight	Phi group	Time	Tumbling barrel speed (15rpm)
%originallength	φ=0		Water volume (5l)
%originalwidth	φ=-3.5 to -3.0		Artifact class (unretouched flakes)
Elongation	φ=-4.0 to -3.5		Sediment lithology (Ham Yard)
Refinement	φ=-4.5 to -4.0		Sediment mass (2kg)
Artifact condition	φ=-5.0 to -4.5		Artifact lithology (flint)
(Ashton 1998)	φ=-5.5 to -5.0		
	φ=-6.0 to -5.5		
	φ=-6.0 to 0		

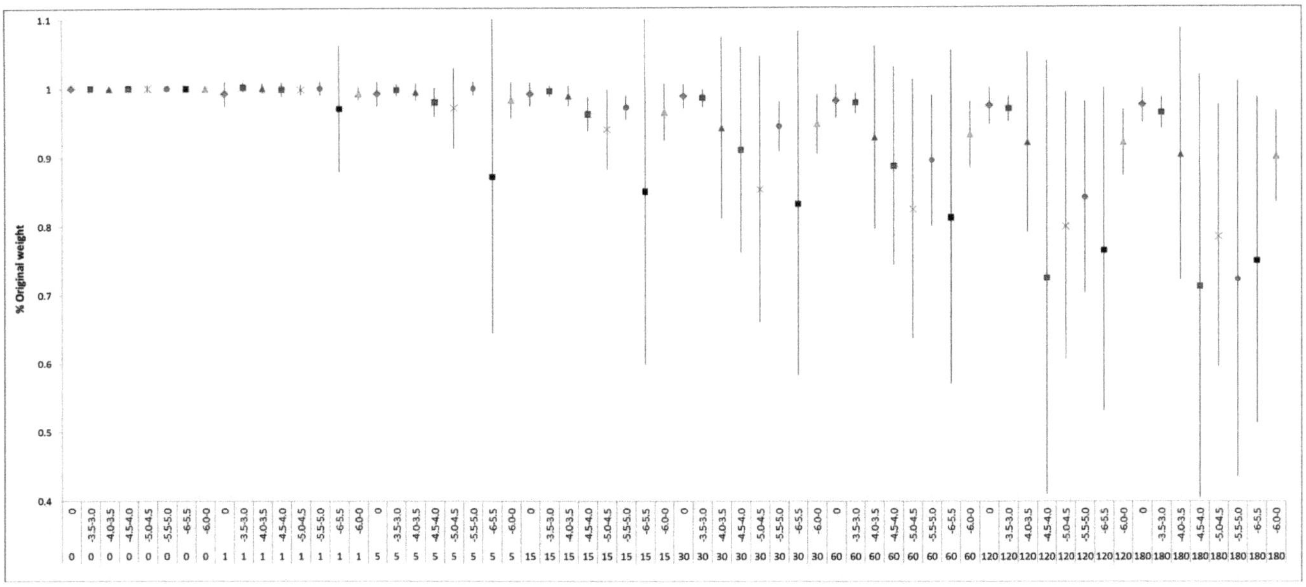

Figure 141: Impact of time and sediment size on %originalweight. *n*=10 for each group. Error bars represent 1 SD in either direction.

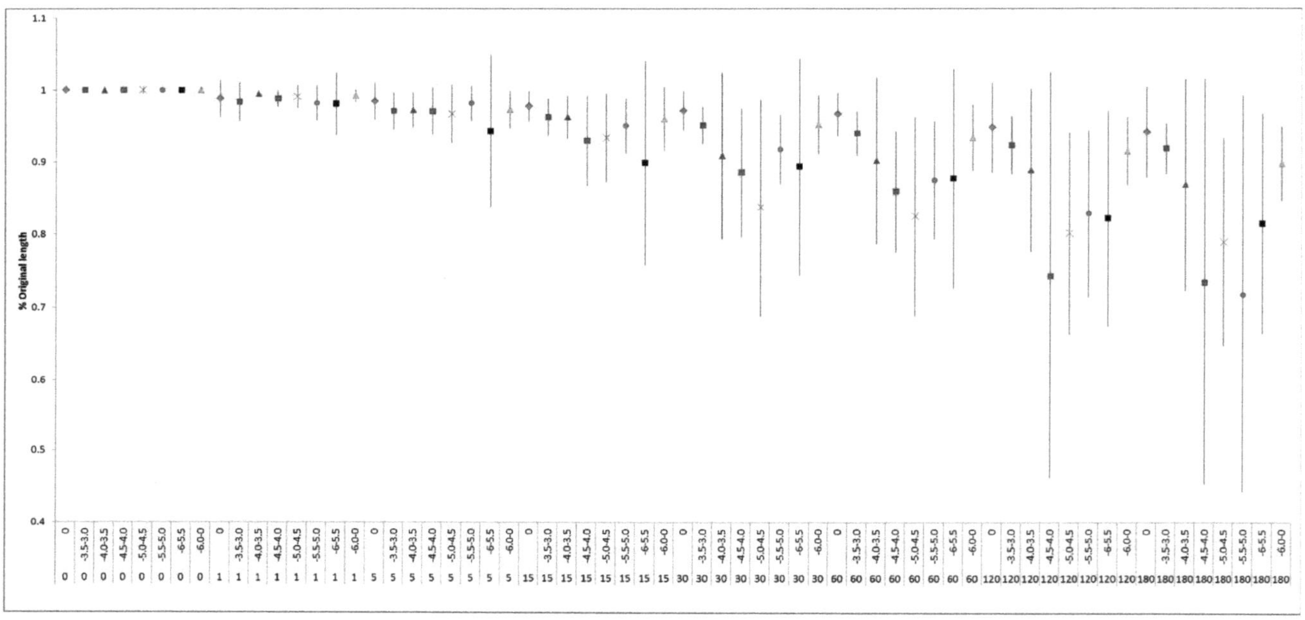

Figure 142: Impact of time and sediment size on %originallength. *n*=10 for each group. Error bars represent 1 SD in either direction.

that sediment size significantly affected %originallength[3] ($p<0.05$; Figure 150), post-hoc tests did not uncover any significant differences in %originallength between the phi groups (Figure 152).

Artifact condition was also influenced by increasing sediment size, with larger sediments producing abraded conditions more rapidly (Figure 148 and Figure 146). These differences however, were small. All flakes were ranked as heavily abraded by 60 minutes of tumbling except the control group that showed no increase in the artifact condition above fresh. The largest sediment size (φ=-6.0 to -5.5) appears to abrade slightly faster than the other sediment categories (Figure 146; Figure 147).

Statistical analysis also showed a significant interaction effect between time and sediment size on %originalweight, %originallength, %originalwidth, and refinement ($p<0.001$; Figure 154). Contrasts showed that this interaction effect was significant between 1 and 15 minutes, and then 60 to 120 minutes for weight, between 1 and 15 minutes for width, and between 5 and 30 minutes for length (Appendix 10-time*sediment size).

[3] Tukey's *post-hoc* comparisons control the family-wise error rate for Type I errors by using a stricter α than in the ANOVA. This means a Type II error occurred in the *post-hoc* results, indicating that impact of sediment size on %originallength is not as strongly correlated as other factors.

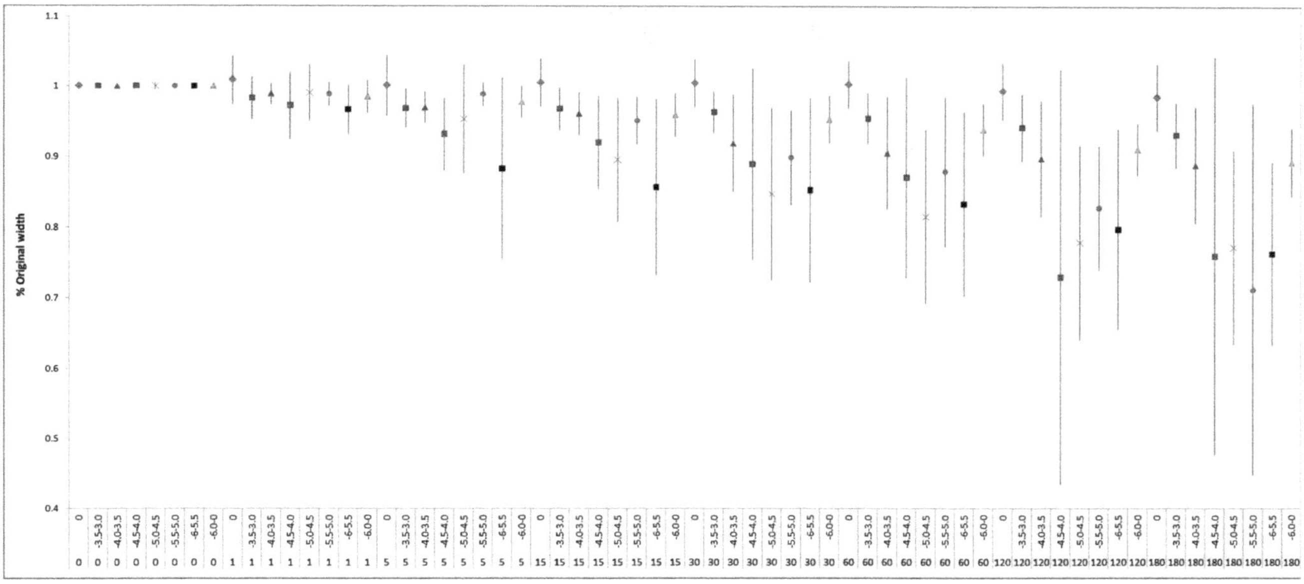

Figure 143: Impact of time and sediment size on %originalwidth. *n*=10 for each group. Error bars represent 1 SD in either direction.

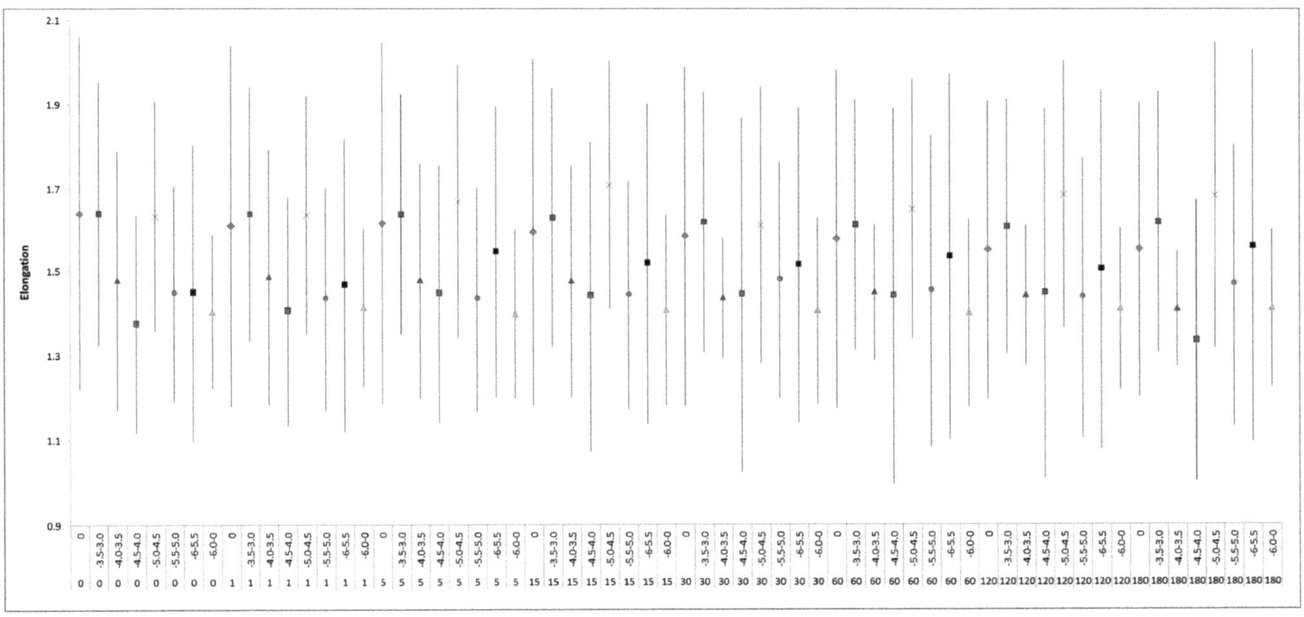

Figure 144: Impact of time and sediment size on elongation. *n*=10 for each group. Error bars represent 1 SD in either direction.

The results of this study lead to three main conclusions:

1. Time played a significant role on the effects of tumbling on both primary metrics and condition. Contrasts revealed that, often, these effects were non-significant beyond 120 minutes. This suggests that in spite of the long abrasion time between the two time-steps, abrasion may be non-linear and begin to slow down after 120 minutes. However, this conclusion is tentative since there is only one observed time-step beyond 120 minutes.
2. If %originalweight is taken as a proxy for abrasion, then larger sediment size significantly increases the effects and rate of artifact abrasion, as seen between sediments sizes of φ≥-5.5 and φ≤-3.5. However, the response to changes in sediment sizes was surprisingly insensitive. Sediment sizes up to φ=-3.5 were indistinguishable from control groups (φ=0). This suggests that small sediments have no effect on the abrasion process or that they operate below the level of background abrasion caused by the tumbling barrel. It was also impossible to statistically distinguish the differences between sediment sizes from φ=-3.5 to -5.5. In the case of the natural aggregate sediment with sediment sizes comprised between φ=0 and φ=-6.0 (Figure 41), the abrasion effect on artifacts was similar to that of sediment sizes from φ=-3.5 to -5.5. This can be explained by the fact that, in spite of having larger sediment sizes within it, the aggregate sediment also contains smaller sediments (from φ≤-3.5) that may provide cushioning effects against larger particles (Kodama 1994b; Kuenen 1956). This impact of sediment size on artifact %originalweight does not

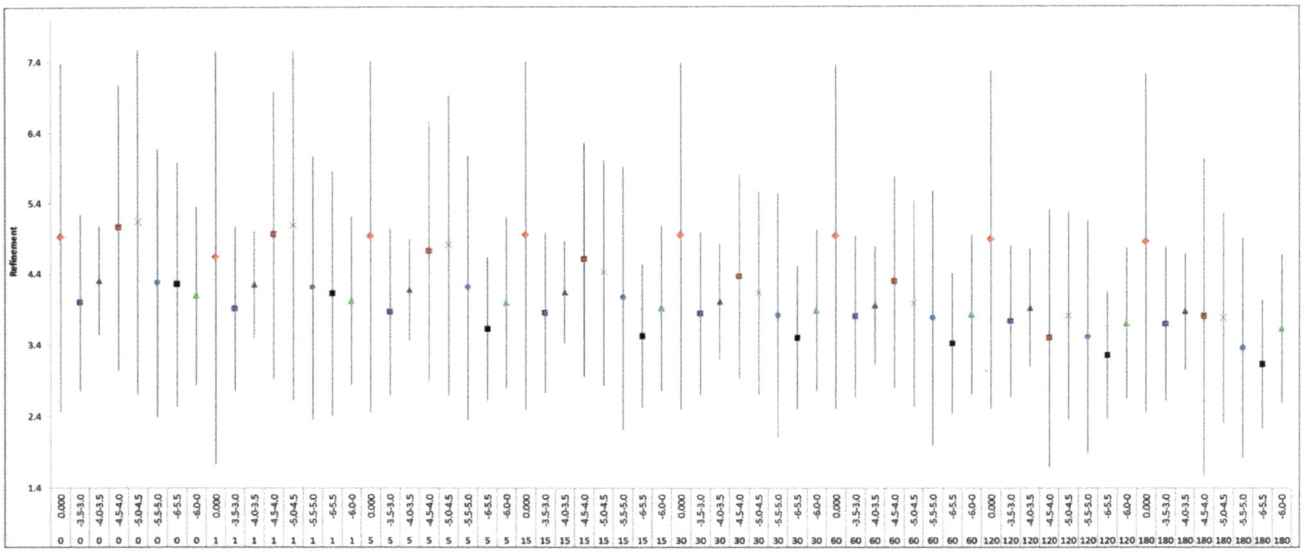

Figure 145: Impact of time and sediment size on refinement. *n*=10 for each group. Error bars represent 1 SD in either direction.

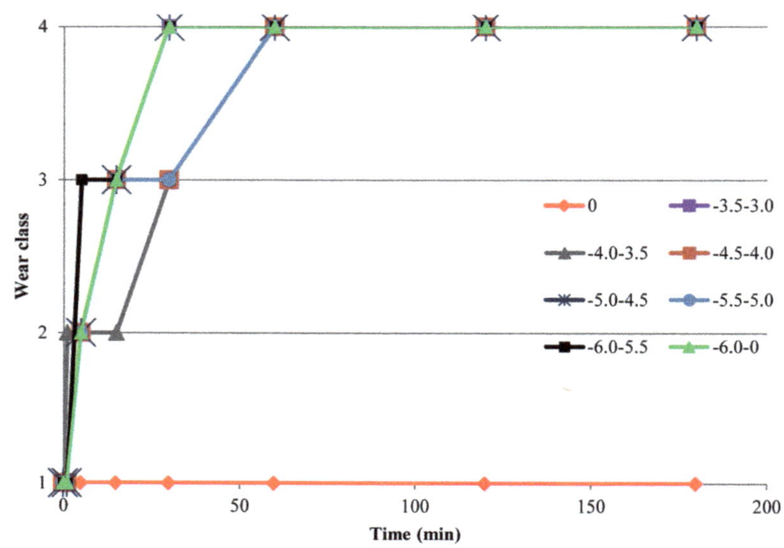

Figure 146: Impact of time and sediment size on artifact condition. *n*=10 for each group.

lead to significant differences in artifact conditions, which show no large variations in the way that they abrade over time.

3. The results of the sediments tested in this study suggest that abrasion time had a significant effect on overall artifact morphology. In the most extreme phi group (φ=-6.0 to -5.5), artifacts retained an average of 75% of their original weight after 3 hours of tumbling. The non-significant change to elongation and refinement suggests that while they are getting smaller, artifacts largely retain their morphology irrespective of the initial starting shape (i.e., scaled reduction in size and shape).

5.5. Sediment lithology, time, and artifact abrasion

The aim of the following analysis is to investigate the impact of natural sediment lithologies on the abrasion of lithic artifacts over time. A test was devised to assess the impact of sediment lithology on artifact abrasion, measured through primary metrics using datasets ₣, #, ₩, Ŧ (Appendix 4), comparing groups of 10 flint flakes abraded in Trent Valley sediments, 10 flint flakes in Ham Yard sediments, and 10 flint flakes in Kennet fluvial sediments, and a 10 flint flake control sample.

Formally stated, the hypotheses for this section were as follows:

> H_0=There is no significant difference in changes to primary metrics over time between artifacts tumbled in different sediment lithologies.
>
> H_1=There is a significant difference in changes to primary metrics over time between artifacts tumbled in different sediment lithologies.

A two-way mixed-design ANOVA was used to assess the significance of time and artifact lithology on primary metrics. Planned contrasts were performed to test the significance of differences in primary metric values

Figure 147: An example artifact tumbled in different gravel sizes after 180 minutes. ***From top left to right: φ=0 (control), φ=-3.5 to -3.0, φ=-4.0 to -3.5, φ=-4.5 to -4.0, φ=-5.0 to -4.5, φ=-5.5 to -5.0, φ=-6.0 to -5.5, φ=-6.0 to 0.***

between successive time-steps. Where significant, post-hoc Tukey tests explored individual relationships between the sediment lithology types in this experiment (see Section 3.5.8). Since artifact condition categories did not vary across time-steps, they were not statistically treated.

In this experiment, six dependent variables (%original-weight, %originallength, %originalwidth, elongation, refinement and artifact condition (Figure 155) were measured on four groups of flint flakes (₣, #, ₩, ₮; Appendix 4) that were tumbled in different gravel lithologies that were sifted into a φ=-5.0 to -4.5 size category (Figure 155). The two independent variables tested were time and lithology. A bootstrapped one-way between subjects ANOVA showed that mean differences of the starting weight, length, width, thickness, elongation, and refinement of the flakes were statistically similar (Appendix 11).

Because time was measured over the same artifacts, it was considered a within-subjects factor (random factor). Sediment size varied between artifacts and was therefore considered a between-subjects factor (fixed factor). Tumbling barrel speed (15rpm), water volume (5l), artifact class (unretouched flakes), artifact lithology (flint), and sediment mass (2kg) were kept constant to reduce the effect of additional variables.

Figure 156 to Figure 161 shows the changes to artifact primary metrics and conditions over time when abraded in different sediment lithologies. Figure 163 reports the summary statistics while Figures 164 to 169 present the results of the mixed ANOVA and the post-hoc tests.

Figure 164 shows the results of the mixed-design ANOVA, testing time. Time had a significant effect on %originalweight, refinement, and %originalwidth ($p<0.05$; Figure 157 and Figure 158), but not on %originallength or elongation ($p>0.05$; Figure 159 and Figure 160). Contrasts revealed that time was significant across %originalweight, %originallength, %originalwidth, and refinement except for at 30 minutes.

Results also report a significant effect of sediment lithology on %originalweight, %originallength, and %originalwidth, but not on elongation or refinement ($p<0.05$; Figure 165). Post-hoc Tukey tests showed that the effect of the Trent Valley, Ham Yard and Kennet sediments on weight, length, and width loss of artifacts (Figure 166, Figure 167, Figure 168) did not differ. However, the control group appears to have a significantly different effect on weight, length and width loss compared to Kennet artifact lithologies, while having a significantly different effect on artifact width loss compared to Ham Yard sediments (Figure 168 and Figure 158).

Abrasion condition revealed that all artifacts were similarly heavily abraded (class 4) in less than 60 minutes in all sediment lithologies, even though a slower abrasion is visible before 60 minutes when using the Trent Valley sediments (Figure 163, Figure 161, Figure 162).

Figure 148: Descriptive statistics of changes to primary metrics and condition over time between flakes tumbled in different sediment sizes.

	Phi group	T=0	T=1	T=5	T=15	T=30	T=60	T=120	T=180
Weight	φ=0	1±0	0.992 ±0.017	0.992 ±0.017	0.991 ±0.017	0.988 ±0.018	0.981 0.024	0.974 ±0.026	0.975 ±0.025
	φ=-3.5 to -3.0	1±0	1.002 ±0.007	0.998 ±0.008	0.996 ±0.007	0.986 ±0.013	0.979 ±0.015	0.970 ±0.018	0.964 ±0.022
	φ=-4.0 to -3.5	1±0	1.000 ±0.007	0.995 ±0.012	0.989 ±0.015	0.943 ±0.131	0.929 ±0.132	0.921 ±0.130	0.904 ±0.181
	φ=-4.5 to -4.0	1±0	0.999 ±0.010	0.98 ±0.021	0.962 ±0.025	0.911 ±0.149	0.887 ±0.144	0.725 ±0.314	0.712 ±0.306
	φ=-5.0 to -4.5	1±0	0.998 ±0.007	0.972 ±0.058	0.941 ±0.058	0.853 ±0.193	0.824 ±0.188	0.800 ±0.195	0.785 ±0.191
	φ=-5.5 to -5.0	1±0	1.000 ±0.010	1.000 ±0.010	0.972 ±0.017	0.945 ±0.036	0.895 ±0.095	0.841 ±0.139	0.722 ±0.286
	φ=-6.0 to -5.5	1±0	0.971 ±0.091	0.8723 ±0.228	0.851 ±0.252	0.832 ±0.249	0.812 ±0.242	0.765 ±0.234	0.750 ±0.236
	φ=-6.0 to 0	1±0	0.993 ±0.0096	0.983 ±0.025	0.966 ±0.041	0.948 ±0.043	0.933 ±0.047	0.921 ±0.047	0.901 ±0.065
Length	φ=0	1±0	0.988 ±0.025	0.984 ±0.026	0.977 ±0.021	0.971 ±0.027	0.966 ±0.030	0.948 ±0.062	0.941 ±0.062
	φ=-3.5 to -3.0	1±0	0.983 ±0.027	0.971 ±0.026	0.962 ±0.026	0.951 ±0.026	0.940 ±0.031	0.924 ±0.040	0.919 ±0.035
	φ=-4.0 to -3.5	1±0	0.995 ±0.004	0.972 ±0.024	0.962 ±0.030	0.909 ±0.115	0.903 ±0.116	0.890 ±0.112	0.869 ±0.145
	φ=-4.5 to -4.0	1±0	0.988 ±0.011	0.970 ±0.033	0.929 ±0.062	0.886 ±0.089	0.860 ±0.083	0.744 ±0.280	0.735 ±0.280
	φ=-5.0 to -4.5	1±0	0.991 ±0.016	0.967 ±0.040	0.934 ±0.061	0.838 ±0.149	0.826 ±0.137	0.802 ±0.138	0.791 ±0.143
	φ=-5.5 to -5.0	1±0	0.981 ±0.024	0.981 ±0.024	0.950 ±0.038	0.918 ±0.048	0.875 ±0.081	0.830 ±0.114	0.718 ±0.274
	φ=-6.0 to -5.5	1±0	0.981 ±0.044	0.943 ±0.105	0.899 ±0.141	0.894 ±0.149	0.878 ±0.151	0.823 ±0.148	0.816 ±0.151
	φ=-6.0 to 0	1±0	0.992 ±0.009	0.973 ±0.026	0.960 ±0.044	0.952 ±0.041	0.934 ±0.046	0.915 ±0.047	0.898 ±0.052
Width	φ=0	1±0	1.009 ±0.035	1.001 ±0.043	1.005 ±0.034	1.005 ±0.034	1.003 ±0.034	0.994 ±0.041	0.985 ±0.048
	φ=-3.5 to -3.0	1±0	0.983 ±0.031	0.968 ±0.028	0.968 ±0.030	0.963 ±0.029	0.954 ±0.036	0.941 ±0.048	0.931 ±0.046
	φ=-4.0 to -3.5	1±0	0.989 ±0.015	0.970 ±0.022	0.961 ±0.030	0.919 ±0.069	0.906 ±0.080	0.898 ±0.0826	0.888 ±0.083
	φ=-4.5 to -4.0	1±0	0.972 ±0.048	0.932 ±0.051	0.920 ±0.066	0.890 ±0.136	0.871 ±0.142	0.729 ±0.296	0.760 ±0.283
	φ=-5.0 to -4.5	1±0	0.990 ±0.040	0.953 ±0.077	0.895 ±0.088	0.847 ±0.122	0.815 ±0.123	0.779 ±0.138	0.772 ±0.137
	φ=-5.5 to -5.0	1±0	0.989 ±0.017	0.989 ±0.017	0.951 ±0.034	0.899 ±0.067	0.879 ±0.106	0.827 ±0.088	0.712 ±0.264
	φ=-6.0 to -5.5	1±0	0.967 ±0.035	0.883 ±0.130	0.857 ±0.125	0.853 ±0.131	0.833 ±0.131	0.798 ±0.142	0.763 ±0.130
	φ=-6.0 to 0	1±0	0.985 ±0.023	0.978 ±0.023	0.959 ±0.031	0.953 ±0.034	0.939 ±0.037	0.911 ±0.037	0.893 ±0.048
Elongation	φ=0	1.640 ±0.422	1.610 ±0.429	1.616 ±0.432	1.595 ±0.414	1.585 ±0.405	1.579 ±0.405	1.553 ±0.357	1.554 ±0.352
	φ=-3.5 to -3.0	1.640 ±0.314	1.639 ±0.303	1.639 ±0.287	1.630 ±0.309	1.62 ±0.312	1.614 ±0.300	1.61 ±0.304	1.621 ±0.311
	φ=-4.0 to -3.5	1.479 ±0.309	1.487 ±0.305	1.479 ±0.281	1.478 ±0.277	1.438 ±0.144	1.452 ±0.162	1.445 ±0.168	1.412 ±0.137
	φ=-4.5 to -4.0	1.377 ±0.261	1.407 ±0.273	1.448 ±0.307	1.442 ±0.369	1.446 ±0.423	1.443 ±0.449	1.450 ±0.441	1.338 ±0.337

Figure 148: ***Continued.***

	***Phi* group**	**T=0**	**T=1**	**T=5**	**T=15**	**T=30**	**T=60**	**T=120**	**T=180**
	φ=-5.0 to -4.5	1.633 ±0.275	1.637 ±0.285	1.669 ±0.326	1.709 ±0.296	1.612 ±0.331	1.652 ±0.310	1.686 ±0.319	1.684 ±0.364
	φ=-5.5 to -5.0	1.448 ±0.257	1.436 ±0.267	1.436 ±0.267	1.445 ±0.274	1.481 ±0.283	1.456 ±0.372	1.440 ±0.335	1.4702 ±0.337
	φ=-6.0 to -5.5	1.45 ±0.353	1.468 ±0.349	1.548 ±0.348	1.52 ±0.382	1.517 ±0.376	1.538 ±0.437	1.507 ±0.428	1.563 ±0.468
	φ=-6.0 to 0	1.404 ±0.184	1.415 ±0.190	1.399 ±0.201	1.409 ±0.228	1.408 ±0.223	1.403 ±0.225	1.413 ±0.20	1.414 ±0.190
Refinement	φ=0	4.922 ±2.454	4.641 ±2.915	4.933 ±2.471	4.948 ±2.452	4.941 ±2.440	4.932 ±2.423	4.881 ±2.378	4.845 ±2.379
	φ=-3.5 to -3.0	4.001 ±1.237	3.916 ±1.149	3.866 ±1.177	3.852 ±1.121	3.84 ±1.141	3.803 ±1.132	3.734 ±1.064	3.695 ±1.070
	φ=-4.0 to -3.5	4.308 ±0.770	4.252 ±0.751	4.176 ±0.717	4.143 ±0.715	4.009 ±0.809	3.956 ±0.826	3.919 ±0.823	3.871 ±0.810
	φ=-4.5 to -4.0	5.065 ±2.009	4.960 ±2.037	4.726 ±1.833	4.608 ±1.645	4.363 ±1.432	4.293 ±1.485	3.503 ±1.798	3.801 ±2.230
	φ=-5.0 to -4.5	5.139 ±2.431	5.090 ±2.460	4.812 ±2.121	4.426 ±1.589	4.133 ±1.423	3.986 ±1.439	3.809 ±1.450	3.780 ±1.472
	φ=-5.5 to -5.0	4.283 ±1.891	4.215 ±1.858	4.215 ±1.858	4.069 ±1.849	3.816 ±1.713	3.781 ±1.794	3.516 ±1.631	3.359 ±1.537
	φ=-6.0 to -5.5	4.262 ±1.726	4.129 ±1.721	3.628 ±1.004	3.527 ±1.045	3.504 ±0.996	3.422 ±0.981	3.258 ±0.893	3.133 ±0.901
	φ=-6.0 to 0	4.098 ±1.252	4.023 ±1.182	3.998 ±1.200	3.918 ±1.158	3.886 ±1.129	3.827 ±1.119	3.705 ±1.053	3.628 ±1.032
Artifact condition	φ=0	1	1	1	1	1	1	1	1
	φ=-3.5 to -3.0	1	1	2	3	4	4	4	4
	φ=-4.0 to -3.5	1	2	2	2	3	4	4	4
	φ=-4.5 to -4.0	1	1	2	3	3	4	4	4
	φ=-5.0 to -4.5	1	1	2	3	4	4	4	4
	φ=-5.5 to -5.0	1	1	2	3	3	4	4	4
	φ=-6.0 to -5.5	1	1	3	3	4	4	4	4
	φ=-6.0 to 0	1	1	2	3	4	4	4	4

Figure 149: Statistical analysis of the effect of time on primary metrics (Greenhouse-Geisser correction).

	Type III Sum of Squares	**df**	**df error**	**Mean square**	**F**	**Sig.**
%originalweight	1.581	2.586	186.178	.611	34.504	**.000**
%originallength	1.515	2.426	174.668	.624	43.019	**.000**
%originalwidth	1.314	2.514	180.990	.534	38.141	**.000**
Elongation	.39	3.505	245.281	.011	.466	.736
Refinement	40.434	2.796	195.569	14.463	23.627	**.000**
Artifact condition	N/A	N/A	N/A	N/A	N/A	N/A

Figure 150: Statistical analysis of the effect of sediment size on primary metrics (Greenhouse-Geisser correction).

	Type III Sum of Squares	df	df error	Mean square	F	Sig.
%originalweight	.204	7	72	.029	2.753	**.014**
%originallength	.093	7	72	.013	2.543	**.021**
%originalwidth	.198	7	72	.028	5.847	**.000**
Elongation	5.269	7	70	.753	1.008	.433
Refinement	94.429	7	70	13.490	.702	.670
Artifact condition	N/A	N/A	N/A	N/A	N/A	N/A

Figure 151: Tukey post-hoc comparisons for %originalweight.

Phi group	*n*	Grouping	
-6.0-5.5	10	A	
-5.5-5.0	10	A	B
-5.0-4.5	10	A	B
-4.5-4.0	10	A	B
-4.0-3.5	10	A	B
-3.5-3.0	10		B
-6.0-0	10	A	B
0	10		B

There was a significant effect of the interaction of time and sediment lithology on %originalweight and refinement of artifacts (Figure 169). However, contrasts revealed that these were only significant on %originalweight between 5 and 15 minutes (Appendix 12-time*sediment lithology).

The results of this section lead to four main conclusions:

1. As in previous experiments, time had a significant effect on all metrics except for %originallength and elongation, for all randomly selected artifacts (i.e., random shapes).
2. Results showed that the various sediment lithologies used in this experiment (i.e., Ham Yard, Trent Valley,

Figure 152: Tukey post-hoc comparisons for %originallength.

Phi group	*n*	Grouping
-6.0-5.5	10	A
-5.5-5.0	10	A
-5.0-4.5	10	A
-4.5-4.0	10	A
-4.0-3.5	10	A
-3.5-3.0	10	A
-6.0-0	10	A
0	10	A

Figure 153: Tukey post-hoc comparisons for %originalwidth.

Phi group	*n*	Grouping		
-6.0-5.5	10	A		
-5.5-5.0	10	A	B	
-5.0-4.5	10	A	B	
-4.5-4.0	10	A	B	
-4.0-3.5	10	A	B	C
-3.5-3.0	10		B	C
-6.0-0	10	A	B	C
0	10			C

Figure 154: Statistical analysis of the effect of the interaction of sediment size and time on primary metrics (Greenhouse-Geisser correction).

	Type III Sum of Squares	df	df error	Mean square	F	Sig.
%originalweight	.835	18.101	186.178	.046	2.605	**.001**
%originallength	.528	16.982	174.668	.031	2.140	**.007**
%originalwidth	.613	17.596	180.990	.035	2.490	**.001**
Elongation	.391	24.538	245.381	.016	.675	.876
Refinement	25.145	19.569	195.693	1.285	2.099	**.006**
Artifact condition	N/A	N/A	N/A	N/A	N/A	N/A

Figure 155: List of experimental conditions for the sediment lithology experiment.

Dependent variables	Independent variables		Controlled variables
	Fixed	*Random*	
%originalweight	Sediment	Time	Tumbling barrel speed (15rpm)
%originallength	Lithology		Water volume (5l)
%originalwidth	(Ham Yard,		Artifact class (unretouched flakes)
Elongation	Trent Valley,		*Phi* group (φ=-5.0 to 4.5)
Refinement	Kennet, Control)		Sediment mass (2kg)
Artifact condition			Artifact lithology (flint)
(Ashton 1998)			

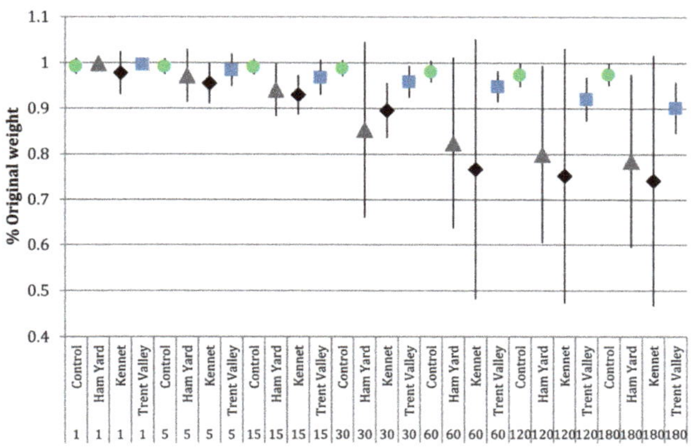

Figure 156: Impact of time and sediment lithology on %originalweight. *n*=10 for each group. Error bars represent 1 SD in either direction.

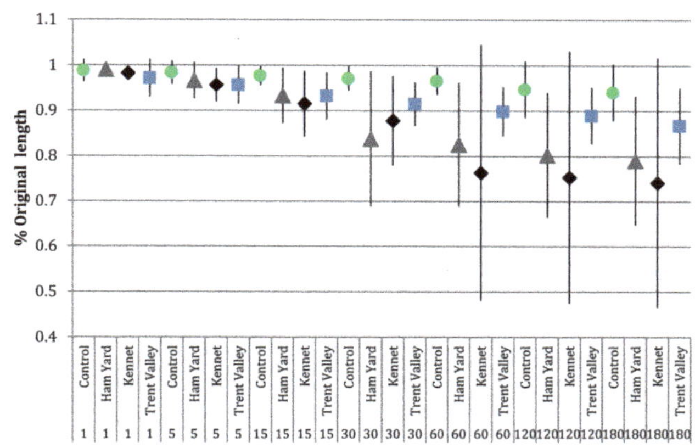

Figure 157: Impact of time and sediment lithology on %originallength. *n*=10 for each group. Error bars represent 1 SD in either direction.

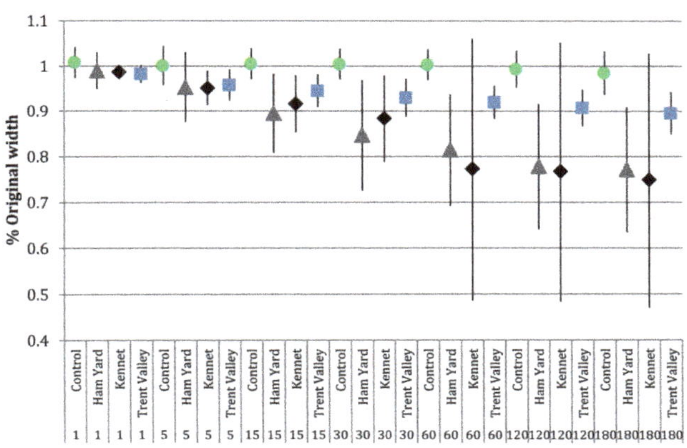

Figure 158: Impact of time and sediment lithology on %originalwidth. *n*=10 for each group. Error Bars represent 1 SD in either direction.

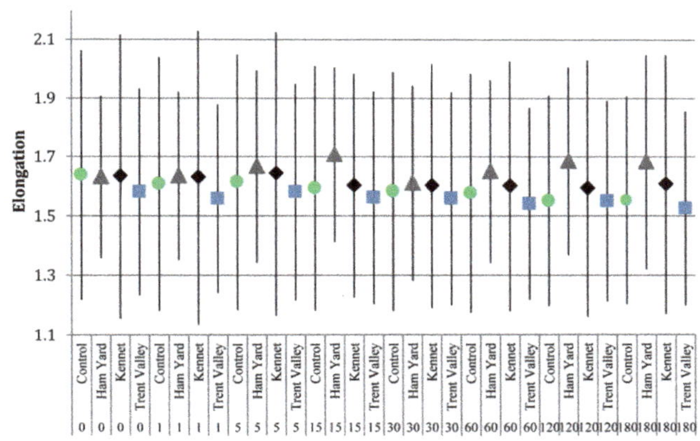

Figure 159: Impact of time and sediment lithology on elongation. *n*=10 for each group. Error bars represent 1 SD in either direction.

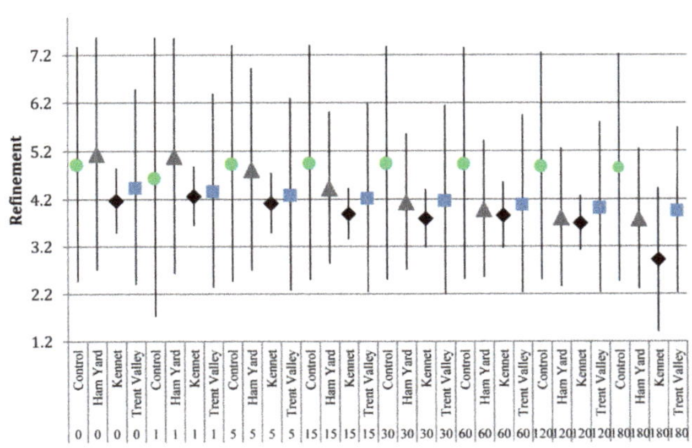

Figure 160: Impact of time and sediment lithology on refinement. *n*=10 for each group. Error bars represent 1 SD in either direction.

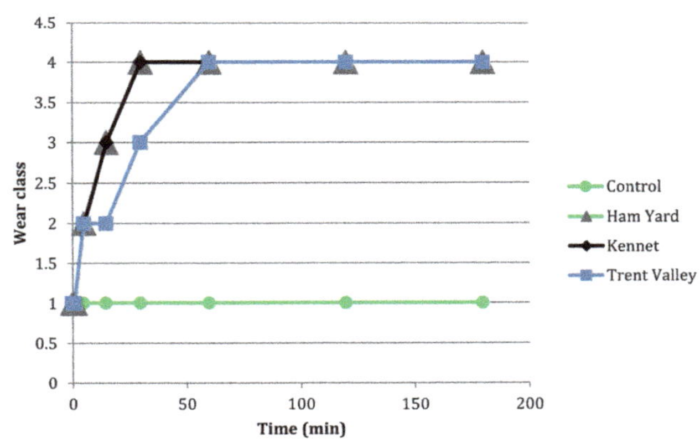

Figure 161: Impact of time and sediment lithology on artifact condition. *n*=10 for each group.

Kennet and control) affect artifacts %originalweight, %originalength, and %originalwidth in a significantly different manner ($p<0.05$; Figure 157). Post-hoc tests revealed that the effects of the Kennet sediments on artifacts primary metrics were significantly different from the control, Ham Yard, and Trent Valley sediments. This result is in line with other experimental observations of polylithologic pebble populations, which showed that the abrasion of sediments of comparable durability remained the same but attrition rates of less durable rock types increased whereas rates of more durable types decreased (Abbott and Peterson 1978). This suggests that Kennet gravels have a significantly different or harder lithological component than flint as a constituent. The artifact condition data however suggests that different sediment lithologies affect the

Figure 162: Artifacts tumbled in different sediment lithologies after 180 min. ***From top left to right: None (control), Ham Yard, Trent Valley, Kennet.***

appearance of abrasion more or less dramatically in spite of no discernible changes to morphology.

3. The results reported no significant effect of sediment type on elongation and refinement, which indicates that the loss of material, if any, is proportional between the various metrics leading to scaled reduction, similarly to what is reported for time (Figure 165).
4. The data suggested that there is an overall significant interaction between sediment lithology and time for weight and refinement. The interaction of time and sediment lithology on weight was only significant at one of the seven time-steps, suggesting that overall, time does not moderate the effect of sediment lithologies on abrasion. It is unclear why this affected refinement without affecting width (as refinement is the ratio of width/thickness and thickness being an essentially static measurement).

5.6. Artifact class, time, and artifact abrasion

The purpose of this analysis was to explore the effect of artifact class on the abrasion of lithic artifacts over time, focusing on flake artifacts and flake handaxes (core tool artifacts). If initial size and artifact morphology plays a role in the abrasion of artifacts (Harding et al. 1987; Lewin and Brewer 2002), then significant differences to abrasion are expected between lithic flakes and handaxes. A test was devised to assess the impact of artifact type on artifact abrasion, measured through primary metrics, using subsets of 10 flint flakes and 10 handaxes (zł + Handaxes; see Appendix 4).

Formally stated, the hypotheses for this section were as follows:

> H_0=There is no significant difference in changes to primary metrics over time between handaxes and flakes.
>
> H_1=There is a significant difference in changes to primary metrics over time between handaxes and flakes.

This analysis used a two-way mixed-design ANOVA to compare the means of the two groups. In this test, time was treated as the random factor while artifact class was considered a fixed factor. Planned contrasts were used to assess the significance of each time-step when compared to the previous for each primary metric (see Section 3.5.8). Since artifact condition categories did not vary across time-steps, they were not statistically treated.

In this experiment, six dependent variables (%originalweight, %originallength, %originalwidth, elongation, refinement, and artifact condition; Figure 170) were measured on flint flakes (zł; Appendix 4) and handaxes. The two independent variables tested were time and artifact type. A bootstrapped Welch's t-test showed that flakes and handaxes were statistically different in terms of their initial mass, length, width, thickness, and refinement, but not in elongation ($p>0.05$; Appendix 13).

Time was measured over the same artifacts and was therefore a random factor. Sediment size varied between artifacts and was therefore considered a fixed factor. Tumbling barrel speed (15rpm), water volume (5l), artifact lithology (flint), sediment lithology (Ham Yard), and sediment mass (2kg) were kept constant to reduce their effect during this experiment.

Figure 171 through Figure 177 illustrate the changes to artifact primary metrics and conditions between flakes and handaxes. Figure 178 reports the summary statistics while Figure 179 to Figure 181 present the results of the mixed ANOVA.

Results showed a significant effect of time on the reduction of artifact weight, length, width, and refinement ($p<0.001$; Figure 179, Figure 171, Figure 172, Figure 173, Figure 175), but no significant effect of time on elongation was reported ($p>0.05$; Figure 179 and Figure 174). Contrasts suggest that these effects were significant for all time-steps (Appendix 14).

Results also showed a statistically significant effect of artifact type on all dependent variables except for %originallength and elongation ($p<0.05$; Figure 180, Figure 171 to 175), with flakes consistently losing more from each significant category over time, except for elongation.

Figure 163: Descriptive statistics of changes to primary metrics and condition over time between flakes tumbled in different sediment lithologies.

		T=0	T=1	T=5	T=15	T=30	T=60	T=120	T=180
%originalweight	*Control*	1±0	0.992 ±0.017	0.992 ±0.017	0.991 ±0.017	0.988 ±0.018	0.981 ±0.024	0.974 ±0.026	0.975 ±0.025
	Ham Yard	1±0	0.998 ±0.007	0.972 ±0.058	0.941 ±0.058	0.853 ±0.193	0.824 ±0.188	0.800 ±0.195	0.785 ±0.191
	Kennet	1±0	0.978 ±0.047	0.955 ±0.044	0.930 ±0.043	0.896 ±0.060	0.767 ±0.285	0.753 ±0.279	0.742 ±0.275
	Trent Valley	1±0	0.997 ±0.014	0.985 ±0.036	0.969 ±0.038	0.959 ±0.035	0.949 ±0.034	0.921 ±0.048	0.902 ±0.056
%originallength	*Control*	1±0	0.988 ±0.025	0.984 ±0.026	0.977 ±0.021	0.971 ±0.027	0.966 ±0.030	0.948 ±0.062	0.941 ±0.062
	Ham Yard	1±0	0.991 ±0.016	0.967 ±0.040	0.934 ±0.061	0.838 ±0.149	0.826 ±0.137	0.802 ±0.138	0.791 ±0.143
	Kennet	1±0	0.983 ±0.014	0.957 ±0.037	0.916 ±0.072	0.878 ±0.099	0.763 ±0.282	0.753 ±0.279	0.741 ±0.276
	Trent Valley	1±0	0.972 ±0.041	0.957 ±0.042	0.933 ±0.052	0.915 ±0.048	0.899 ±0.054	0.890 ±0.062	0.868 ±0.084
%originalwidth	*Control*	1±0	1.009 ±0.035	1.001 ±0.043	1.005 ±0.034	1.005 ±0.034	1.003 ±0.034	0.994 ±0.041	0.985 ±0.048
	Ham Yard	1±0	0.990 ±0.040	0.953 ±0.077	0.895 ±0.088	0.847 ±0.122	0.815 ±0.123	0.779 ±0.138	0.772 ±0.137
	Kennet	1±0	0.987 ±0.011	0.952 ±0.038	0.916 ±0.063	0.884 ±0.096	0.773 ±0.287	0.768 ±0.284	0.749 ±0.279
	Trent Valley	1±0	0.983 ±0.020	0.958 ±0.035	0.945 ±0.036	0.930 ±0.042	0.920 ±0.037	0.908 ±0.041	0.896 ±0.047
Elongation	*Control*	1.640 ±0.422	1.610 ±0.429	1.616 ±0.432	1.595 ±0.414	1.585 ±0.405	1.579 ±0.405	1.553 ±0.357	1.554 ±0.352
	Ham Yard	1.633 ±0.275	1.637 ±0.285	1.669 ±0.326	1.709 ±0.296	1.612 ±0.331	1.652 ±0.310	1.686 ±0.319	1.684 ±0.364
	Kennet	1.6354 ±0.481	1.632 ±0.497	1.645 ±0.481	1.604 ±0.379	1.603 ±0.414	1.602 ±0.424	1.594 ±0.435	1.608 ±0.440
	Trent Valley	1.583 ±0.350	1.559 ±0.319	1.583 ±0.367	1.563 ±0.360	1.560 ±0.361	1.543 ±0.326	1.552 ±0.341	1.528 ±0.329
Refinement	*Control*	4.922 ±2.454	4.641 ±2.915	4.933 ±2.471	4.948 ±2.452	4.941 ±2.440	4.932 ±2.423	4.881 ±2.378	4.845 ±2.379
	Ham Yard	5.139 ±2.431	5.090 ±2.460	4.812 ±2.120	4.426 ±1.589	4.133 ±1.423	3.986 ±1.439	3.809 ±1.450	3.780 ±1.472
	Kennet	4.170 ±0.684	4.262 ±0.617	4.115 ±0.633	3.890 ±0.545	3.792 ±0.617	3.860 ±0.700	3.698 ±0.584	2.918 ±1.518
	Trent Valley	4.443 ±2.042	4.367 ±2.026	4.286 ±2.013	4.222 ±1.978	4.171 ±1.983	4.088 ±1.863	4.017 ±1.784	3.956 ±1.730
Artifact condition	*Control*	1	1	1	1	1	1	1	1
	Ham Yard	1	1	2	3	4	4	4	4
	Kennet	1	1	2	3	4	4	4	4
	Trent Valley	1	1	2	2	3	4	4	4

Contrasts suggest that these significant interactions took place over all time-steps (Appendix 14-time).

Artifact conditon showed important differences in artifact abrasion rates with flakes becoming heavily abraded by 30 minutes, while handaxes did not become heavily abraded until 120 minutes (Figure 178, Figure 176, Figure 177).

Figure 181 indicates that the interaction between time and artifact class was significant for %originalweight, %originalwidth, and refinement. Contrasts revealed, however, that these differences were all non-significant between time-steps (Appendix 14-time*artifact class).

Figure 164: Statistical analysis of the effect of time on primary metrics (Greenhouse-Geisser correction).

	Type III Sum of Squares	df	df error	Mean square	F	Sig.
%originalweight	.775	1.418	51.059	.547	15.638	**.000**
%originallength	.835	1.395	50.208	.599	17.202	.093
%originalwidth	.704	1.335	48.045	.527	15.699	**.000**
Elongation	.060	2.742	95.958	.019	.857	.458
Refinement	19.610	2.605	91.192	7.526	12.088	**.000**
Artifact condition	N/A	N/A	N/A	N/A	N/A	N/A

Figure 165: Statistical analysis of the effect of sediment lithology on primary metrics (Greenhouse-Geisser correction).

	Type III Sum of Squares	df	df error	Mean square	F	Sig.
%originalweight	.104	3	36	.035	4.559	**.008**
%originallength	.073	3	36	.024	3.882	**.017**
%originalwidth	.130	3	36	.043	6.776	**.001**
Elongation	.437	3	35	.146	.136	.938
Refinement	43.494	3	35	14.498	.536	.661
Artifact condition	N/A	N/A	N/A	N/A	N/A	N/A

Figure 166: Tukey post-hoc comparisons for %originalweight.

Sediment lithology	*n*	Grouping	
Kennet	10	A	
Ham Yard	10	A	B
Trent Valley	10	A	B
Control	10		B

Figure 167: Tukey post-hoc comparisons for %originallength.

Sediment lithology	*n*	Grouping	
Kennet	10	A	
Ham Yard	10	A	B
Trent Valley	10	A	B
Control	10		B

The results of this experiment can be summarized as follows:

1. Time had a significant effect on artifact %originalweight, %originallength, %originalwidth, refinement and artifact condition for at least 3 hours (Figure 179). Time, however, did not have a significant effect on artifact elongation ($p>0.05$; Figure 179), indicating that changes to primary metrics were largely scaled across length and width.
2. Artifact class does play a role in the abrasion of artifacts' primary metrics, with flakes abrading significantly quicker than handaxes in all measurement categories except for length and elongation. This suggests that while initial artifact class plays a role in the rate of overall abrasion of artifacts over time, the way in which it affects length is the same, or at least comparable.

Figure 168: Tukey post-hoc comparisons for %originalwidth.

Sediment lithology	*n*	Grouping	
Kennet	10	A	
Ham Yard	10	A	
Trent Valley	10	A	B
Control	10		B

3. Artifact class has a major effect upon the rate of artifact abrasion with flakes abrading much more rapidly than handaxes. This is likely due to the fact that flakes were smaller in all primary metrics with more acute edges and therefore more susceptible to damage. It is also possible that the bifacially flaked edges of the handaxes were more stable than the flakes, making them abrade less rapidly.

Figure 169: Statistical analysis of the effect of time and sediment lithology on primary metrics (Greenhouse-Geisser correction).

	Type III Sum of Squares	df	df error	Mean square	F	Sig.
%originalweight	.104	3	36	.035	4.559	**.008**
%originallength	.305	4.18	50.208	.073	2.092	.093
%originalwidth	.355	4.004	48.045	.089	2.639	.055
Elongation	.123	8.225	95.958	.015	.580	.796
Refinement	16.370	7.816	91.192	2.094	3.363	**.002**
Artifact condition	N/A	N/A	N/A	N/A	N/A	N/A

Figure 170: List of experimental conditions for the artifact class experiment.

Dependent Variables	Independent Variables		Controlled Variables
	Fixed	*Random*	
%originalweight	Artifact Class	Time	Tumbling barrel speed (15rpm)
%originallength	(Handaxes, flakes)		Water volume (5l)
%originalwidth			Sediment Lithology (Ham yard)
Elongation			*Phi* Group (φ=0 to -6)
Refinement			Sediment Mass (2kg)
Artifact condition			Artifact lithology (flint)
(Ashton 1998)			

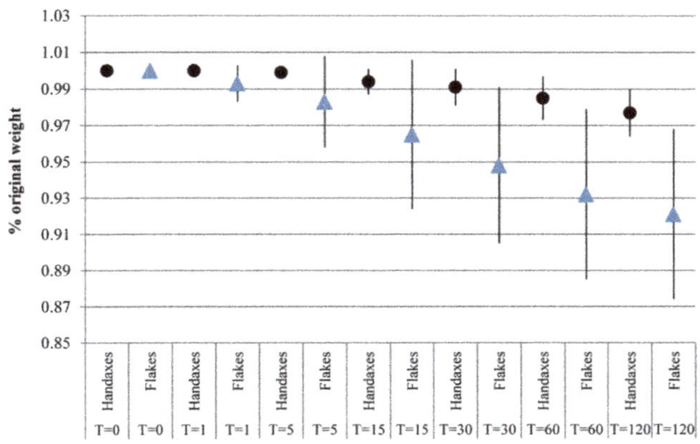

Figure 171: Impact of time on %originalweight for flakes (n=10) and handaxes (n=10). Error bars represent 1 SD in either direction.

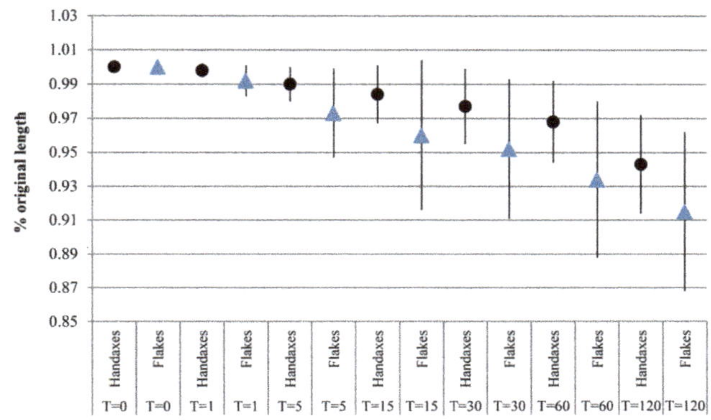

Figure 172: Impact of time on %originallength for flakes (n=10) and handaxes (n=10). Error bars represent 1 SD in either direction.

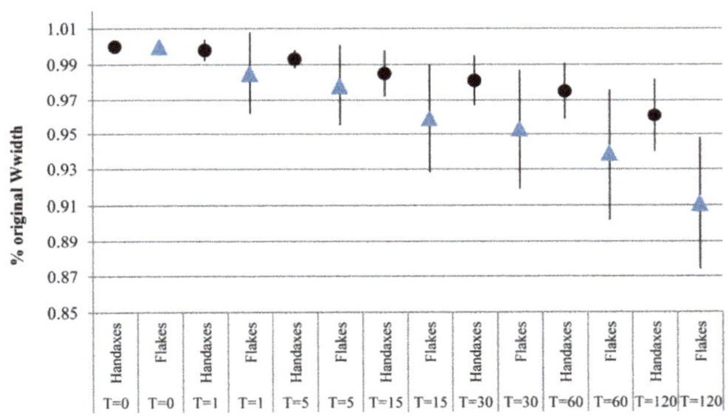

Figure 173: Impact of time and flakes (n=10) and handaxes (n=10) on %originalwidth. Error bars represent 1 SD in either direction.

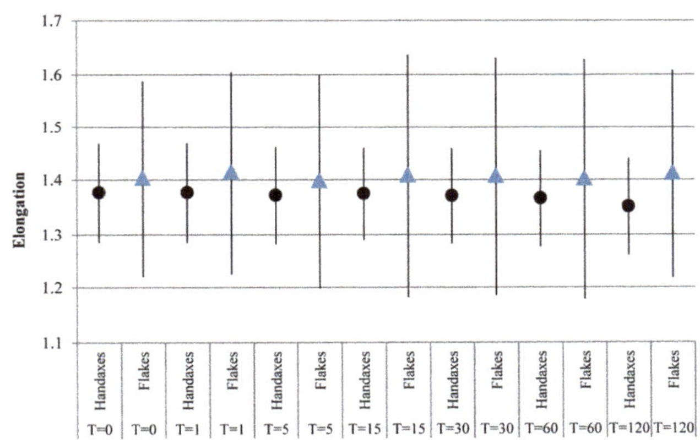

Figure 174: Impact of time on elongation for flakes (n=10) and handaxes (n=10). Error bars represent 1 SD in either direction.

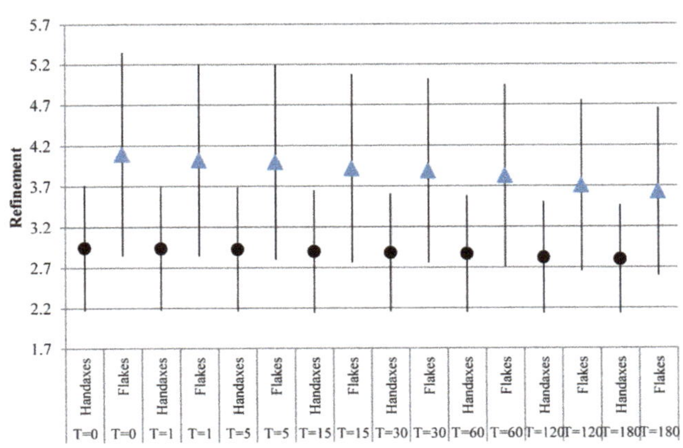

Figure 175: Impact of time on refinement for flakes (n=10) and handaxes (n=10). Error bars represent 1 SD in either direction.

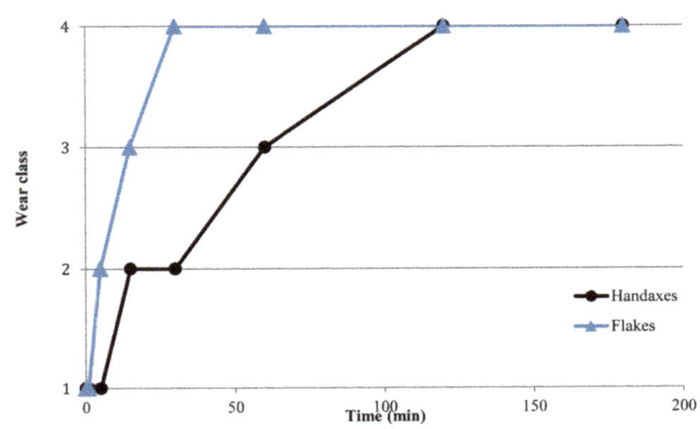

Figure 176: Impact of time on artifact condition for flakes (n=10) and cordiform handaxes (n=10).

Figure 177: Different artifact classes used in this experiment after 180 min. ***Left is a flake while right is a handaxe.***

4. The fact that interaction of time and artifact class had an overall significant effect on %originalweight, %originalwidth, and refinement but not between individual time-steps, is likely due to a steady and non-drastic loss throughout the experiment, which ultimately results in a significant loss compared to the initial metrics.

5.7. Chapter summary

Figure 182 summarizes the significance of each of the dependent variables at each time-step from the tumbling barrel experiments above. The data show that time has a significant effect on %originalweight, %originallength, %originalwidth and refinement, but not elongation, presumably because of scaled reductions in length and width. This suggests that increased abrasion time tends not to affect the overall planiform morphology of the artifact (length/width), but does affect the lateral view morphology. Contrasts revealed that, in some experiments, these effects were non-significant between 120 minutes and 180 minutes, suggesting that abrasion may begin to slow down after 2 hours.

The analysis in Section 5.2 on sediment mass revealed that there is a correlation between the number of sedimentary impacts (using sediment load as a proxy) and the abrasion of lithic artifacts. In the early stages of abrasion, this is mediated by time, suggesting that increased impacting in the early stages of abrasion have a greater effect in the earlier than in the later stages in the abrasion process.

The analysis in Section 5.3 (artifact lithology) demonstrated that three different cryptocrystalline raw materials did not differ in how they morphologically abraded, even though their abrasion condition varied considerably. Patinated flint artifacts abraded most rapidly followed by flint and quartzite flakes. This does not support the prediction that hardness or grain size affects actual changes to artifact morphology, but that it is relevant to physical condition.

The data analysis in Section 5.4 relates sediment size to the abrasion of artifacts. It shows that larger gravel sizes (above φ=-5.5) abrade artifacts more rapidly and dramatically than smaller sediment sizes (less than φ=-3.5). This is in support

Figure 178: Descriptive statistics of changes to primary metrics and condition over time between handaxes and flakes.

		T=0	T=1	T=5	T=15	T=30	T=60	T=120	T=180
%originalweight	*Handaxes*	1±0	1±0.001	0.999±0.001	0.994±0.007	0.991±0.01	0.985±0.012	0.977±0.013	0.972±0.015
	Flakes	1±0	0.993±0.01	0.983±0.025	0.965±0.041	0.948±0.043	0.932±0.047	0.921±0.047	0.901±0.065
%originallength	*Handaxes*	1±0	0.998±0.004	0.99±0.01	0.984±0.017	0.977±0.022	0.968±0.024	0.943±0.029	0.936±0.029
	Flakes	1±0	0.992±0.009	0.973±0.026	0.96±0.044	0.952±0.041	0.934±0.046	0.915±0.047	0.898±0.052
%originalwidth	*Handaxes*	1±0	0.998±0.006	0.993±0.005	0.985±0.013	0.981±0.014	0.975±0.016	0.961±0.021	0.954±0.024
	Flakes	1±0	0.985±0.023	0.978±0.023	0.959±0.031	0.953±0.034	0.939±0.037	0.911±0.037	0.893±0.048
Elongation	*Handaxes*	1.377±0.092	1.377±0.092	1.372±0.09	1.375±0.085	1.371±0.088	1.366±0.089	1.351±0.09	1.352±0.099
	Flakes	1.404±0.184	1.415±0.19	1.399±0.201	1.409±0.228	1.408±0.223	1.403±0.225	1.413±0.195	1.414±0.19
Refinement	*Handaxes*	2.943±0.773	2.933±0.759	2.923±0.764	2.896±0.752	2.88±0.727	2.864±0.718	2.819±0.687	2.796±0.668
	Flakes	4.098±1.252	4.023±1.182	3.998±1.2	3.918±1.158	3.886±1.129	3.827±1.119	3.705±1.053	3.628±1.032
Condition	*Handaxes*	1	1	1	2	2	3	4	4
	Flakes	1	1	2	3	4	4	4	4

Figure 179: Statistical analysis of the effect of time on primary metrics (Greenhouse-Geisser correction).

	Type III Sum of Squares	df	df error	Mean square	F	Sig.
%originalweight	.058	2.405	43.283	.024	22.297	**.000**
%originallength	.094	2.454	44.173	.038	40.073	**.000**
%originalwidth	.075	1.993	35.883	.005	47.969	**.000**
Elongation	.003	2.271	40.870	.002	.274	.789
Refinement	1.597	1.737	31.273	.233	7.548	**.003**
Artifact condition	N/A	N/A	N/A	N/A	N/A	N/A

Figure 180: Statistical analysis of the effect of artifact class on primary metrics (Greenhouse-Geisser correction).

	Type III Sum of Squares	df	df error	Mean square	F	Sig.
%originalweight	.054	1	18	.054	12.623	**.002**
%originallength	.021	1	18	.021	4.246	.054
%originalwidth	.038	1	18	.038	11.215	**.004**
Elongation	.066	1	18	.066	.348	.563
Refinement	40.289	1	18	40.289	5.509	**.031**
Artifact condition	N/A	N/A	N/A	N/A	N/A	N/A

Figure 181: Statistical analysis of the effect of time and artifact class on primary metrics (Greenhouse-Geisser correction).

	Type III Sum of Squares	df	df error	Mean square	F	Sig.
%originalweight	.016	2.405	43.283	.007	6.187	**.003**
%originallength	.004	2.454	44.173	.001	1.500	.232
%originalwidth	.010	1.993	35.883	.005	6.253	**.005**
Elongation	.007	2.271	40.870	.003	.546	.605
Refinement	.405	1.737	31.273	.233	7.548	**.003**
Artifact condition	N/A	N/A	N/A	N/A	N/A	N/A

of the prediction made in Chapter 1. However, the presence of smaller sediment sizes in real world aggregate sediments appears to dampen the effect of larger sediments during abrasion (see Figure 41 for specific breakdown of sediment sizes), leading to only extremely coarse gravels having a significantly greater effect on artifact abrasion compared to smaller sediments.

The experiment in Section 5.5 shows that sediment lithology affects the abrasion (primary metrics as well as condition) of lithic artifacts, depending on their polyclastic composition.

The analysis in Section 5.6 shows that artifact class plays an important role in the abrasion of artifacts, with handaxes being abraded much more slowly than flakes due to their more resistant morphology.

The significance, synthesis, and limitations of the tumbling barrel results and will be discussed in more detail in Section 7.3.

Figure 182: Summary of the effects of time for all primary metrics across all five experiments. 'X' denotes $p<0.05$ while '0' denotes ($p>0.05$).

		0 and 1	1 and 5	5 and 15	15 and 30	30 and 60	60 and 120	120 and 180
Time and weight contrasts	*Sediment mass*	N/A	X	0	X	X	X	N/A
	Artifact lithology	N/A	X	X	X	X	X	X
	Sediment size	N/A	X	X	X	X	X	X
	Sediment lithology	N/A	X	X	X	0	X	X
	Artifact class	N/A	0	0	X	X	X	X
Time and length contrasts	*Sediment mass*	N/A	X	X	X	X	X	N/A
	Artifact lithology	N/A	X	X	X	X	X	0
	Sediment size	N/A	X	X	X	X	X	0
	Sediment lithology	N/A	X	X	X	0	X	X
	Artifact class	N/A	X	0	X	X	X	X
Time and width contrasts	*Sediment mass*	N/A	0	0	0	X	X	N/A
	Artifact lithology	N/A	X	X	X	X	0	0
	Sediment size	N/A	X	X	X	X	X	0
	Sediment lithology	N/A	X	X	X	0	X	X
	Artifact class	N/A	X	X	X	X	X	X
Time and elongation contrasts	*Sediment mass*	0	X	0	0	0	0	N/A
	Artifact lithology	0	0	0	0	0	0	0
	Sediment size	0	0	0	0	0	0	0
	Sediment lithology	0	0	0	0	0	0	0
	Artifact class	0	0	0	0	0	0	0
Time and refinement contrasts	*Sediment mass*	0	0	0	0	0	X	N/A
	Artifact lithology	X	0	X	X	0	0	0
	Sediment size	X	0	X	X	X	X	0
	Sediment lithology	0	0	X	X	0	X	0
	Artifact class	X	X	X	X	X	X	X

Chapter 6

Flume experiments

6.1. Introduction

The last chapter (Chapter 5) examined the influence of macroscopic fluvial wear on gross artifact morphology. The experiments in this chapter focus on the changes to lithic artifacts in the smaller sediment sizes that did not elicit the gross morphological changes in the tumbling barrel (see Section 7.3 for that experiment), focusing specifically on the development of surface micro-abrasion, ridge wear, and edge damage.

This study used an annular flume, scanning electron microscope (SEM) and an energy-dispersive X-ray spectrometer to study the effects of time, sediment size, and transport mode (*in situ* v *ex situ* wear) on the type and amount of surface micro-abrasion, ridge wear, and edge micro-flaking on replica lithic artifacts (refer to Section 3.4 for methods). This chapter is divided into three parts, each examining one of these three dependent variables univariately.

6.2. Sediment size, time, mobility, and surface micro-abrasion

This section assesses the impact of time, sediment size, and artifact transport mode (*in situ* v *ex situ*) on the development of flaked artifact surface micro-abrasion.

Three separate hypotheses (A, B, and C) were tested here to explore the effect of sediment size, time, and mobility on lithic surface micro-abrasion:

H_{A0}=Sediment size does not significantly change the rate of flaked lithic surface micro-abrasion.

H_{A1}=Sediment size significantly changes the rate of flaked lithic surface micro-abrasion.

H_{B0}=Increasing time does not significantly increase the quantity of flaked lithic surface micro-abrasion.

H_{B1}=Increasing time significantly increases the quantity of flaked lithic surface micro-abrasion.

H_{C0}=Changing mobility does not significantly change the rate of flaked lithic surface micro-abrasion.

H_{C1}=Changing mobility significantly changes the rate of flaked lithic surface micro-abrasion.

An ordinal logistic regression (logit; OLR; see Section 3.5.9) was conducted to provide statistical support for the observations of developing flaked lithic surface micro-abrasion, using sediment size, time, and mobility as predictors.

In this experiment, artifact surface micro-abrasion was measured on nine *microscopic observation areas* (2x2mm, 20mm apart) across 27 different artifacts (see Section 3.4.1). Artifact surfaces were ranked according to a four-point scale (Figure 183) where (1) indicated no damage (100% of original surface preserved; Figure 184b), (2) indicated slightly abraded (less than 33% of original surface removed; Figure 184c), (3) indicated moderately abraded (between 33% and 66% of original surface removed; Figure 184d), and (4) indicated very abraded (66% or more of original surface removed; Figure 184e).

Figure 183: List of experimental conditions for the surface micro-abrasion experiment.

Dependent variables	Independent variables	Controlled variables
<u>Artifact surface micro-abrasion:</u> 1. No damage: (100% of original surface preserved) 2. Slightly abraded: (less than 33% of original surface removed) 3. Moderately abraded: (between 33% and 66% of original surface removed) 4. Very abraded: (66% or more of original surface removed)	**<u>Time:</u>** 24h, 72h, 96h, 168h **<u>Sediment sizes:</u>** • Control (φ=0) • Medium sands • (φ=2.25 to 2) • Very fine/fine gravels • (φ=-1.75 to -2.25) • Medium gravels • (φ=-3 to -3.25) • Medium/coarse gravels (φ=-3.5 to -4.25) **<u>Transport mode:</u>** • Fixed • Unfixed	**<u>Transportation mode of abrasive sediment</u>**: Saltation **<u>Lithology</u>**: Flint **<u>Artifact class:</u>** Unretouched handaxe thinning flake

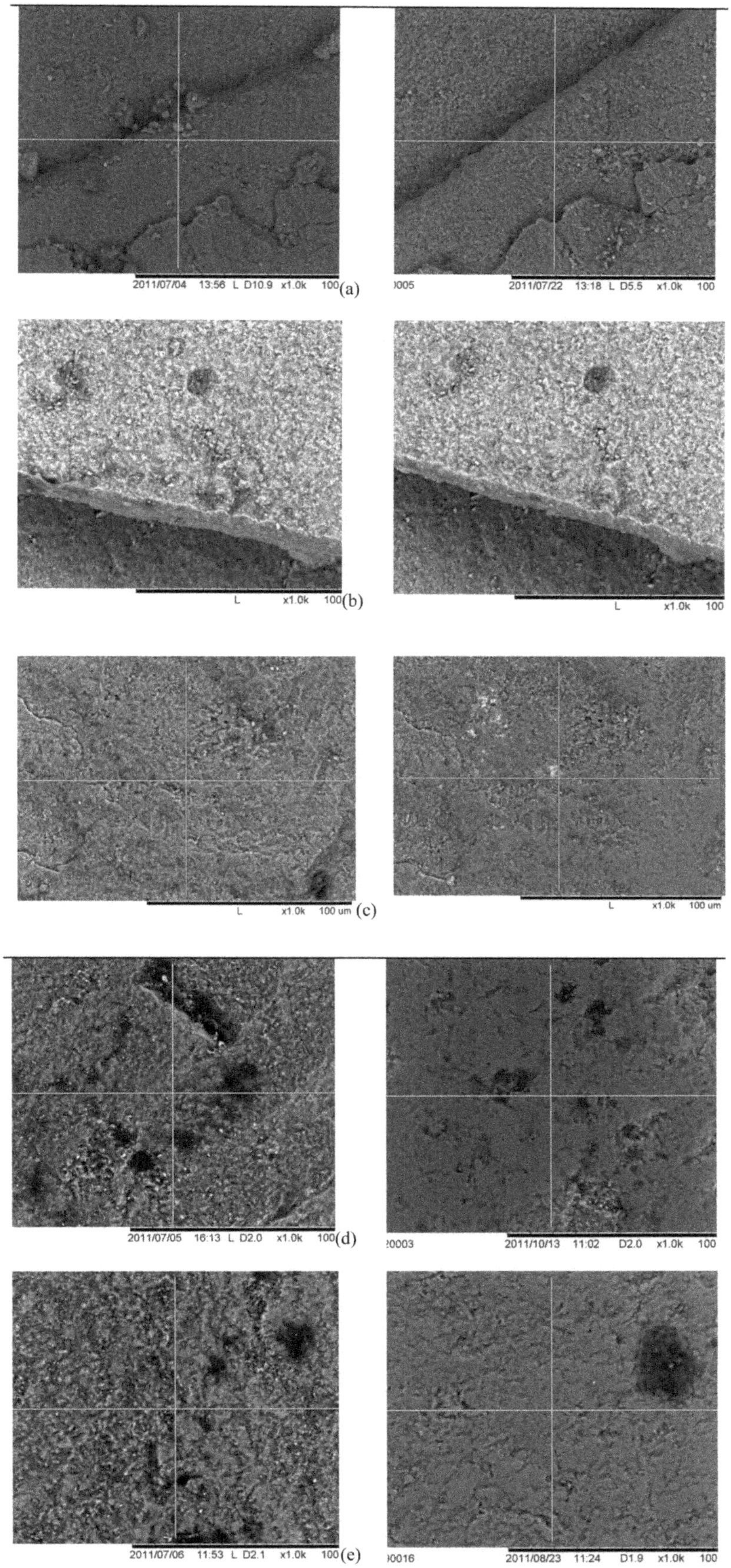

Figure 184: Examples of the four different categories of surface micro-abrasion (1000x) - (a) a control flake (F5M3) (b) a flake illustrating no change after abrasion (F1M10) (c) a flake surface illustrating slightly abraded category (F4M2_ridge) (d) A flake illustrating the moderately abraded category (F12M1) (e) a flake illustrating the very abraded category (F15M4). The left hand column shows the pre-trial photographs while the right hand column represents the post-trial photograph. All micro-graphs were taken with SEM. All artifacts were >99% silica. (Reproduced from Chu *et al.* 2013 with permission from Elsevier.)

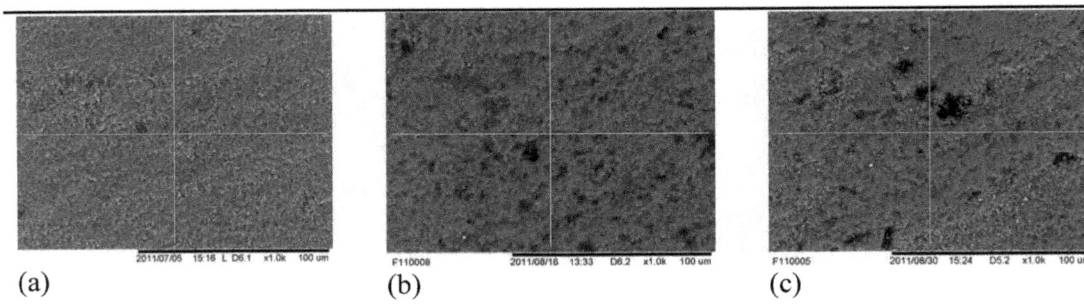

Figure 185: Microscopic observation area F11M6 showing changes to surface micro-abrasion over time (a) t=0 (b) t=96 (c) t=168 hours. (Reproduced from Chu *et al.* 2013 with permission from Elsevier.)

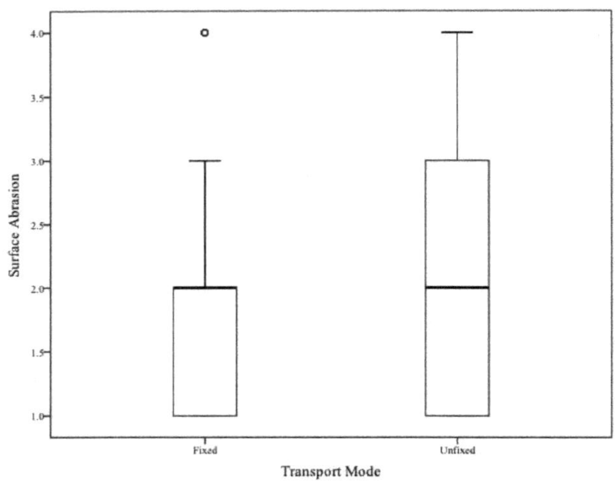

Figure 186: Impact of transport mode on surface micro-abrasion. (Reproduced from Chu *et al.* 2013 with permission from Elsevier.)

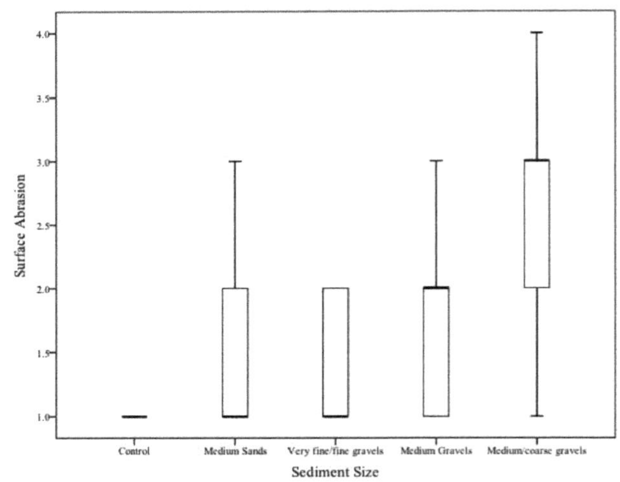

Figure 187: Impact of sediment size on surface micro-abrasion; $\chi 2(6)=171.589, p<0.001$ (ordinal logistic regression). (Redrawn from Chu *et al.* 2013 with permission from Elsevier.)

The independent variables tested in this experiment were time, sediment size, and mobility, which measured if the artifact was mobile or stationary. Time was analyzed as a continuous variable, while sediment size and mobility were treated as fixed conditions (Figure 185 and Figure 189). Artifact lithology, artifact class, and transportation mode of the sediment were kept constant to minimize error.

OLR is a linear regression, and as a result, supposes that the relationship between the dependent variable (surface micro-abrasion) and the independent variables (time, sediment size, and transport mode) is statistically the same across different levels. This assumption is verified using the test of parallel lines, which if significant indicates that OLR is unsuitable for the analysis. The results of the test of parallel lines for the first analysis were $\chi^2(12)=10.745$, $p=0.551$, indicating that the null hypothesis was upheld and that the data was suitable for testing.

Figure 186, Figure 187 and Figure 188 illustrate differences to surface abrasion under different sedimentary conditions. Figure 189 shows the summary statistics for surface abrasion and Figure 190 presents the results of the ordinal logistic regression.

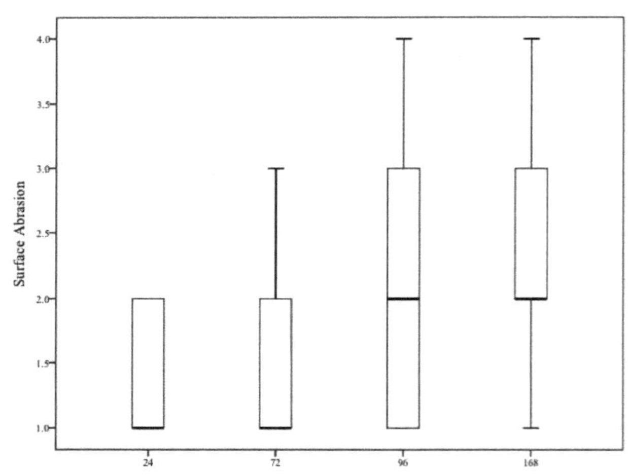

Figure 188: Impact of time on surface micro-abrasion; $\chi 2(6)=171.589, p<0.001$ (ordinal logistic regression). (Redrawn after from Chu *et al.* 2013 with permission from Elsevier.)

Pre-trial SEM micro-graphs of microscopic observation areas showed that artifacts had homogeneous cryptocrystalline surfaces comprised of overlapping 'scales' (Luedtke 1992) of 100% silicon with occasional traces of calcium (>2%) attributed to minor genetic

Figure 189: Descriptive statistics of the conditions of microscopic observation areas in the surface abrasion experiment. Values are *n* and % in parentheses. (Redrawn from Chu *et al.* 2013 with permission from Elsevier.)

Gravel Size (Wentworth class)	Mobility	Time (hours)	No damage	Slightly abraded	Moderately abraded	Very abraded
Medium sand	Fixed	24	14 (100)	0 (0)	0 (0)	0 (0)
Medium sand	Fixed	72	3 (100)	0 (0)	0 (0)	0 (0)
Medium sand	Fixed	96	3 (33)	6 (67)	0 (0)	0 (0)
Medium sand	Fixed	168	9 (36)	12 (48)	4 (16)	0 (0)
Very fine/fine gravel	Fixed	24	2 (22)	7 (78)	0 (0)	0 (0)
Very fine/fine gravel	Fixed	72	7 (22)	2 (22)	0 (0)	0 (0)
Very fine/fine gravel	Fixed	96	7 (0)	0 (0)	0 (0)	0 (0)
Very fine/fine gravel	Fixed	168	7 (100)	0 (0)	0 (0)	0 (0)
Medium gravel	Fixed	24	7 (0)	0 (0)	0 (0)	0 (0)
Medium gravel	Fixed	72	4 (44)	5 (56)	0 (0)	0 (0)
Medium gravel	Fixed	96	2 (25)	3 (38)	3 (38)	0 (0)
Medium gravel	Fixed	168	1 (13)	7 (88)	0 (0)	0 (0)
Medium/coarse gravel	Fixed	24	7 (78)	2 (22)	0 (0)	0 (0)
Medium/coarse gravel	Fixed	72	0 (0)	8 (89)	1 (11)	0 (0)
Medium/coarse gravel	Fixed	96	0 (0)	1 (11)	4 (44)	4 (44)
Medium/coarse gravel	Fixed	168	0 (0)	2 (22)	4 (44)	3 (33)
Very fine/fine gravel	Unfixed	72	15 (83)	3 (17)	0 (0)	0 (0)
Medium gravel	Unfixed	72	3 (33)	6 (67)	0 (0)	0 (0)
Medium/coarse gravel	Unfixed	72	0 (0)	6 (67)	3 (33)	0 (0)
Medium/coarse gravel	Unfixed	96	2 (25)	6 (75)	0 (0)	0 (0)
Medium/coarse gravel	Unfixed	168	0 (0)	0 (0)	1 (11)	8 (89)

inclusions (Figure 184). These findings were consistent with previous SEM analyses of East Anglian flint (Luedtke 1992). Comparisons with post-trial micro-graphs revealed that flume treatment resulted in wear to lithic surfaces over time, characterized by ablation with the removal of asperities or microscopic high-points from the flint surface (Stachowiak and Batchelor 2005: 487). There was no pitting, displacement, or cracking shown in any of the post-trial photographs (Figure 184).

Of the total 204 microscopic observation areas, 39% (n=80) showed no wear: the control group (n=24) microscopic observation areas showed no change. The descriptive statistics are detailed in Figure 189. Of the various wear categories, 34% of the microscopic observation areas (n=74) were slightly abraded, 11% (n=26) were moderately abraded, and 7% (n=15) were heavily abraded (Figure 184). The results of the OLR were significant, indicating that the predictors were capable of discriminating the level of abrasion among microscopic observation areas (χ^2 (6)= 171.589, p<.001). Pseudo-R^2 values of R^2=0.54 (n=204; Cox and Snell), and 0.60 (n=204; Nagelkerke) indicated a moderately strong relationship between prediction and grouping (Figure 190).

To break down individual factors, time in the flume significantly affected surface abrasion (p<0.001; Figure 190, Figure 185, Figure 188). For smaller size categories (medium sands, very fine/fine gravels), time had a significant effect on the number of microscopic observation areas exhibiting wear, but not on the level of observed surface micro-abrasion. For larger sediment size categories (medium/coarse gravels, medium gravels), time increased both the frequency and intensity of the surface micro-abrasion (Figure 186).

The level of wear was significantly affected by sediment size (p<0.001; Figure 190 and Figure 187), with 85% of microscopic observation areas abraded in medium/coarse gravels exhibiting a wearing away of the surface, while only 23% of those abraded in medium sands exhibited post-trial wear and none in the medium sands (Figure 189). The

Figure 190: Ordinal logistic regression results of the effect of time, sediment size, and transport on surface micro-abrasion; R2=0.54 (Cox and Snell) and 0.60 (Nagelkerke). Model χ2(6)=171.589, (p<0.001). Note: sediment classes are separated because they were not analyzed as continuous variables, rather they were all compared to medium/coarse gravels.

	Log Odds Ratio	Std. Error	Wald	df	Sig.	95%Confidence Interval	
						Lower Bound	*Upper Bound*
Time	-.024	.004	39.442	1	**.000**	.016	.031
Transport mode	-.378	.353	1.144	1	.285	-1.070	.315
Control	-24.381	.000	.	1	.	-24.381	-24.381
Medium sands	-3.557	.525	45.977	1	**.000**	-4.585	-2.529
Very fine/fine gravels	-3.339	.499	44.761	1	**.000**	-4.318	-2.361
Medium gravels	-2.241	.446	25.274	1	**.000**	-3.115	-1.367
Medium/coarse gravels	0			0			

negative log-ordered coefficient levels when comparing medium coarse gravels to all other sizes indicate that larger sediment sizes are more likely to assign higher wear ratings than those found in smaller sediment.

Flake transportation, however, did not affect the amount or intensity of surface micro-abrasion among microscopic observation areas ($p>0.05$; Figure 190 and Figure 186).

The results of this analysis support hypotheses that lithic surfaces show microscopic evidence of surface-wear as a result of fluvial abrasion in a controlled laboratory setting (Burroni *et al.* 2002; Fernandes *et al.* 2007; Keeley 1980; Lenoble 2005; Shackley 1974; Stapert 1976). This was characterized by ablative abrasion that formed within 24 hours of exposure to sediments as small as medium sands.

The pseudo-R^2 values suggest that the factors analyzed in this experiment were capable of predicting at least 50% of the variance found in this model. This suggests that these factors are relatively powerful predictors of surface micro-abrasion levels but also indicate that other uncontrolled factors were at play. These might include the artifact shape, sediment lithology, sediment shape, and the amount of abrasive particles in motion (Figure 17).

This analysis also showed that surface micro-abrasion is sensitive to sediment size. All sediment size categories used in this experiment elicited wear in post-trial micrographs. This indicates that the sediment size threshold for inducing surface micro-abrasion is below medium sands. Still, its abrasion was less common in smaller sediments, suggesting that increasing sediment size can increase the frequency of abrasion expression. In addition, increasing sediment size (even on ¼ *phi* scale) significantly predicts higher wear ratings. It should also be noted that the log ordered coefficients reduce relatively linearly in each category, suggesting that both time and sediment size exhibit an asymptotic curvilinear function indicating that their expression tails off though increasing sediment size.

The results also indicated that time played a part in the abrasion of lithic surfaces regardless of sediment size ($p<0.05$; Figure 190). The longer artifacts were left in the flume, the more likely they were to exhibit surface micro-abrasion as well as increased levels. In the largest sediment size category of this experiment, near or total saturation was often achieved (i.e., the entire original surface was removed) within 168 hours of abrasion.

Though this has yet to be tested on archaeological materials, geomorphological studies have suggested that proveniencing sediments with micro-surface textures can often provide some limited provenience information (Whalley 2003; Whalley *et al.* 1982).

Transport mode did not significantly affect surface abrasion—it was impossible to distinguish whether the abrasion was made while the artifact was *in situ,* or through transport.

The results of this experiment are discussed along with previous experiments and their application to the archaeological record further in Section 7.4.

6.3. Sediment size, time, mobility, and flaked lithic ridge wear

The aim of this section is to assess the impact of time, sediment size, and artifact transport mode (*in situ* v *ex situ*) on flake artifact ridge widths.

Three separate hypotheses (A, B, and C) were tested here to explore the effect of sediment size, time, and mobility on lithic ridge wear:

H_{A0}=Sediment size does not significantly affect flaked lithic ridge wear.

H_{A1}=Sediment size significantly affects flaked lithic ridge wear.

H_{B0}=Time does not significantly affect flaked lithic ridge wear.

Figure 191: List of experimental conditions for the lithic ridge wear analysis.

Dependent variables	Independent variables	Controlled variables
<u>Artifact ridge width:</u> (µm)	<u>Time:</u> 24 h, 72 h, 96 h, 168 h <u>Sediment sizes:</u> Control (φ=0) Medium sands (φ=2.25 to 2) Very fine/fine gravels (φ=-1.75 to -2.25) Medium gravels (φ=-3 to -3.25) Medium/coarse gravels (φ=-3.5 to -4.25) <u>Transport mode</u>: fixed, unfixed	**Transportation mode of abrasive sediment:** saltating **Lithology:** flint **Artifact class:** unretouched thinning flake

Figure 192: Descriptive statistics for the artifact ridge experiment.

Time (h)	Sediment size (*Phi*)	Transport mode	Mean (µm)	Std. Deviation (µm)	*n*
24	*Medium sands*	Fixed	44.29	.	1
		Total	44.29	.	1
	Very fine/fine gravels	Fixed	48.00	24.042	2
		Total	48.00	24.042	2
	Medium gravels	Fixed	118.20	21.788	5
		Total	118.20	21.788	5
	Medium/coarse gravels	Fixed	6.00	.	1
		Total	6.00	.	1
	Total (all sediment types)	Fixed	81.92	48.139	9
		Total	81.92	48.139	9
72	*Control*	Fixed	8.80	6.181	5
		Total	8.80	6.181	5
	Very fine/fine gravels	Fixed	51.00	6.928	3
		Unfixed	35.00	20.785	3
		Total	43.00	16.395	6
	Medium gravels	Fixed	73.00	37.639	4
		Unfixed	57.25	51.801	8
		Total	62.50	46.412	12
	Medium/coarse gravels	Fixed	116.00	.	1
		Unfixed	145.00	2.828	2
		Total	135.33	16.862	3
	Total (all sediment types)	Fixed	46.54	40.109	13
		Unfixed	65.62	54.491	13
		Total	56.08	47.876	26
96	*Medium sands*	Fixed	90.33	35.774	9
		Total	90.33	35.774	9
	Very fine/fine gravels	Fixed	64.80	29.141	5
		Total	64.80	29.141	5
	Medium gravels	Fixed	93.43	33.877	7
		Total	93.43	33.877	7
	Medium/coarse gravels	Fixed	196.67	26.690	3
		Unfixed	118.00	2.828	2
		Total	165.20	47.061	5
	Total (all sediment types)	Fixed	99.21	49.914	24
		Unfixed	118.00	2.828	2
		Total	100.65	48.151	26
168	*Medium sands*	Fixed	37.92	10.291	13
		Total	37.92	10.291	13
	Medium gravels	Fixed	184.29	97.672	7
		Total	184.29	97.672	7
	Medium/coarse gravels	Fixed	210.11	99.524	9
		Unfixed	163.67	25.384	3
		Total	198.50	88.103	12
	Total (all sediment types)	Fixed	126.69	107.908	29
		Unfixed	163.67	25.384	3
		Total	130.16	103.339	32
Total (all times)	*Control*	Fixed	8.80	6.181	5
		Total	8.80	6.181	5
	Medium sands	Fixed	58.71	34.598	23
		Total	58.71	34.598	23
	Very fine/fine gravels	Fixed	57.30	22.716	10
		Unfixed	35.00	20.785	3
		Total	52.15	23.551	13
	Medium gravels	Fixed	122.91	71.606	23
		Unfixed	57.25	51.801	8
		Total	105.97	72.383	31
	Medium/coarse gravels	Fixed	185.93	97.483	14
		Unfixed	145.29	25.191	7
		Total	172.38	82.175	21
	Total (all sediment types)	Fixed	98.63	80.691	75
		Unfixed	87.78	60.594	18
		Total	96.53	77.033	93

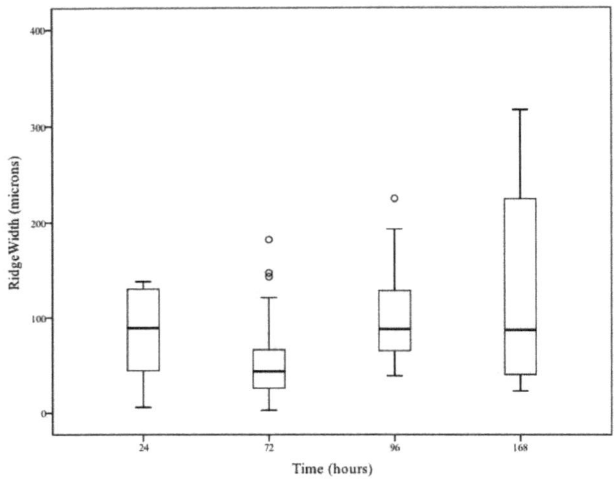

Figure 193: Impact of time on ridge widths. F(3, 93)=3.05, *p*=0.019. (Redrawn from Chu *et al.* 2013 with permission from Elsevier.)

H_{B1}=Time significantly affects flaked lithic ridge wear.

H_{C0}=Mobility does not significantly affect flaked lithic ridge wear.

H_{C1}=Mobility significantly affects flaked lithic ridge wear.

The hypotheses in this experiment are that, if time, sediment size, and/or transport mode affect the abrasion of artifacts over time, then the mean ridge widths between groups should be statistically different from each other.

This experiment tests ridge widths univariately with a three-way robust Analysis of Variance (ANOVA; see Section 3.5.7 for details).

In this experiment, artifact ridge width was measured, where applicable, before and after flume abrasion on 93 microscopic observation areas in total, distributed across 26 artifacts (Figure 192). Artifacts were then subjected to saltation in the flume where time, sediment size, and transport mode varied (Figure 191). To reduce the effect of additional variables, the artifact class, artifact lithology, and mode of abrasive transport was kept the same.

The homogeneity of variance (homoscedasticity) between the different levels of examination was verified using Levene's test. The variances for ridge width were significantly different ($F(2, 12)=5.12$, $p<0.001$), however this was likely due to the conservativeness of the test and the analysis could be performed with confidence.

Figure 192 shows the descriptive statistics for ridge wear under different abrasion conditions. Figure 193, Figure 194, and Figure 195 show ridge widening under different times, sediment sizes, and transport modes. Figure 196 shows the results of the ANOVA. Figure 197 and Figure 198 report the results of the ANOVA's *post-hoc* tests.

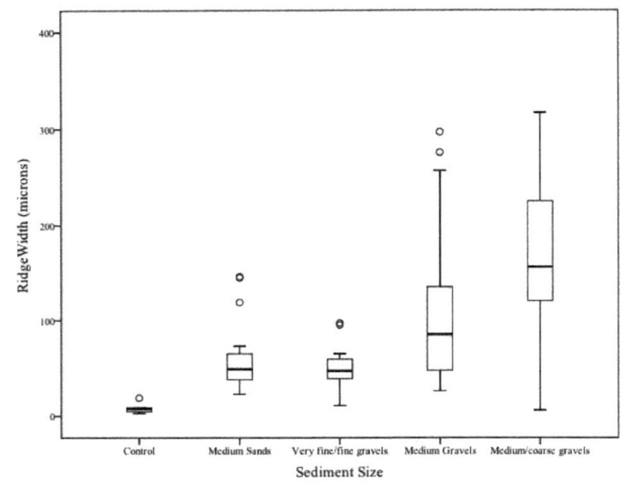

Figure 194: Impact of sediment size on ridge widths. F(4, 93)=6.02, *p*<0.001. (Redrawn from Chu *et al.* 2013 with permission from Elsevier.)

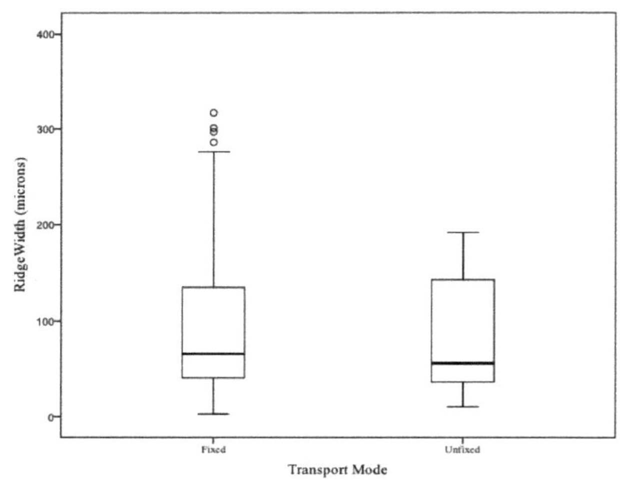

Figure 195: Impact of transport mode on ridge widths. F(1, 93)=4.12, *p*=0.046. (Redrawn from Chu *et al.* 2013 with permission from Elsevier.)

Like surface micro-abrasion, ridge wear was also characterized by ablation, resulting in a general smoothing and widening of the ridge (Figure 196 and Figure 194). The results of the robust 3-way ANOVAs are reported in Figure 196. These results illustrate that there was a significant effect of time (p=0.019; Figure 193), sediment size (p<0.001; Figure 194), and transport mode (p=0.046; Figure 195) and the interaction of time and sediment size (p=0.001; Figure 196) on ridge widths. Two-way interactions showed a significant interaction between time and sediment size on ridge width (p=0.001) but for neither the interaction of time and transport mode nor the interaction of sediment size and transport mode at the p<0.05 level (Figure 196). An $R2$ value of 0.656 indicates that 2/3 of the variance in wear scores is explained by the effects of time, sediment size, and transport mode (Figure 196).

Post-hoc comparisons using the Hochberg GT2 correction indicated that the mean ridge width of all sediment sizes

Figure 196: Statistical analysis of the effect of time, sediment size, and mobility and interactions on lithic ridge wear.

Source	Type III Sum of Squares	Df	Mean Square	F	Sig.
Corrected model	358316.612[a]	19	18858.769	7.337	**.000**
Intercept	237991.919	1	237991.919	92.597	**.000**
Time (hours)	27051.910	3	9017.303	3.508	**.019**
Sediment size	61984.481	4	15496.120	6.029	**.000**
Transport mode	10608.320	1	10608.320	4.127	**.046**
Time * sediment size	72263.660	7	10323.380	4.017	**.001**
Time * transport mode	5007.648	2	2503.824	.974	.382
Sediment size*transport mode	1155.592	2	577.796	.225	.799
Error	187624.188	73	2570.194		
Total	1412518.604	93			
Corrected total	545940.800	92			

R Squared=0.656 (Adjusted R Squared=0.567)

Figure 197: Bootstrapped Hochberg's GT2 post-hoc comparisons showing homologous subsets for ridge widths at different times. Cells without values are statistically distinguishable from each other.

Time (h)	*n*	Grouping	
24	9	A	
72	26	A	
96	26	A	B
168	32		B

Figure 198: Bootstrapped Hochberg's GT2 post-hoc comparisons showing homologous subsets for ridge widths in different sediment sizes. Cells without values are statistically distinguishable from each other.

Sediment Size (*Phi*)	*n*	Grouping		
Control	5	A		
Medium sands	13	A	B	
Very fine/fine gravels	23	A	B	
Medium gravels	31		B	
Medium/coarse gravels	21			C

(i.e., total in Figure 192) at 24 hours (μ=81.72, σ=48.613; Figure 192) and 72 hours (μ=56.08, σ=47.87; Figure 192) was significantly different from ridge width at 168 hours (μ=130.16, σ=103.33; Figure 192) at the p<0.05 level. However, the ridge width at 96 hours (μ=100.65, σ=48.151; Figure 192) did not significantly differ from the ridge widths at 24, 72, or 168 hours at the p<0.05 level (Figure 192 and Figure 197).

Hochberg's GT2 *post-hoc* comparisons for sediment size (Figure 198) indicated that the mean ridge width for the controls (μ=8.8, σ=6.18; Figure 192) was significantly different from medium gravels (μ=105.97, σ=72.383), and medium/coarse gravel (μ=172.38, σ=82.175; Figure 192) sediment sizes. It also revealed that the ridge width at medium/coarse gravels significantly differed from the control, medium sands and very fine/fine gravels. However, the ridge width in medium gravels did not significantly differ from the ridge widths at medium sands, very fine/fine gravels, or medium/coarse gravels. Overall, ridge width increases with increasing gravel size (Figure 192).

Taken together, these results suggest that time spent in the flume does have an influence on the width of ridges. Specifically, the results illustrate that, when replica flint artifacts are abraded in a flume, their ridges increase in width (Chambers 2003; Shackley 1974). However, it should be noted that the time left in the flume must be large to see a significant effect as 96 hours in the flume does not significantly affect wear differently from 24 hours or 168 hours.

These results show that all sediment size classes result in the ridge attrition of replica flint artifacts. The higher the sediment size in which they are abraded, the wider their ridges become. Sediment size must differ in size to elicit different ridge wear levels. The data obtained in this experiment suggests that for sediment sizes less than medium/coarse gravels on the *phi* scale, sediments should differ by more than 1 *phi* scale group in order for artifacts to exhibit statistically different ridge wear levels. However, not all ridges are necessarily affected, suggesting that a better predictor of wear is the dispersion of ridge wear values.

Additionally, artifacts that are abraded *in situ* have a slightly larger average ridge width than those allowed to abrade in transit. This suggests that the abrasion process of mobile artifacts is reduced, possibly because of decreased clast-artifact impacts and/or diminished impact force.

The significant interaction between sediment size and time show that time mediates the abrasion of ridges between sediment sizes. Contrasts revealed that this is largely insignificant at most time-steps. Those time-steps where the interactions were significant, time did not illustrate any clear patterns. This supposes that time moderates abrasion of ridges in different sediment sizes; larger sediment sizes abrade faster than smaller ones.

6.4. Sediment size, time, mobility, and edge damage

Many artifacts exhibited edge damage in the form of small, continuous bifacial feather terminating flakes (>1mm)

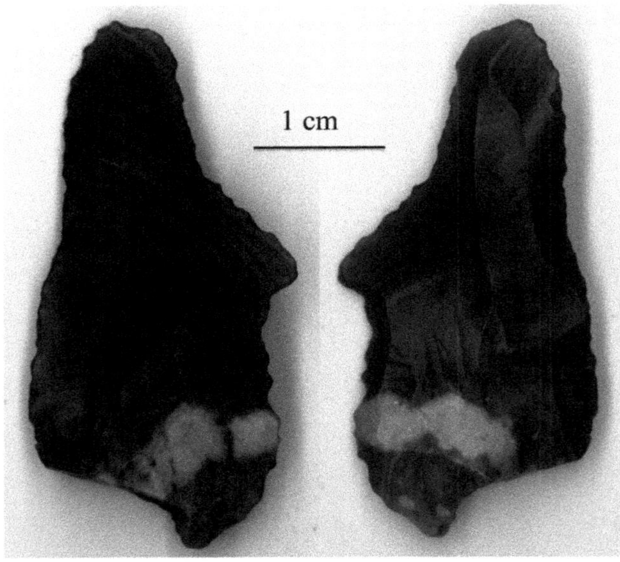

Figure 199: The ventral (left) and dorsal (right) side of an artifact damaged in the flume. (Redrawn from Chu *et al.* 2013 with permission from Elsevier.)

around the margins of the edges (*micro-flaking*; Figure 199). The aim of this section is to assess the impact of time, sediment size, and artifact transport mode (*in situ* v *ex situ*) on flaked artifact edge damage.

Three separate hypotheses (A, B, and C) were tested here to explore the effect of sediment size, time, and mobility on lithic edge damage:

> H_{A0}=Sediment size does not significantly affect flaked lithic edge damage.
>
> H_{A1}=Sediment size significantly affects flaked lithic edge damage.
>
> H_{B0}=Time does not significantly affect flaked lithic edge damage.
>
> H_{B1}=Time significantly affects flaked lithic edge damage.
>
> H_{B0}=Mobility does not significantly affect flaked lithic edge damage.
>
> H_{B1}=Mobility significantly affects flaked lithic edge damage.

Two tests were used to assess if sediment size, time, and mobility affect artifact edge damage over time. The first tested the influence of time and transport mode (Figure 200) with an ordinal logistic regression model (see Section 3.5.9). Sediment size was not included in this model as a predictor, because it completely predicted edge damage categories, rendering it unsuitable for such a model (i.e., that only one of the sediment sizes was associated with only one value of edge damage—a perfect statistical fit).

As a result, sediment size is analyzed separately using the Kruskall-Wallis test (refer to Section 3.5.10).

In this experiment, edge damage was measured on 27 different artifacts (Figure 200; see Section 3.4.1 for experimental design). Edge damage was ranked in one of three categories: (1) No damage, (2) Light damage, defined by continuous edge damage (all along the entire edge) with flake scars no larger than 1mm in width, and (3) Damaged, defined by continuous damage with flake scars between 1 and 2mm in width. Scar size was the most significant and diagnostic aspect of the edge damage.

The independent variables tested were (i) time, that was divided over four time-steps (24 hours, 72 hours, 96 hours, and 168 hours); (ii) Sediment size, which tested four different sediment sizes (medium sands, very fine/fine sands, medium gravels, medium/coarse gravels) and a control (no sediment), and (iii) transport mode (fixed, unfixed) that tested the effect of the artifact's transport with the mobile sediment versus its fixation to the flume floor.

Ordinal logistic regression supposes that the relationship between the dependent variables and the independent variable is statistically the same (otherwise known as the proportional odds assumption). It presumes that the relationship between the predictor and the outcome

Figure 200: List of experimental conditions for the edge damage analysis.

Dependent variables	Independent variables	Controlled variables
Edge damage: none, light damage, heavy damage	**Time:** 24 h, 72 h, 96 h, 168 h **Sediment sizes:** • Control (φ=0) • Medium sands • (φ=2.25 to 2) • Very fine/fine gravels • (φ=-1.75 to -2.25) • Medium gravels • (φ=-3 to -3.25) • Medium/coarse gravels • (φ=-3.5 to -4.25) **Transport mode:** fixed, unfixed	**Mode of transportation of abrasive sediment:** saltating **Lithology:** flint **Artifact class:** unretouched thinning flake

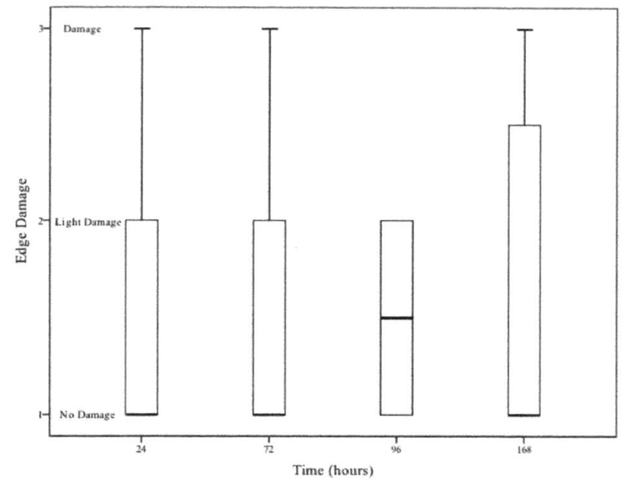

Figure 201: Boxplot expressing edge damage by time.

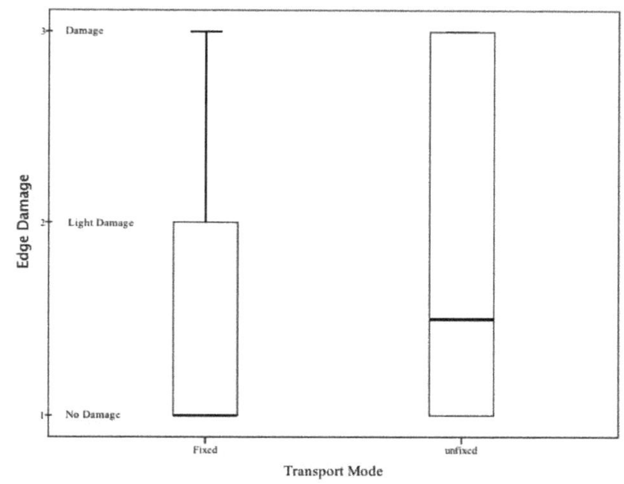

Figure 202: Boxplot expressing edge damage by transport mode. (Redrawn from Chu *et al.* 2013 with permission from Elsevier.)

variables is similar across variables (i.e., that the relationship between all pairs of groups is the same). This assumption is verified using the test of parallel lines, which if significant, rejects the null hypothesis that the dependent and independent variables are statistically similar. The results of this test were $\chi^2(2)=2.485$, $p=0.289$, indicating that the null hypothesis was upheld and that the data was suitable for testing.

Figure 201, Figure 202 and Figure 203 all report the effects of time, transport mode, and sediment size on artifact edge damage. Figure 204 reports the results of the ordinal logistic regression.

Three control flakes (2 fixed, 1 unfixed) exhibited no damage after being in the flume for 72 hours, indicating that the flume itself had no discernible effect on the artifacts over that time interval.

The OLR analyzing the influence of time and transport mode showed that neither time or transport mode had an effect on the appearance or severity of artifact edge damage ($p>0.05$; Figure 201, Figure 202, Figure 204).

Edge damage was however significantly affected by the sediment size (Figure 203), $H(3)=19.933$, $p<0.001$. There was a complete separation of data with artifacts abraded in very fine/fine gravels or smaller showing no signs of edge damage (Figure 203). Among medium gravels, all edge damage was lightly damaged while 67% of those abraded in medium/coarse gravels exhibited heavy damage. A Mann-Whitney test showed that medium gravels and medium/coarse gravels differed significantly ($U=25$, $z=-3.48$, $p<0.05$ $r=0.55$).

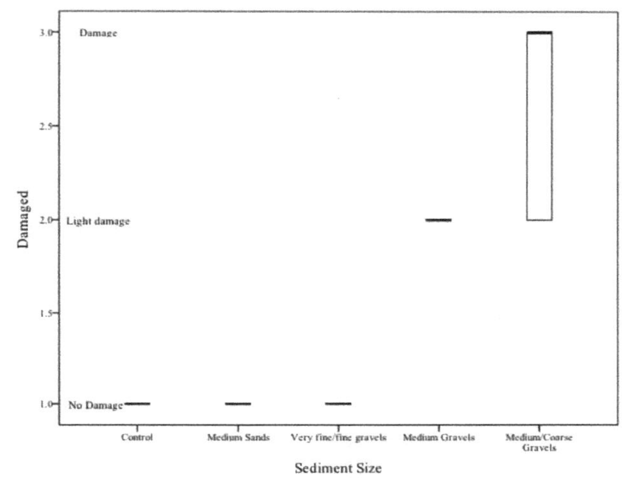

Figure 203: Boxplot expressing edge damage by sediment size. $H(3)=19.933$, $p<0.001$. (Redrawn from Chu *et al.* 2013 with permission from Elsevier.)

The results of this analysis suggest that edge damage of thinning flakes is not significantly affected by either time or its entrainment within a given sediment. This means that edge damage of artifacts can occur in less time than 24 hours although it is hard to know what this means in real world terms. The results do provide evidence however that edge damage is strongly mediated by sediment size. Specifically, they demonstrate that for the feathered edges, edge damage is not elicited in sediment sizes below very fine/fine gravels. They are however, sensitive to small changes in sediment size, with medium gravels showing significantly different levels of micro-flaking from medium/coarse gravels.

Figure 204: Effect of time and transport mode on edge damage. ***Note:*** ***R2*****=0.02 (Cox and Snell) and 0.02 (Nagelkerke). Model $\chi2(2)=0.68$, $p=0.727$(*n.s.*).**

	Estimate (B)	Std. Error	Wald	df	Sig.	95% Confidence Interval	
						Lower Bound	*Upper Bound*
Time	.003	.007	.117	1	.732	-.012	.017
Transport Mode	-.700	.878	.637	1	.425	-2.420	1.020

6.5. Chapter summary

The above statistical analyses on the data collected during the flume experiment provides support for the hypothesis that lithic tools transported in fine-grained (φ>-4.25) fluvial sediments show signs of microscopic surficial wear expressed in four ways:

1. Fine-grained fluvial abrasion of flint tool surfaces and ridges is characterized by ablation wear (Shackley 1975).
2. Flint lithic surface abrasion occurs in sediments as small as medium sands. The frequency and intensity of this abrasion increases as a function of time and sediment size. Significant differences can occur over experimental periods as short as days and between sediments that differ by as little as ¼ *phi* scale. Lithic surfaces examined in this experiment became fully/nearly fully abraded in medium/coarse sediments within 168 hours.
3. Flint ridge widths significantly widened as an effect of time in the flume. There were significant effects at the ¼ *phi* scales tested in this experiment though they appeared to be less sensitive to time than surface abrasion. When artifacts were allowed to move with the bedload, their ridges were slightly less abraded than those that were abraded *in situ*.
4. Artifact edges were commonly damaged in the flume. This was not related to either transport mode or time in the flume, but was highly predicted by sediment size. Artifacts that were abraded in very fine/fine gravels or less did not exhibit any change to edge morphology. Increasing sediment size directly correlated to the size of the continuous, bilateral micro-flaking along the margins and extents of the artifact.

The significance, synthesis, and limitations of the flume results will be discussed in more detail in Section 7.4.

Chapter 7

Discussion

7.1. Introduction

This chapter contextualizes the experimental results of this book and applies them to five archaeological case studies. Therefore, the chapter is divided into two main parts. The first part, Sections 7.2 to 7.4, interprets the experimental results from Chapters 4, 5, and 6, synthesizing them with previous experimental findings, discussing their application to Pleistocene contexts, and outlining their limitations. The second part (Section 7.5) applies the findings of these experiments to the northern European Lower and Middle Palaeolithic sites, findspots, and landscapes of Happisburgh 3 (Ashton and Lewis 2012), Pakefield (Parfitt *et al.* 2005), Happisburgh 1 (Ashton *et al.* 2008), the Trent Valley (Howard *et al.* 2008), and Caours (Locht *et al.* 2010). The aim of this chapter is to better understand fluvial assemblages and site formation processes during the Lower and Middle Palaeolithic of Northern Europe and their implications for interpretations of early hominin behavior.

7.2. Fluvial experiments

This section discusses the results of a field experiment examining the effects of fluvial processes on the horizontal displacement of replica Palaeolithic assemblages in the River Glenderamackin, UK (refer to Section 3.2 for methodology and Chapter 4 for results). The experiment focused on the interrelationships and effects on dispersal and artifact primary metrics (maximum length, maximum width, maximum thickness, and weight) of river discharge, animal trampling, artifact trapping, artifact burial, and insertion location on artifact recovery, transport distances, orientations, dip, micro-flaking, and breakages. Here, the results will be contextualized within the framework of other fluvial experiments described in Section 2.3.2 and summarized in Figure 205. Fundamentally, this study found that flakes deposited in a relatively high-energy fluvial environment demonstrated characteristic dispersal patterns related to the size and shape of the artifacts, breakage patterns, and their insertion contexts.

7.2.1. Scatter locations and horizontal dispersal

A primary aim of this project was to assess the role of insertion locations on the horizontal dispersal of lithic artifacts in the River Glenderamackin. The results of a mixed model and a follow up two-sample t-test showed that scatter insertion locations indeed played a statistically significant role in the horizontal dispersal of artifacts ($p<0.05$; refer to Section 4.4.1 and see Figure 87). The main difference found was between artifacts deposited on the banks of the channel (which were exposed to floodwaters on multiple occasions) and those that were deposited within the channel with the former moving on the order of centimeters and meters, while the latter were more widely dispersed, transported as far as 250m.

Previous experiments studying discard location have also suggested that insertion location plays a role in the dispersal of artifacts (Harding *et al.* 1987; Hosfield and Chambers 2005c; Petraglia and Nash 1987; Schick 1986). Schick (1986) showed that lithic assemblage disturbances were highly variable according to site location (e.g., floodplain, river bank, channel bed). Harding *et al.* (1987) reported similar results, noting that artifacts placed on an active gravel bar moved more readily than those placed 'in slacker water' (p.118).

Many previous studies provide data allowing for direct comparison of horizontal displacement of artifacts. The results of this experiment are within these reported ranges (Figure 206). Among artifacts deposited on the riverbanks (Stations E and H), Petraglia and Nash (1987) reported that scatters were buried '*in situ*'. Schick (1986) also reported seven experiments (Sites 3a-3b, 14-15, 25, 28, 34) in which artifacts emplaced on the banks of rivers were 'minimally disturbed' (i.e., artifacts stayed mostly in place with a maximum transport of 7m). However, Schick (1986) described other bankside experiments (Sites 13, 20-22) as more heavily modified, reporting that transport values increased to as far as 19m in one case. Schick attributed these differences to the fact that these scatters were located lower and closer to the active channel than other scatters (1987).

Among scatters placed within the channel (either directly or on a bar), previous experiments showed greater transport distances. Schick (1986) found that such sites were scoured, truncated, or experienced 'major disturbance,' resulting in transport of up to 90m (Sites 1c, 3a-3b, 19, 23, 24, 26-27, 36; Figure 206). Petraglia and Nash (1987) reported similar results (Stations C, D) with in-channel artifacts dispersed up to an average of 33.1m. Hosfield and Chambers (2005c) reported a maximum transport distance of 84.95m while Harding *et al.* (1987) showed variable disturbance with some artifacts not moving at all, while others were transported up to 150m. Only Isaac (1967) reported minor horizontal movement to his in-channel artifact scatters, though a likely reason for this and the variable distances of Harding *et al.* (1987) was their use of heavier, less mobile handaxes compared to the predominantly flake-based assemblages of this and other experiments. Transport distances in previous experiments were lower than this

Figure 205: Comparison of this volume's field experiments to previous experiments.

	Aim(s)	Site	Assemblage	Duration	Dependent Variable(s)	Independent Variable(s)	Result(s)	Limitation(s)
Isaac 1967	• Discriminate between anthropogenic and fluvial site patterning	5 locations on a braided ephemeral stream; Lake Magadi, Kenya	100 cast concrete bifaces and a number of lava flakes	2 years	• Transport distance • Artifact orientation	• Time • Bed lithology • Artifact class	• Bifaces transported and tilt upstream • Bifaces imbricate to flow • Flakes removed in sand/ hard beds	• Small sample size • Not flint artifacts • Single channel form
Schick 1993: 203; Schick 1986 (Flume work)	• Observe fluvial forces on archaeological materials	Flume; 54ft long, 2ft wide, slope (0.5%)	*Uns.*	54ft	• Entrainment mode • Transport rate	• Time • Flow velocity (<2.5ft/sec) • Rate of velocity change • River bed substrate (bare, medium sands) • Artifact size and shape	• Smaller artifacts move first/further • Flow surges affect whole assemblages • Artifact dimensions and sediment substrate affect transport • Artifact orientations bias • Artifact interactions retard erosion	• No quantification of results
Bunn et al. 1980; Schick 1987; Schick 1993; Schick 1986	• Examine the behavior of artifacts in different hydrological environments	Locations along a braided ephemeral stream, Koobi Fora, Kenya	>10k cobbles, cores, and various flakes and fragments	4 years	• Transport distance • Dip-and-strike	• Hydrological events • Hydrological settings • Artifact class • Artifact size (a-axis)	• Smaller artifacts removed • Rivers 'stretch' sites downstream • Hydrology affects site preservation	• Results applicable to only one hydrological regimen
Petraglia and Nash 1987	• Understand the spatial impact of different channel settings • Improve methodological rigor of previous experiments	5 locations along Jemez Floodplain, New Mexico	544 chert/flint/quartz flakes, cores, cobbles	2 years	• Transport distance	• Time • Channel type/ setting • Artifact weight • Raw materials • Orientation	• Tempo/magnitude/ duration of fluvial events affects preservation • Size and movement correlations not straight-forward • Artifacts rework with other assemblages • Sites stabilize rapidly	• Only one hydrological regimen • No empirical application of hydrological regimen to effects
Harding et al. 1987	• Determine if handaxe concentrations are fluvial products • Determine handaxe travel distances • Relate abrasion to transport distances • Relate artifact weight loss to time in a fluvial system	Locations on a gravel-bedded river, Afon Ystwyth, Wales	60 flint bifaces	1 year	• Transport distance • Abrasion • Artifact weight	• Hydrological discharge • Hydrological setting • Handaxe form • Handaxe weight • Handaxe size (A, B, and C-axis)	• Inverse relationship between distance travelled and artifact size • Discard location important • Surface polish/edge damage • Weight loss related to transport distance	• Only handaxes • Small sample size • Results not quantified • Surface features not examined before deposition

Figure 205: *Continued.*

	Aim(s)	Site	Assemblage	Duration	Dependent Variable(s)	Independent Variable(s)	Result(s)	Limitation(s)
Dennell 2004	• Explore lithic artifacts movement	Hillside stream gully in the Soan Valley, Pabbi Hills, Pakistan	148 stones	1 year	• Transport distance	• Artifact long axis	• Fluvial movement causes minimal flaking • Artifacts are rapidly moved downstream and lost in gully systems	• Limited application to lithic materials
Chambers 2004; Hosfield and Chambers 2004; 2005a; 2005b; Hosfield *et al.* 2000	• Understand transportation and transformation of lithic artifacts in a gravel-bedded river	13 locations on a gravel-bedded river, Afon Ystwyth, Wales	563 flint flakes 83 flint and greensand chert handaxes	3 years (max length of 16 months)	• Transport distance • Abrasion	• Time • Hydrological setting • Artifact size (A, B, and C-axis)	• Flake scatters stayed together in the beginning (>10m) • Micro-flaking on artifacts • Discrete scatters mixed • Flakes vertically dispersed • Handaxes ridge widened • Handaxes develop cones of percussion	• Painted clast recovery system • Limited handaxe transport
Lenoble 2005 Experiment 7	• Understand assemblages modifications in an ephemeral stream	Lot-et-Garonne, France; ephemeral channel 50cm wide	37 flakes	26 months	• Transport distance	• Artifact long axis	• Partial assemblage movement • Torrential flows move all artifacts • Traps cause secondary concentrations	• Painted clast recovery system • Limited to upland sites
Current Study	• Determine effect of artifact dimensions, river discharge, trampling, trapping, and burial and insertion location on horizontal space, artifact orientation, and edge damage	9 locations along a gravel-bedded Glenderamackin River, Cumbria, UK	456 flint flakes, cores, and bifaces	18 months	• Transport distance • Edge damage (micro-flaking, snapping) • Orientation • Dip	• Hydrological discharge • Artifact dimensions • River discharge • Trampling • Trapping • Burial • Depositional location	• Width had a significant impact on the recovery of transported artifacts • Depositional location had an impact on artifact transport • Burial effects transport distance • Shorter flakes are transported farther • Wider flakes are transported farther • Artifact orientation stayed random • Animal trampling does not affect the dip of artifacts. • Short transport distances (10 m>) induced micro-flaking • Large animal trampling resulted in the transverse snapping of artifacts	• Painted clast recovery system

Figure 206: Horizontal displacement of previous fluvial experiments for comparison to the present study. Some experiments are not included as they are only casually reported (e.g., Dennell 2004). Where multiple observations on scatter movements were made, maximum movements are reported. Refer to Section 2.3.2 for specific experiment details.

Reference	Experiment	Principle context	Mean transport distance (m)	Min transport distance (m)	Max transport distance (m)	Major effects
Isaac 1967	Plots 1, 4, 5	Broad sand/silt channel	*Uns.*	*Uns.*	*Uns.*	'Minor change'
	Plot 2	Narrow sandy channel	*Uns.*	*Uns.*	*Uns.*	'Trivial movement'
	Plot 3	Narrow hardbed channel	*Uns.*	*Uns.*	75	Distribution close to scatter
Schick 1986	Site 1a	Delta margin	*Uns.*	*Uns.*	*1.3 (time 3)*	Some artifacts removed, then buried in vegetation
	Site 1c	Sand channel	*Uns.*	*Uns.*	10	Widely dispersed
	Sites 3a-3b	Large (60m) channel bank	*Uns.*	*Uns.*	*3.5*	Little movement/burial
	Site 12	Channel braid island	*Uns.*	*Uns.*	*Uns.*	Little movement/burial
	Site 13-15	Channel levee	*Uns.*	*Uns.*	<19	Minimal disturbance but severe winnowing to site 13
	Site 19	Channel sand bar	*Uns.*	*Uns.*	*Uns.*	Site truncated
	Site 20	Medium (20m) channel bank	*Uns.*	1	9	Major disturbance
	Site 21	Medium (20m) channel bank	*Uns.*	*Uns.*	*Uns.*	Heavily disturbed
	Site 22	Low-lying channel bank	*Uns.*	*Uns.*	*Uns.*	Major disturbance; transport of several meters
	Site 23	Channel bar	*Uns.*	*Uns*	23	Major disturbance (2 artifacts recovered)
	Site 24	Low-water channel	*Uns.*	10	>20	Major disturbance
	Site 25	Levee	*Uns.*	*Uns.*	*Uns.*	Minor disturbance
	Sites 26-27	Small sand channel	*Uns.*	*Uns.*	*Uns.*	Distributed several meters downstream
	Site 28	Medium (20m) channel bank	*Uns.*	*Uns.*	7	Minor disturbance
	Site 34	Flood basin	*Uns.*	*Uns.*	*Uns.*	Scour/burial
	Site 36	Shallow drainage channel	*Uns.*	*Uns.*	90	Major disturbance
	Site 37	Small channel	*Uns.*	*Uns.*	*Uns.*	Scour/burial
Petraglia and Nash 1987	Station A	Aeolian sediments *control*	11.0	*Uns.*	*Uns.*	N/A
	Station B	Hillslope	13.3	*Uns.*	*Uns.*	Minor disturbance
	Station C	Ephemeral stream	N/A	*Uns.*	*Uns.*	Major disturbance (none recovered)
	Station D	Sandy lag deposit	33.1	*Uns.*	*Uns.*	Major disturbance
	Station E	Floodplain/alluvial fan	*Buried*	*Uns.*	*Uns.*	Minor disturbance
	Station F	Small ephemeral stream	10.4	*Uns.*	*Uns.*	Minor disturbance/burial
	Station G	Anthill	3.8	*Uns.*	*Uns.*	Minor disturbance
	Station H	Floodplain	*Uns.*	*Uns.*	*Uns.*	Buried
Harding *et al* 1987		Gravel bedded river	*Uns.*	0	150.0	Variable disturbance
Lenoble 2005	Experiment 7	Ephemeral stream	*Uns.*	5	20	Truncation and dispersal
Hosfield and Chambers 2005c	Scatters 1, 3, 6, 7, 8, 10	Semi-stable floodplain and point bar complex	*Uns.*	0.05	34.53	Variable dispersal
	Scatters 11 and 12	Point/midstream bar complex	*Uns.*	0.07	84.95	Variable dispersal
Current Study	Scatters 1-6 *Apr 2012*	Erosion riverbank	.50	.01	4.42	Minor disturbance
	Scatters 7-9 *Sep 2012*	Gravel point bar	27.88	.08	250.10	Major disturbance

study's maximum transport distance of 250m, however this is likely the result of variable artifact recovery as where reported, river discharge was always higher in other experiments (see Figure 208).

The results of this experiment and others suggest the following:

1. Channel morphology affects the probability that artifacts will be widely dispersed. All of the experiments that showed minimal disturbance to bankside artifacts were conducted in meandering river systems of varying discharge levels. Schick (1986), whose experiments were conducted in braided channels was the only author to report disturbance among bankside artifacts. This is unsurprising given that braided channels are wide and shallow, characterized by flashy regimes with highly erodible banks (Easterbrook 1999: 126). Meandering channels are less flashy, characterized by deep and narrow channels whose banks are erosion resistant (Easterbrook 1999: 126). This means that artifacts placed on the banks of meandering systems are likely to be topographically higher relative to water levels and subjected to lower seasonal and annual variations in discharge (Figure 207). This interpretation is supported by Schick's (1986) own observations that differences between insertion heights were the likely reason for variable transport of Scatters 13–15.
2. Bankside vegetation has a 'binding effect' on artifacts, inhibiting their transport. Riparian flora is a well-known inhibitor of sediment transport as it increases the friction co-efficient of the underlying sediments. This is robustly supported by personal observations and Petraglia and Nash's study (1987), which also reported similar results in their vegetated bankside regime.
3. Artifacts placed in highly active areas (i.e., channel bars) were therefore more likely to be transported than those discarded in riparian environments (riverbanks), but the story is likely to be more complicated. Gravel-bars certainly are active erosion zones, but they are also depositional areas. The nature of this balance between aggradation and deposition is complicated, as Harding *et al.* (1987) note, and depends on slope, particle size, and channel morphologies (Florsheim 1985; Laronne and Duncan 1992). It is conceivable, therefore, that bars can be locations where artifacts are stored and/or buried, while also acting as source areas for artifact transport.

The implications of these results for Pleistocene archaeology are that in warmer phases, where vegetated meandering channels predominate, artifacts discarded on the riverbanks may remain there for long periods of time as a result of lower flood frequencies and the 'binding effects' of vegetation (Bridgland 2000). These phases are also more likely to attract/support a hominin presence. Conversely, in cooler periods, when broad, braided rivers predominate, artifacts may be more susceptable to erosion and incorporation into the active channel. These findings must be interpreted with caution, however, since the overall time-scales of these experiments are inevitably short and it is unclear if and when the channel would have eroded the banks and incorporated their sediments into the active channel. However, if Late Glacial and Holocene rivers give any indication, radiocarbon dating suggests that lateral channel migration can take centuries or even millenia (Cleveringa *et al.* 1998; Huben 2001; 2003). This suggests that if not buried, artifacts may have been exposed to a suite of other taphonomic modifications such as trampling or weathering and soil movement, without themselves being fluvially displaced.

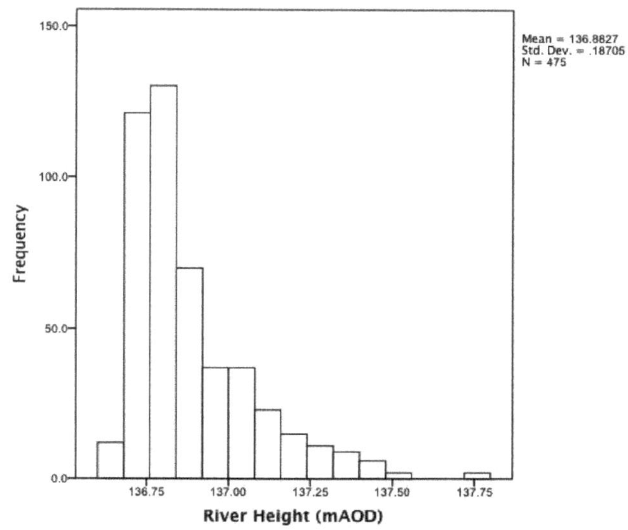

Figure 207: Daily river heights of the Glenderamackin from the Threlkeld gauging station for the duration of this project.

7.2.2. Maximum and average river discharge

Chapter 4 tested the hypothesis that maximum or mean river discharge significantly affected the horizontal transport of artifacts. The results of a mixed model showed no statistically significant association between the maximum or mean river discharge and artifact transport distances (refer to Section 4.4.1). However, there are some problems with interpreting these results. Most of the artifacts that formed the basis of these tests did not move considerable distances. There was only one monitoring period between April 2011 and September 2011 (Scatters 7–9; n=155) when artifacts were significantly transported. They were also transported out of the project area before the subsequent visit, yielding no further transport distance values. Therefore, these results should be treated with caution.

Comparing discharge values of the River Glenderamackin to those of other rivers observed in earlier experiments, daily mean flow and average flows were lower than those previously reported, though the high flows for this study of 30m^3/s were comparable to those of Petraglia and Nash (1987; Figure 208). An estimate of river flashiness, calculated as a ratio of the two values also showed that the Glenderamackin was the least flashy of all the experimental rivers.

Figure 208: Reported maximum and average river discharge flows for previous fluvial field experiments. **Values reported by Harding* et al. *(1986). †Values obtained from the USGS website (http://waterdata.usgs.gov/nwis) from the gauging station (Jemez River below Jemez Canyon Dam, 08329000).*

	Mean daily flow (m^3/s)	Max discharge (m^3/s)	'Flashiness' (Mean daily flow/high flow)
Harding *et al.* 1986*	5.5	90	0.06
Petraglia and Nash 1987†	3.07 (108ft^3/s)	33.74 (1180ft^3/s)	0.09
Hosfield and Chambers 2005c	5.81	75.62	0.07
Current study	1.50	30	0.05

Synthesizing this study with others is difficult because other authors do not report associations with fluvial output and differences in assemblage composition and artifact recovery rates make straight comparisons untenable. Understanding the impact of fluvial output, therefore, remains an outstanding question for future research. Nevertheless, this study does identify a new minimum fluvial output regime for transporting artifacts. This suggests that the exposure of artifacts to active flows in smaller channels such as the Glenderamackin can still have deleterious effects on archaeological assemblages.

7.2.3. Artifact dimensions

Artifact size is commonly used as an indicator of fluvial disturbance among archaeological assemblages. The results of a mixed model (refer to Section 4.4.1) and a linear regression (refer to Section 4.4.2) showed that within the gravel bed, artifact size appears to have an influence on transport distance. The results of both of these statistical analyses demonstrated that length and width were positively correlated to transport distance ($p<0.05$).

Interestingly, while previous experiments have reported correlations between either artifact dimensions or size, class, and horizontal displacement (Harding *et al.* 1987; Isaac 1967; Petraglia and Nash 1987; Schick 1986: 79), others have not (Hosfield and Chambers 2004). This study agrees with the former, particularly Schick, who found that shorter flakes moved over larger distances.

Though clast dimensions are intimately related to each other, results of this study confirms what Hosfield and Chambers (2005c) have pointed out, namely that a-, b-, and c-axes are better predictors of transport distance than weight (cf. Wilcock 1997). This study also suggests that the shorter length and width of artifacts may play some role in retarding the suggested saltation transport of artifacts (cf. Chambers 2004).

The differences between this study's results and those of Hosfield and Chambers (2005c) are probably in part due to their reliance on flakes in the Afon Ystwyth, an extremely active river bar setting. Harding *et al.* (1987) used that same setting, but the transported assemblage was comprised of handaxes. This study observed axes as well as flakes but the output regimen of the Glenderamackin was never above 30m^{3}s^{-1}, while the Afon Ystwyth between 2000–2003 (Hosfield and Chambers 2005c) reached a maximum of 75.62m^{3}s^{-1} (Figure 208). Hosfield and Chambers' (2005c) finding of an insignificant relationship between artifact length, width and thickness, and horizontal displacement may simply be the result of an output regimen too extreme to discriminate between such small ranges of artifact sizes.

Hosfield and Chambers (2005c) also point out that the periodic 'trapping' of artifacts, or the hampering of mobility by their being temporarily lodged in the bedload, may impact their transport distances. This experiment however, also saw trapped artifacts, but did not report the same findings; trapping did not have any statistical influence on the movement of artifacts in terms of their size and transport distances. Figure 209 shows a comparison of the bedloads and discharges for the two experimental sites. Given that the shapes, sizes, and lithology of the river clasts were similar, it is likely that trapping was the result of either the mean daily or maximum daily flow that was in both cases significantly higher at the Llanafan site. This

Figure 209: Comparison of the fluvial characteristics of Hosfield and Chambers (2005c) Llanafan site and the Glenderamackin from this volume. Descriptions of the Llanafan bedload are from Hosfield and Chambers (2005c) while the discharges were taken from the Environment Agency (http://www.nwl.ac.uk/ih/nrfa).

	Hosfield and Chambers 2005c (June 2002–March 2003)	This study (April–Sept 2011)
Mean daily flow (m^3/s)	5.81	1.50
High flow (m^3/s)	75.62	30
Median gravel size (mm)	*c.*16–32mm	*c.*16–32mm
River clast shape	Disc- and blade-like	Oblate, bladed, and prolate
River bed lithology	Palaeozoic shales and gritstones	Palaeozoic, greywacke, arenite, siltstone and mudstones

higher output may have resulted in higher bedload mobility thereby trapping artifacts more readily although perhaps only temporarily.

This suggests that dispersed knapping scatters found within gravel-bedded channels may be identified by the absence of smaller flakes, however given the weak association and large transport distances, disturbance is probably more accurately identified by other methods such as scatter density and or edge micro-flaking.

7.2.4. Artifact burial

Artifacts in this experiment were commonly buried either partially or completely by fluvial sediments. The results from a mixed model (refer to Section 4.4.1) suggested this statistically inhibited their transport distances ($p<0.05$).

Partial and total artifact burial is a finding common to all comparable experiments (Dennell 2004; Isaac 1967; Harding *et al.* 1987; Hosfield and Chambers 2005b; Petraglia and Nash 1987; Schick 1986). Some authors have drawn conclusions concerning the importance of burial in the preservation of *in situ* sites. Nash and Petraglia, for example, noted, 'Burial is a particularly important factor for preservation of sites in fluvial situations. It protects or at least retards the transport process in a streambed' (Petraglia and Nash 1987: 125). Harding *et al.* (1987), on the other hand, hypothesized that the nature and timing of burial and reburial was likely episodic, with cycles of transport–burial–re-exposure repeating over time.

This study agrees with both; burial in this experiment undoubtedly hindered the transport of artifacts as has been noted in other similar experiments (Petraglia and Nash 1987; Isaac 1967). However, the burial, most likely, was only temporary. The slumping alluvial banks on which Scatters 1–6 were placed were actively eroding in 'avalanches' which were absorbed in the stream *en masse*. There is no doubt that, had this experiment continued for a longer time, some experimental scatters would have been subject to the same erosion.

7.2.5. Micro-flaking and transport distance

This experiment showed that fluvial transport within the gravel bed typically resulted in damage to artifacts in the form of small (>1mm in maximum dimensions), irregular, continuous bifacial retouch around the perimeter of flakes, called micro-flaking (Harding *et al.* 1986; Hosfield and Chambers 2005c). These developed on flakes that were transported between 1 and 250m downstream (Figure 84). The results of a binary ordinal regression showed that the expression of micro-flaking was highly correlated with increasing transport distance, with artifacts being micro-flaked in as little as 7m. This analysis also showed that the dimensions of the flakes did not predict whether artifacts were micro-flaked or not (refer to Section 4.4.6). Finally a small limited dataset (n=3) showed that micro-flaking affected the primary metrics of artifacts, but there was no clear relationship between this change and transport distance, suggesting that transport modes and unexplored aspects of their morphology such as ridges and edge angles may also play roles.

These results are well supported by laboratory-based work (Tringham *et al.* 1974; Shackley 1974; Chambers 2003; Burroni *et al.* 2002) and other field studies (Harding *et al.* 1987; Dennell 2004; Hosfield and Chambers 2005c). Hosfield and Chambers (2004) recorded the development of micro-flaking in a gravel-bedded river after a minimum horizontal displacement of 5m. Harding *et al.* (1987) reported edge damage developing by at least 38m. Data from this study suggested that artifacts were micro-flaked unambiguously within 7m of transport.

Dennell (2004: 246), Petraglia and Nash (1987), Schick (1986), and Isaac (1967) did not report on micro-flaking; in fact, they did not report any damage, presumably because it was not observed. This suggests that micro-flaking development may also reflect the nature of transport in gravel-bedded rivers. Research in this book (Chapter 6) suggests that bedload grain size is an important factor in the formation of edge micro-flaking and that in sandy substrates, like the ones utilized by Dennell (2004: 246), Petraglia and Nash (1987), Schick (1986) and Isaac (1967), micro-flaking simply did not occur. It also suggests that edge damage is not necessarily only the result of saltation transport as Chambers has suggested (2004). All this implies that making inferences about artifact transport modes from micro-flaking in gravel-bedded rivers may not be so straightforward.

Hosfield and Chambers (2005c) suggest that micro-flaking is reminiscent of intentional retouch. However, the near-complete continuous bifacial nature of the micro-flaking in this experiment, and that of Hosfield and Chambers, and the distinct nature of the small scars (even along the striking platform) is inconsistent with commonly observed archaeological patterns of anthropogenic flake retouch (Tringham *et al.* 1974). It is also unlikely that taphonomic processes such as trampling, solifluction, or plowing could be the cause of the type of micro-flaking observed (see Figure 11 for references).

Hosfield and Chambers (2004) note the presence of larger 'notches' (<15mm) removed from artifacts. This was not observed in this experiment and may again reflect the high fluvial output of the river in which their experiments were conducted. Nonetheless, these notches were found on pieces that were also micro-flaked, indicating that interpreting such notches as anthropogenic in fluvially derived pieces should be treated with caution.

This study did not find any significant association between micro-flaking and flake size, suggesting that micro-flaking is a powerful and indiscriminant indicator of fluvial modification. However, when comparing the weight loss for the three recovered flakes in this experiment to the smallest handaxe category from Harding *et al.*'s (1987)

Figure 210: A comparison of weight loss from fluvial transport, contrasting Harding *et al.* (1987) and the current study.

Reference	Range of original maximum length (mm)	Mean distance moved	Mean weight loss (g)	Weight loss/meter (g/m)
Harding *et al.* 1987	71.4–86.5	20.5	.03	.001
Current Study	43.77–43.66	184.28	2.15	.01

study, this study showed a ten times greater weight loss per meter traveled (Figure 210). Given that the bed lithologies of the two experiments are similar (Figure 209), this shows that artifact type (i.e., handaxe or flake) may play a role in the micro-flaking of artifacts and further work needs to explore the role of edge angles and edge margin thickness in micro-flaking expression.

In short, continuous bifacial micro-flaking is a good marker of archaeological site disturbance in gravel-bedded rivers—a marker that is extremely sensitive to small transport distances or even small-grained *in situ* abrasion, and one that can be useful in identifying low-energy fluvial disturbance to Palaeolithic sites.

7.2.6. Artifact displacement and trampling

Isaac (1967: 198) indicated that a main goal for future fluvial field experiments was to '[determine] the effects of trampling on concentrations of artifacts under alluvial circumstances.' The trampling of Scatters 3–6 between April 2011 and September 2011 allowed a unique opportunity to test the hypothesis that artifact horizontal transport would be greater among non-trampled scatters. Because cattle trampled Scatters 3 and 4, and sheep trampled Scatters 5 and 6 (animal size categories 3 and 4 respectively; Brain 1983; Smith *et al.* 2003), this study also tested the hypothesis that animal size would increase the horizontal transport of artifacts by virtue of their increased weight and strength. Contrary to these expectations, the results of a mixed model showed no statistically significant impact of animal trampling on transport distance ($p>0.05$; refer to Section 4.4.1)

The means and ranges were lower in this study than when compared to previous studies (Figure 211; Eren *et al.* 2010; Pargeter and Bradfield 2012; Nielsen 1991). When confronted with the same results, Eren *et al.* (2010) suggested (comparing their data to Nielsen's (1991)), that because horizontal displacement was not related to either sediment saturation, horizontal displacement, or animal size, it was the result of their shorter trampling times (approximately 3.5 minutes compared to upwards of 80 passes). However, given that the trampling times in this experiment were considerably longer, it is suggested that *intensity* may be a better proxy for trampling disturbance because this experiment was longer in duration and likely did not experience the same rapidly successive trampling 'passes' that the other experiments did and which could be argued to be more 'artificial.'

This study also hypothesized that animal trampling would impact the vertical distribution of lithic artifacts. When measured after the trampling period, the mean scatter height changed between -3 and 1cm. However, when these values were compared to previous values for the same artifacts, this was only significant for one of the cattle trampling scatters (Scatter 3; refer to Section 4.3.3).

This can be attributed to the weights of the animals. The sheep did not exhibit enough downward pressure to push the artifacts deeper into the ground. However, this would fail to explain why Scatter 4, which was also trampled by cattle, did not experience significant vertical movement. Given that the scatters were deposited in comparable settings (they were located within 10m of each other), were subjected to trampling by the same animals, and that previous studies have indicated that artifact size likely does not impact vertical displacement (Eren *et al.* 2010 and references therein), the difference may have to do with the local topography of the area and the cattle's preferred areas of activity.

When compared to previous experiments, our trampled scatters showed similar vertical displacement values (Figure 212). They were least like the results of Eren *et al.*'s (2010) 'wet sediment' experiments, suggesting that the fluvial context tested in this experiment was more analogous to the dry sediments used in Neilsen (1991) and Pargeter and Bradfield (2012). Scatters 4, 5, and 6 were also within the range of previous experiments with a mean vertical increase of 1cm. This increase of height is possible due to artifacts being kicked onto higher topographies as a result of the trampling (Eren *et al.* 2010) or that sediments

Figure 211: Mean and range of artifact horizontal displacement for this study (between April 2011 and Sept 2011) and those with comparable data (Nielsen 1991; Eren *et al.* 2010). * *Figures are based on a maximum x or y value of movement in a grid. † Based on one value.*

Reference	Mean (cm)	Range (cm)
Nielsen 1991 (TR-II)	23.9	126.0
Nielsen 1991 (TR-III)	19.2	122.0
Eren *et al.* 2010 'buffalo dry'	17.9	101.8
Eren *et al.* 2010 'buffalo wet'	17.8	71.7
Eren *et al.* 2010 'goat dry'	11.5	60.3
Eren *et al.* 2010 'goat wet'	13.5	44.4
Pargeter and Bradfield 2012*	24.1	118.4
Current study (Scatter 1)	51.5	271
Current study (Scatter 2)†	107	107
Current study (Scatter 3) cattle	6	71
Current study (Scatter 4) cattle	9	106
Current study (Scatter 5) sheep	9	110
Current study (Scatter 6) sheep	26	94

Figure 212: **Mean and range of artifact vertical displacement for this study and those with comparable data.**

Reference	Mean (cm)
Nielsen 1991	1.5
Eren *et al.* 2010 'buffalo dry'	1.1
Eren *et al.* 2010 'buffalo wet'	-5.8
Eren *et al.* 2010 'goat dry'	0.6
Eren *et al.* 2010 'goat wet'	-5.3
Pargeter and Bradfield 2012 'backed tools surface'	-1.5
Pargeter and Bradfield 2012 'bladelets surface'	-1.4
Current study (Scatter 3) cattle	-2
Current study (Scatter 4) cattle	1
Current study (Scatter 5) sheep	1
Current study (Scatter 6) sheep	1

were 'raised' due to lateral displacement in response to an animal footprint.

7.2.7. Trampling and artifact dip

Section 4.4.5 tested the hypothesis that artifact dips in riparian settings would vary under different trampling conditions. The results of Rayleigh's, Kuiper's, and Watson's tests showed that dips measured between untrampled artifacts and those that were trampled by either sheep or cattle, did not significantly differ ($p>0.05$).

Eren *et al.* (2010) found similar results when measuring the dips of artifacts that were trampled by buffalo and goats in dry, medium-grained silt/coarse-grained sand. However, they did find significant alteration to dips when artifacts were trampled in a 'muddy' substrate. Another recent study by Benito-Calvo *et al.* (2011) examined the effects of trampling by humans, also in a dry 'clayey/ sandy matrix.' The authors not only found no change in artifact dips but also concluded that, if anything, trampling stabilized the initial artifact inclinations rather than altering them. In fluvial settings this may have implications for the fluvial reworking of artifacts; if artifacts are trampled post-discard, they may less susceptible to short-term artifact entrainment.

Given this, the results of the current study suggest that meandering river riparian zones, even ones that are frequently inundated, perform similarly to non-muddy sediments when recording the effects of trampling dip. This may be because vegetation stabilizes the underlying root structures of riparian zones making it more like dry sediments (Simon and Collison 2002).

Another reason for these different results may be different trampling intensities. Eren *et al.*'s (2010) study reported continuous trampling by a minimum of 16 animals for approximately 3.5 minutes. While the intensity of trampling in this experiment was unknown, the larger project area suggests it was lower. This study's results nevertheless support the conclusion that trampling does not affect the natural background dips of artifacts in a non-waterlogged meandering riparian setting.

Interestingly, the range of this experiment's dip values and other values are consistent with the sedimentological literature that shows that the most disk-like and ellipsoidal particles generally vary between 10 and 30° in a fluvial setting (Potter and Pettijohn 1977: 42). Contrastingly, Eren *et al.*'s (2010) non-fluvial trampled values fall

Figure 213: A compilation of reported dips from field experiments for comparison to the current study.

Reference	Context	Dip (°)	Dip Max
Isaac 1967 (Plot 4)	Sand/silt bed braided channel	*Uns*	60
Eren *et al.* 2010 'buffalo dry'	Medium silt fine sand	10±8	42
Eren *et al.* 2010 'buffalo wet'	Medium silt fine sand	44±24	90
Eren *et al.* 2010 'goat dry'	Medium silt fine sand	6±4	19
Eren *et al.* 2010 'goat wet'	Medium silt fine sand	41±29	90
Schick 1986 (flooded workshop site)	Stream channel bank	15-30	*Uns.*
Schick 1986 ((Site 1c)	Sand channel	2-30	*Uns.*
Schick 1986 (Site 13)	Channel levee	*Uns.*	40
Schick 1986 (Site 15)	Channel levee	5-12	*Uns.*
Schick 1986 (Site 19)	Channel sand bar	15-34, mean=27	34
Schick 1986 (Site 24)	Low-water channel	Mean=25.6	45
Schick 1986 (Sites 26-27)	Small sand channel	Mean=29	*Uns.*
Schick 1986 (Site 28)	Medium (20m) channel bank	Mean=13.7	*Uns.*
Schick 1986 (Site 34)	Flood basin	Upstream dip	*Uns.*
Current study (Scatter 1)	Erosion riverbank	8±10	36
Current study (Scatter 2)	Erosion riverbank	21±8	33
Current study (Scatter 3) cattle	Erosion riverbank	8±13	39
Current study (Scatter 4) cattle	Erosion riverbank	20±13	44
Current study (Scatter 5) sheep	Erosion riverbank	10±16	38
Current study (Scatter 6) sheep	Erosion riverbank	19±16	75

Figure 214: A compilation of trampling experiments for comparison to the present study (modified from Eren *et al.* 2010; table 1).

Reference	Principal Substrate	Flake lithology	Trampling agent	Duration
Stockton 1973	Sand	Red glass	Humans	1 day
Flenniken and Haggerty 1979	Silt loam	Obsidian	Humans	1000 crossings
	Alluvial sand	Obsidian	Humans	1000 crossings
	Clay and gravel	Obsidian	Humans	1000 crossings
	Basalt gravel	Obsidian	Humans	1000 crossings
	Obsidian flakes	Obsidian	Humans	1000 crossings
Villa and Courtin 1983	Loose, well-sorted silty sand	Flint	Humans	16, 22, 32, 36 days
Goerke 1983	Hard packed sand with small pebbles	Chert and chalcedony	Elephant	25 crossings
Gifford *et al.* 1985	Compact sandy silt	Obsidian	Humans	2 hours
	Unconsolidated medium fine sand	Obsidian	Humans	2 hours
Pryor 1988	Sandy soil	Obsidian	Humans	2 hours
	Loamy soil	Obsidian	Humans	2 hours
Nielsen 1991	Muddy gravel	Obsidian	Humans	1500 crossings
	Muddy gravel	Obsidian	Humans	800 crossings
	Muddy gravel	Obsidian	Humans	800 crossings
Shea and Klenck 1993	Sandy soil	Brandon flint	Humans	15 min
	Sandy soil	Brandon flint	Humans	30 min
	Sandy soil	Brandon flint	Humans	45 min
McBrearty *et al.* 1998	Low density sand	Chert	Humans	1 hours
	High density sand	Chert	Humans	1 hours
	Low density sand	Chert	Humans	1 hours
	High density sand	Chert	Humans	1 hours
	Low density loam	Chert	Humans	1 hours
	High density loam	Chert	Humans	1 hours
	Low density loam	Obsidian	Humans	1 hours
	High density loam	Obsidian	Humans	1 hours
Eren *et al.* 2010	Medium silt, fine sand	Limestone	Water buffalo	3 min, 24 s
	Medium silt, fine sand	Limestone	Water buffalo	3 min, 24 s
	Medium silt, fine sand	Limestone	Goats	3 min 36 s
	Medium silt, fine sand	Limestone	Goats	3 min 36 s
Current study	Vegetated river-cut cliffs	Brandon flint	Cattle	4 weeks
	Vegetated river-cut cliffs	Brandon flint	Sheep	4 weeks

outside of that range (Figure 213). This may indicate that fluvial processes can have an overriding effect on changes to artifact dip caused by trampling. Though it may be tempting to suggest that archaeological assemblages in fluvial contexts with artifact dips similar to those found in this study may suggest fluvial modification (especially if they are imbricated), their dip is also dependent on their maximum projection plain dipping up-current and therefore the appropriate stereograms should be constructed to demonstrate accordingly.

The results of these experiments suggest that while trampling may not affect the overall dip of artifacts, dip may be a powerful predictor of fluvial action. Further work is needed to confirm this and a major issue that still needs clarification is how the dip values of this and other experiments relates to the natural background of artifact dips, particularly in freshly knapped scatters. Finally, as clast dips tend to imbricate relative to the stream, future work should also incorporate the appropriate orientations in stereograms as further evidence.

7.2.8. Trampling and artifact damage

The results presented in Section 4.4.7 showed that among bankside artifacts, damaged artifacts were only found in scatters that were subject to trampling. 14% of artifacts that were trampled exhibited breakages as opposed to none of the artifacts that were not trampled. These breakages were characterized by either impact notching or snap fractures (refer to Section 4.2). Patterns of artifact snapping on artifacts trampled by cattle (21%) and sheep (3%) differed statistically significantly from each other suggesting that breakage patterns were influenced by animal size.

Comparing this data to previous experiments, the values were broadly similar (Figure 214). Eren *et al.* (2010) reported similar breakage patterns (Figure 215) in similar

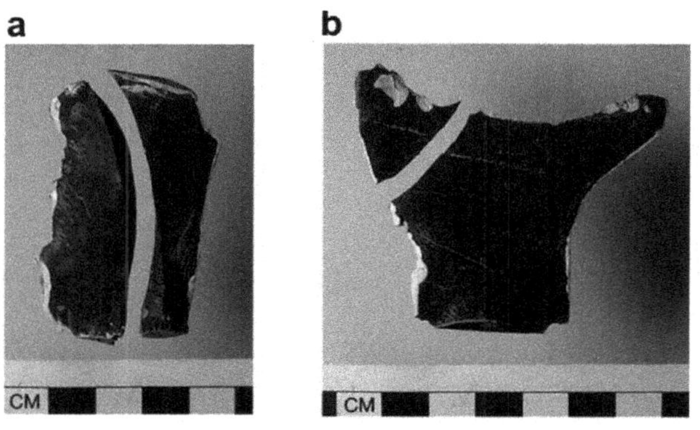

Figure 215: Examples of artifact breakage during the Eren *et al.* 2010 experiment. One artifact was split down its axis of percussion (a), while another had its distal end broken off (b) (Eren *et al.* 2010; figure 8 with permission from Elsevier).

frequencies. With supplementary data from Eren *et al.*'s (2010) article, Fisher's exact tests were performed to see what the effects of animal size and sediment saturation were. Eren *et al.*'s (2010) data showed that the number of artifacts broken in saturated (between 3–4% wet mass) sediments (17%) and those broken in dry sediments (3%; p=0.02) differed significantly, but between water buffalo (10%) and goats (10%; p>0.05) it did not differ significantly. When separated into wet and dry substrates, breakage patterns by different animals were also not statistically significantly different.

The reason this study found differences in breakage patterns for trampling by animals of different sizes and Eren *et al.* (2010) did not, is not clear, especially considering that the sizes of animals used in both experiments were broadly similar in size (cattle: 450–1000kg and sheep: 20–200kg v. water buffalo: 250–550kg and goat: 25–95kg; size estimates from Nowak 1999).

There are three possible reasons for this:

1. This study used flint, while Eren *et al.* (2010) used a siliceous limestone. However, this is an unlikely reason as the author noted that the limestone was particularly 'tough' and produced overall breakage patterns similar to this study.
2. The amount of trampling that artifacts in this study underwent is unknown. Differences in this study's breakage patterns may reflect varying intensity.
3. Water buffalo have adapted splayed hooves so that they do not sink into mud. This larger footprint may distribute weight more evenly than cattle hooves that are more compact and may result in fewer breakages.

Animal trampling is a well-researched part of experimental archaeological research (Boot 1987; Gifford-Gonzalez *et al.* 1985; Lopinot and Ray 2007; McBrearty *et al.* 1998; Nielsen 1991; Shea and Klenck 1993). This study, as well as Eren *et al.* (2010), show that the effects of animal trampling were less intense than those proposed by earlier studies, that suggested artifact trampling was severe enough to duplicate retouch or even produce identifiable tool types. Most likely the different findings are the result of the fine alluvial substrates used in this and Eren *et al.*'s (2010) study, as opposed to the substrates containing gravels used in other studies (Goerke 1983; Flenniken and Haggarty 1979; Nielsen 1991; Figure 213). Another possibility is that in other studies, humans trampled the artifacts, whereas in Eren *et al.* (2010) and this study, even-toed ungulates (*Artiodactyla*) trampled the artifacts (Figure 213).

It is tempting to assume that all snapped artifacts in alluvial settings are indicators of trampling, but other processes such as sediment consolidation (Eren *et al.* 2011; Warren 1905) and anthropogenic breaks can produce similar effects. These causes of snapping need to be further explored and differences in the breakage patterns should be examined more systematically.

The Pleistocene implications are that trampling impacts may be more marked on gravel substrates common in cooler climate braided river-type floodplain settings. However, breakages in the form of impact notches and snap fractures may be useful diagnostics in identifying trampling among fine-grained vegetated sediments.

It is clear from this experiment that there are a large number of variables at stake here unexplored by this book. Given that vegetation may bind artifacts to the floodplain, we might be dealing with years or decades of trampling in various substrates by a variety of animal sizes as large as mammoths and numbers by herds of animals. Therefore, many more experiments are needed before a true understanding of trampling can be achieved.

7.2.9. Orientations

Previous experiments have suggested that sediments (Hakon Wadell 1935; Potter and Pettijohn 1977: 42–43), lithic artifacts (Isaac 1967; Lenoble and Bertran 2004; Schick 1986: 245), and bones (Aslan and Behrensmeyer 1996; Boaz and Behrensmeyer 1976; Kaufmann *et al.* 2011; Reid and Frostick 1985) readily orient themselves anisotropically in fluvial environments. Chapter 4 tested the hypothesis that the artifacts' long axes orientations were biased in fluvial environments. This was done for all artifacts (see scatter history descriptions in Section 4.3) and for artifacts with distinctly long axes (i.e., $l/w \geq 1.5$; see Section 4.4.4) using Rayleigh's, Watson's, and Kuiper's tests. The results of these analyses indicated that in both cases, the artifacts' long-axis orientation did not develop anisotropic patterns.

Comparing the data from this experiment to those of previous experiments (Figure 216), Isaac (1967) found that handaxes oriented themselves transverse to the current, with a small number of clustered artifacts (n=3) orienting themselves parallel to the flow. Schick found that artifacts in different contexts were oriented either transverse or parallel to the flow, with smaller artifacts tending to be more in line with the stream flow, sometimes without being

Figure 216: Reported artifact imbrications from field experiments for comparison to the current study.

Reference	Experiment	Context	Mean orientation from stream vector flow
Isaac 1967	Plot 4	Sand/silt braided channel	Mostly transverse (i.e., 90); though some clustered pieces (n=3) were parallel
	Site 1a	Small sand channel delta	Mostly parallel with stream flow
	Site 13	Channel levee	Larger artifacts oriented transverse to current while smaller ones oriented parallel to the flow.
	Site 15	Channel levee	All distinctive long axes oriented either transverse or parallel to the flow trajectory despite short movement
	Site 19	Channel sand bar	Transverse or parallel to channel axis
	Site 24	Low-water channel	Parallel or perpendicular to channel axis (n=11)
	Site 25	Levee	Transverse or parallel to channel axis
	Sites 26-27	Small sand channel	Transverse or parallel to channel axis
	Site 28	Medium (20m) channel bank	Larger artifacts transverse, smaller artifacts parallel to channel axis
	Site 34	Flood basin	Probably transverse to flow
	Site 36	Shallow drainage channel	Transverse to flow
Hosfield and Chambers 2005c	Scatters 1, 3, 6, 7, 8, 10	Semi-stable floodplain and point bar complex	No clear orientation
	Scatters 11 and 12	Point/midstream bar complex	No clear orientation
Current study	Scatters 1-6	Semi-stable floodplain	No clear orientation
Current study	Scatters 7-9	Point bar	No clear orientation

transported significant distances. Hosfield and Chambers (2005c) found that none of their scatters were clearly imbricated.

The reasons for these variable results may be due to the following:

1. Artifact imbrication may differ between gravel-bedded and sand-bedded channels. Figure 216 shows that experiments conducted in braided, sand-bedded channels readily imbricated (i.e., Isaac 1967; Schick 1986) while this study and Hosfield and Chambers (2005c) showed no such patterning. Hosfield and Chambers' (2005c) suggest that a possible mechanism was 'localized trapping of flakes between larger clasts results in random fabric patterns' (2005c: 14). Given the similar bed lithologies of the two experimental sites, this explaination provides a plausible reason for the lack of imbrication among this study's channel inserted scatters (7–9). However, clast imbrication is also related to packing density, sorting and stream gradient, in addition to particle shape and size and in different gravel-bedded regimes, artifact imbrication may be more likely (Potter and Pettijohn 1977: 43).
2. Among the in-channel scatters (7–9), the nature of a gravel bed may produce turbulent 'eddies' that complicate local flow directions (Charlton 2008: 88).
3. Artifact morphology may play a role in artifact imbrication. Isaac's experiments focused on concrete cast handaxes, while Schick's experiments usually included a portion of long-bone artifacts in their results. Their distinct long axis may have made their imbrication either more likely, or more identifiable. However, given that this study analyzed artifacts with distinct axes (where l/w>1.5), this explanation may not be valid.
4. Among bankside scatters (1–6), underlying vegetation cover may have served as an additional roughness element, 'anchoring' artifacts. Secondly, artifacts were only occasionally submerged. Finally, animal trampling may have served to 'stabilize' their position within sediments (Benito-Calvo *et al.* 2011).
5. Differential recording methods in measuring long axes may have played a role. Though previous fluvial field studies do not describe their long axis recording methods (as yet no standard methodology has been developed for this), recent research on archaeological material suggests that different ways of recording long axes, either by their symmetrical axis or their longest axes, can have powerful impact on orientation results (Domínguez-Rodrigo *et al.* 2012; Benito-Calvo *et al.* 2011).
6. Different presentation methods. Previous studies' results of imbrication are largely anecdotal. Presented results are rarely measured and never statistically tested. Often, they rely on small sample sizes that may be inadequate to draw robust conclusions.

The Palaeolithic implications of these results are simple and potentially fundamental: artifacts within gravel-bedded channels simply may not imbricate as readily as previous studies have suggested for braided sand-bedded channels. Further work is still required to understand the necessary factors to invoke imbrication of artifacts and what effect artifact dimensions have on their orientation.

7.2.10. Artifact recovery

Limited recovery rates of experimentally transported fluvial artifacts have posed a perennial problem in archaeological tracer studies with no apparent solution in sight (Bertran *et al.* 2012; Dennell 2004; Harding *et al.* 1987; Hosfield and Chambers 2004; 2005c; Petraglia and Nash 1987; Schick 1986). This study was no exception. Paired t-tests for individual scatters did not show statistically significant differences between recovered and unrecovered artifacts among bankside artifacts (refer to individual scatter descriptions in Section 4.3). However, among fluvially transported artifacts (Scatters 7–9), unrecovered artifacts (n=48) were statistically significantly smaller by length and width (refer to Section 4.4.3).

This is in agreement with the results of Schick (1986: 67), who reported a similar recovery bias and attributed it to the selective transport of smaller artifacts, based on the long axis, out of the site. This suggests that artifact size distributions are valuable indicators of fluvial disturbance.

7.2.11. Limitations and future work

There are a number of advantages to studying fluvial taphonomy through real world experiments. Firstly, experiments provide a profile of differential preservation of refuse patterns after site abandonment (Schick 1986: 60). Secondly, they allow for observations of modifications to artifacts in real areas and situations, reminding the researcher of the oversimplification and often-erroneous aspects of laboratory settings (Lewin and Brewer 2002). Thirdly, the gathering of real world experimental data has allowed comparisons with previous experiments reviewed above.

The application of this exploration of artifact deposition in the Glenderamackin to the Palaeolithic record has some limitations due to the scope of this experiment. However, some of these may be overcome through further work:

- First, the short duration of tracer studies gives insufficient information about the long-term/long-range effects of artifact transport (Hosfield and Chambers 2004).
- Second, though these results may be applicable to sites from interglacial settings, these experiments only examined a single river type and their application to non-meandering Pleistocene river types (e.g., braided, anastomosing) is therefore limited.
- Third, relationships of artifact transport are based on a maximum transport distance of 250m. These results may not be applicable to more heavily transported datasets, as longer transport periods may generate different results, but may be relevant to locally disturbed sites.
- Fourth, there are undoubtedly more variables involved in artifact fluvial transport that require exploration, notably riparian vegetation cover and bankside erosion rates (Ashton *et al.* 2006).
- Fifth, methods for accurately evaluating artifact dip and orientation are an ongoing field of study. How one accurately defines the major axis of irregular artifacts in the field is currently a matter of debate (i.e., Benito-Calvo and de la Torre 2011; Domínguez-Rodrigo *et al.* 2012). Additionally, artifact imbrication is likely a much more complicated process that may be better explored by limiting analyses to artifacts with more distinct long axes (i.e., $l/w \geq 2.0$) or by excluding heavier artifacts. However, such approaches were not possible in this case, because excluding the necessary data would not have provided sufficiently sized datasets. This limitation can be overcome in future experiments which should additionally focus on understanding the influence of soil movement and the natural variations in micro-topographic relief.
- Finally, this chapter could not overcome the inherent shortcomings of the painted clast recovery system: though considerable efforts were made to account for unrecovered artifacts at scatter sites, conclusions were limited because there was no way to identify all artifacts' final depositional location. Non-recovery was assumed to be the result of transport out of either the scatter or project area, though it was not clear if this was by several centimeters or several kilometers. It is also possible that some artifacts were not recovered because they were buried or simply not recognized. This may eventually be overcome with more sophisticated artifact recovery techniques such as RFID tags and/or radio transmitters and require further examination (see Section 2.3.2 for review).

7.2.12. Section summary

The frequent dispersal of archaeological material in fluvial deposits is of considerable interest to Palaeolithic archaeologists because it obscures anthropogenic spatial patterns. Previous researchers have suggested that fluvial site disturbance can be identified by comparing size distributions of artifacts to the results of experimental fluvial disturbances of 'control' assemblages derived from experimental knapping events (Bertran *et al.* 2012; Dibble *et al.* 1997; Petraglia and Potts 1994; Schick 1986). The absence of smaller artifacts found in some experiments of this kind has generally been interpreted as a possible indication of downstream fining (Isaac 1967; Schick 1986). Other experiments, however, have found no such association, suggesting that artifact dispersal is a highly complicated process influenced by the river's output regimen (Petraglia and Nash 1987), the discard location of the artifacts (Dennell 2004: 250; Harding *et al.* 1987), and the nature of the bedload/rugosity (Hosfield and Chambers 2004).

This fluvial study has demonstrated a significant and robust association between sedimentary factors and artifact transport, most notably, the association of artifact size with transport distance. It has also confirmed the association of transport with edge damage. The study has neither conclusively proven nor falsified previous hypotheses concerning the nature of fluvial site formation processes. It has however highlighted the complex nature of alluvial site formation processes in a repeatable experiment highlighting the transport of replica Palaeolithic material. It examined the variables of scatter location, artifact dimensions, river discharge, orientations, and animal trampling, and demonstrated their relevance to the

Figure 217: Comparison of this volume's coarse-grained experiments and previous experiments.

Reference	Aim(s)	*n*	Equipment	Duration	Sediment	Velocity	Controlled variable(s)	Dependent variable(s)	Independent Variable(s)	Result(s)	Limitation(s)
Boule 1905	• Identify a natural genesis of eoliths	*Uns.*	Industrial cement mixer in Guerville (France)	29 hours	Chalk and Clay	16rpm	• Rotational speed • Artifact Lithology (French chalk flint)	• Concoidal fracture	• N/A	• Torrents produce pseudo-retouched geofacts	• Equipment limitations (cement mixer)
Tringham et al. 1974	• Distinguish fluvial edge damage from usewear	5	Agitated plastic container	1 hour	Water, sand, stones	N/A	• Artifact lithology (European chalk flint)	• Edge damage (bifacial, location)	• Artifact lithology	• Randomly oriented micro-flaking along the perimeter of the artifact	• No attention paid to the microscopic morphology of the artifact
Shackley 1975	• Distinguish artifact abrasion groupings • Assess abrasion for regional archaeological patterns	6	Tumbling barrel (large plastic barrel)	100 hours (10 intermediate time-steps)	Sand to clay	25.53rpm	• Rotational speed • Lithology (European chalk flint)	• Ridge width • Ridge features	• Time • Handaxe shape • Sediment size (sand to clay) • Water volume	• Ridge abrasion related to time and abrasive • Water and handaxe shape insig.	• Small sample • Small sediments • Only saltating transport mode • Ridges vary across artifacts
Hosfield 1999; Hosfield *et al.* 2000	• Create a local ridge abrasion index for the Hampshire basin • Extend Shackley's running time	15	Commercial cement mixer	50 hours (14 intermediate steps)	3kg River Test gravels <95mm 12l water	*Uns.*	• Rotational speed • Artifact lithology (flint) • Artifact class (handaxe)	• Weight • Primary metrics (Roe 1981) • Ridge width	• Time	• Time and ridge wear linearly related. • Edge damage rapid • Abrasion affects form	• Equipment limitations (cement mixer)
Chambers 2003; 2004	• Assess relationships between artifact transport duration and ridge development • Examine the effects on overall morphology	4	12m glass walled recirculating flume	250m (16 intermediate steps) 1000m (31 intermediate steps	*Uns.*	Bedload transport (rolling /saltation/ sliding)		• Ridge width • Micro-flaking	• Time • Artifact Lithology (Flint, Greensand chert) • Bedload transport (rolling/saltation/ sliding)	• Toolstone affects ridge abrasion rates • Artifact morphology affects transport mode • Transport mode affects damage type • Ridge widths vary on same handaxe	• The gravel 'liners' rather than actual gravel • Time laborious • No established relationship of artifact form (just subjective lenticular v. ovate)

Figure 217: ***Continued.***

Reference	Aim(s)	*n*	Equipment	Duration	Sediment	Velocity	Controlled variable(s)	Dependent variable(s)	Independent Variable(s)	Result(s)	Limitation(s)
Grosman *et al.* 2011; 2013	• Understand the effect of fluvial processes on Acheulean handaxes	8	100 liter cylindrical plastic barrel	200 rolls (7 intermediate time-steps)	250g to 2920g basalt cobbles	*Uns.*		• Damage profile (distance between successive profiles) • Volume • Asymmetry • Roughness	• Rolls	• Tip loses most contour • Lost 10% of volume after rolling • Asymmetry increases with rolls • Roughness increases with damage	• Artifact is oriented according to center of mass and 3 eigenvectors • Only looked at pointed handaxes
Current Study	• Assess time, impact frequency, artifact lithology, sediment size, sediment lithology, and artifact class on primary metrics and artifact condition	188	Polyethylene drum (84cm height, *d*=49cm)	3 hours (8 intermediate time-steps)	3 graded English Pleistocene sediments	15rpm	• Rotational speed • Water volume	• Primary metrics • Artifact condition (Ashton 1998)	• Time • Sediment mass • Artifact lithology • Sediment size • Sediment lithology • Artifact class	• Time affects weight, length, width, refinement • Sediment impacts affects abrasion • Artifact lithology does not impact abrasion (except condition) • Sediment size affects abrasion • Sediment lithology affects abrasion • Artifact class affects abrasion	• Abrasion is likely asymptotic • Increasing sedimentary impacts increases abrasion • Artifact lithology influences artifact condition • Large sediments (>φ=-5.5) abrade artifacts more rapidly • Sediment lithology affects the abrasion • Flakes abrade more rapidly than handaxes

transport of flint artifacts. It also identified that in a gravel-bedded regimen, morphological characteristics such as edge condition and assemblage size distributions may provide details about artifact 'life-histories' (Hosfield and Chambers 2004). Which of these previous studies are more useful or appropriate for assessing archaeological sites? It will, and should, vary: quasi-experimental studies such as this one are likely best applied to sites that are most analogous to their study area in terms of sediments, setting and flow regime.

7.3. Coarse-grained experiments (tumbling barrel experiments)

This section discusses the results of the coarse-grained abrasion experiments from Chapter 5 (refer to Section 3.3 for methodology), which were aimed at understanding how artifact primary metrics and conditions are influenced by (1) abrasion time, (2) sediment mass, (3) artifact lithology, (4) sediment lithology, and (5) artifact class. In the course of the discussion, possible mechanisms or pathways will be given and contextualized within the pre-existing literature (discussed in Section 2.4 and summarized in Figure 217). The strengths and limitations of these experiments will also be discussed in addition to their implications for the archaeological record.

7.3.1. Artifact condition and primary metrics

A main hypothesis of this study was the relationship of artifact condition (see Figure 46) and changes in primary metrics with time in the tumbling barrel. The results showed no clear relationships (refer to experiments in Chapter 5). Though both were positively associated with time (i.e., more time in the barrel=increased condition class and reduction in artifact dimensions), their rates of change were consistently different from each other. Both developed at different rates, modulated by the factors explored in this experiment. This experiment presents the first set of data comparing change in artifact conditions and changes to artifact morphology.

1. There are three reasons why the rates of change in artifact conditions and primary metrics differ:
2. Condition and primary metrics follow different abrasion pathways: Primary metrics are changed primarily by fracture wear (Attal and Lavé 2009). A major constituent for evaluating condition is surface features and ridge width that are propagated by erosion wear (Attal and Lavé 2009). The rates of these different types of abrasion are known to vary on lithic materials depending on their hardness (erosion wear) and brittleness (fracture wear; Frings 2008). Flint (and other cryptocrystalline raw materials) is hard (7.0/10; Mohs hardness scale), but also brittle, so that breakage is more likely.
3. Experimental conditions: artifact abrasion patterns and rates of development changed according to sediment size, sediment lithology, artifact lithology, and volume of sediment and artifact class. Fracture wear and erosion wear were modulated by variations in the bedload conditions (Frings 2008).
4. Limitations of the abrasion recording method: the abrasion recording method (see Figure 46) uses a discrete variable (only existing between 1 and 4) that is being used as a proxy for continuous measurements (0 to infinity). In most cases, artifacts were categorized as 'heavily abraded' early on in their time (by 60 min usually) in the tumbling barrel, but in some instances, they did not achieve this ranking throughout the course of this experiment, suggesting that abrasion lacks a simple linear association with time in the tumbling barrel.

The implications of these experiments for the archaeological record is that artifact conditions may not be good predictors for the level of change which they have experienced morphologically, because the two are governed by different abrasion pathways. This re-emphasizes the need for interassemblage comparisons to factor in at least some of the sedimentary factors' impact upon the artifacts when assessing their level of abrasion and degree of potential disturbance.

7.3.2. Artifact shape

The results of Chapter 5 showed that time in the tumbling barrel significantly reduced artifact length and width and refinement (w/t), but that changes to elongation (l/w) were largely insignificant. This suggests that this modification was a scaled reduction across the planiform of both flakes and handaxes. The results suggested that refinement (w/t) was a measurement that significantly changed through time, even though the flakes and handaxes had different starting refinements, mostly because thickness remained a near constant throughout the experiment. Hosfield and Chambers (2004) and Grosman *et al.* (2011; 2013) both found that changes in artifact morphology largely amounted to loss of length, because of abrupt breaks at the tips of their artifacts. This study found no such association.

Hosfield and Chambers and Grosman *et al.*'s findings may be explained by their focus on elongated handaxe morphologies (i.e., lanceolates/triangulates), though this is unlikely given that this study also examined these forms. Instead, it is probably a reflection of the larger gravel sizes they used, which in both cases exceeded the maximum gravel size used in this study (≤64mm) by at least 50% (96mm). This would seem to suggest that such larger clasts would produce higher energy impacts capable of conchoidally fracturing the more fragile tips or fracture them more substantially.

7.3.3. Time

In accordance with previous studies, these results revealed that time spent in the tumbling barrel had a significant impact on the overall reduction of artifact weight and primary metrics (Grosman *et al.* 2011; 2013; Hosfield 1999: 116; Hosfield *et al.* 2000). This experiment's patterns confirmed that artifact abrasion was higher in the early stages of tumbling for all sediment sizes, confirming suspicions that, at least in terms of gross morphology, artifact abrasion is an asymptotic process. This implies that form can change, even where the condition implies limited transport. This is probably a result of their unique shape and a switch from fracture wear (the chipping off of the delicate margins of the handaxe) to erosional wear (the subsequent grinding of the lithic surface; Krumbein 1941; Le Bouteiller *et al.* 2011; Kuenen 1956). This was confirmed by the insignificance of the changes in primary metrics for most artifacts between 2 and 3 hours.

7.3.4. Volume of sediment

This study demonstrated that artifacts tumbled in greater volume of sediment showed significant increases in abrasion of all primary metrics and artifact condition (refer to Section 5.2).

This is in broad agreement with the sedimentological literature (Gölz *et al.* 1995; Lewin and Brewer 2002), even if the underlying mechanisms for this increase are not entirely clear. It is possible that artifacts in greater sediment volumes may be subjected to a greater number of impacts within the tumbling barrel. It is also possible that the increase is due to unknown changes in the abrasion mechanisms (Lewin and Brewer 2002). Attal and Lavé (2009) suggest that decreased sediment volume promotes greater fragmentation of river cobbles, but given the fragility of the flakes tested and the similar weight loss curves of both the high sediment volume and the low sediment volume groups, this prediction was not met for flint artifacts.

7.3.5. Sediment size

Previous hydrological studies have suggested that sediment size has an overall effect on abrasion (Bradley 1970; Frings 2008; Sternberg 1875; Attal and Lavé 2009). This study was no different. The presence of larger particles within the same volume of sediment implies fewer impacts but a higher transfer of force, potentially promoting fracturing (Šolc *et al.* 2012). As expected, artifacts abraded in larger sediments showed greater loss in primary metrics and increased ridge/edge wear (abrasion category) than those abraded in smaller sediment size categories. Though previous studies did not test the effect of sediment size, Hosfield and Chambers (2004) and Grosman *et al.* (2011; 2013) reported artifact tumbling in sediments with particles in excess of 96mm—at least 50% larger than the largest sediments used in this study (64mm). Consequently, Hosfield and Chambers (2004) and Grosman *et al.* (2011) reported weight losses between 10 and 26.47%, whereas in this study comparable handaxes did not lose more than 4% of their original weight, suggesting that gravel size, especially in the higher size categories, plays a critical role in abrasion.

The lower abrasion rates in this study for artifacts tumbled in 'natural' unsorted sediments, when compared to 'artificial' sediments comprised of well-sorted single size classes, suggests, however, that the process of fluvial entrainment is a more complex process. This is potentially explained by the dampening effects of smaller sediments on larger clasts in polyclastic sediments (Abbott and Peterson 1978; Kodama 1994a; Kuenen 1956). Uncontrolled variables in this experiment such as different gravel lithologies may also contribute to this phenomenon. More precise measurements of impact numbers would be needed to confirm the effect of sediment composition on artifact abrasion.

7.3.6. Sediment lithology

The impact of variable aggradation terrace lithologies has remained a challenging problem in understanding the formation of archaeological assemblages (Ashton and Hosfield 2010; Bridgland *et al.* 2006) and no existing study has addressed this concern. The impact of variant lithologies has been explored for the abrasion of river cobbles, showing that increasing durability decreases the rate of abrasion (Abbott and Peterson 1978; Morris and Williams 1999). However, the exact effect of variable lithologies on lithic artifacts is still unknown. The current experiments demonstrated that variant sediment lithologies do play a significant role in the abrasion of archaeological artifacts. One study suggests that in polylithologic samples, the hardest constituents may 'attack' weaker, more brittle stones (Abbott and Peterson 1978), but in this study, flint, the hardest lithology, was present in all samples.

7.3.7. Artifact lithology

These experiments demonstrated that artifacts of different lithologies (flint and quartzite) showed no significant differences in the way they abraded in terms of their primary metrics, but did differ significantly when judged by their abrasion condition (refer to Section 5.3).

Burroni *et al.* (2002) observed similar changes in artifact condition when they abraded different lithic materials. These common findings are likely explained by differences between artifacts' 'durability' (i.e., hardness) and 'grain structure' (i.e., brittleness). Brittleness tends to promote fracture wear, while durability promotes erosion wear (Abbott and Peterson 1978; Attal and Lavé 2009; Kodama 1994a; Morris and Williams 1999). The data suggest, therefore, that the range of brittleness of the materials was constrained enough so that they would not fracture significantly differently (hence the recording of comparable changes to metrics).

This finding is largely in agreement with Bridgland *et al.* (2006: 22), who suggested that lithological 'hardness' plays a role in artifact condition. A potential weakness of this conclusion is that the condition assessment criteria used in this examination (Ashton 1998) were designed for flint and may not be applicable to other forms of chert or quartzite.

7.3.8. Artifact class

Finally, this study has demonstrated that artifact class plays a significant role in the abrasion of artifacts, with flakes abrading significantly quicker than handaxes in all categories except length. This suggests that, while artifact class plays a role in the overall abrasion of artifacts, length loss tends to be similar across artifact categories, perhaps reflecting tip vulnerability in handaxes. The findings of this book, through its fluvial, barrel, and flume experiments, all indicate that length and width may therefore be a predictor of artifact transport in fluvial environments.

This is consistent with other studies (Schick 1986; Chambers 2004) and suggests that entrainment of artifacts is largely affected by rolling or sliding along their long axis transverse to the flow. This may also be how artifacts are moving in the tumbling barrel and may explain the minimal damage to artifact lengths. This mode of transport is consistent with the 'avalanche' mode of tumbling barrel entrainment whereby particles tumble down the surface of the other particles, that are 'moving and jostling' against one another (Lewin and Brewer 2002). Chambers' findings should be treated with caution, however, because it is unclear what the influence of artifact size was on her study. Some of the abrasion differences between handaxes and flakes may be attributed to the size differences. Furthermore, the bifacially flaked edges of handaxes may serve as a 'stabilizer' retarding the fracturing abrasion of the artifact. It would be interesting to compare the abrasion performances of scaled flakes and handaxes in a flume, as well as perform edge analysis to see if edge flaking differences were present.

7.3.9. Significance and application to Pleistocene contexts

The experiments presented here have a wider relevance to Pleistocene sediments/Palaeolithic assemblages:

- First, that for artifacts entrained in coarse-grained fluvial sediments abrasion is likely to be asymptotic, and that, therefore, distinguishing the degree of abrasion becomes harder with time.
- Second, that gravel composition, particularly in larger size categories, impacts on the quantity of abrasion, and that in polyclastic gravels, the abrasive effects of large clasts may be dampened. This suggests that the abrasive impacts of the larger sediments observed by previous experiments (Grosman *et al.* 2011; Hosfield 1999), may be overestimations of the effects of coarse-grained abrasion in the real world.
- Third, that sediment lithology can affect abrasion rates, suggesting that artifact abrasion may be regionally variable.
- Fourth, that artifact abrasion rates vary with artifact lithology, a key factor for future interpretations of assemblages where there is a mixture of raw materials.
- Fifth, that artifact classes can affect their abrasion rates, possibly due to their varying entrainment modes, suggesting that artifacts within the same assemblage can be variably abraded.

7.3.10. Limitations and future work

> *It is ridiculous to suggest that a few laboratory experiments, made under artificial conditions, are equivalent to the action of Nature, whose forces are so varied. (Haward 1913: 349)* In response to J. Ried Moir's sack and letter-press experiments aimed at replicating fluvial and other taphonomic processes (Moir 1914; see O'Connor 2007: 154–167 for discussion).

There are a number of limitations to this study. First, time in the tumbling experiments does not provide accurate estimates of transport distances and may not entirely simulate the complexities of fluvial entrainment processes (Lewin and Brewer 2002). There are also other factors such as particle velocity, artifact initial size, edge angularity/retouch that were not tested here. Additionally, the individual factors tested here were non-exhaustive in their range of values and there is considerable room to test different sediment lithologies, sizes, volumes and their subsequent permutations, as well as variable weathering conditions and artifact classes.

This study does, however, re-state a principle: given the known impacts of fluvial entrainment parameters and their disparate influences on artifact morphology and condition, future studies evaluating artifact transport levels and assemblage derivation should take into consideration the overall sedimentological context of palaeoliths and their local findspots.

Future studies should focus on elaborating the analysis of current velocity and sediment transport modes as abrasion variables because they are likely to play a significant role (Frings 2008). Additionally, the results of this experiment showed that different lithologies (flint and quartzite) and degrees of chemical weathering (artificial patination) are significant factors in understanding the preservation of artifacts, particularly their condition, a phenomenon that needs further exploration.

7.3.11. Section summary

This study applied statistical models to a series of experiments in a tumbling barrel to isolate the roles of the main factors controlling artifact abrasion rates and comminution during high-energy fluvial transport. Taken together, this research indicates the following:

1. Coarse-grained fluvial entrainment of Palaeolithic artifacts affected their overall morphology and abrasion condition, but not as significantly as previous studies have indicated (Grosman *et al.* 2011; 2013; Hosfield 1999; Hosfield *et al.* 2000; Tringham *et al.* 1974). This is probably largely a function of the sediment sizes and methods they used, suggesting that current studies may over-estimate possible degrees of derivation.

2. Artifact planiforms are highly conservative in spite of abrasion. Compared with previous studies, this research showed significantly less modification to artifacts that were heavily abraded and found that, even in cases of heavy abrasion, artifacts were generally always identifiable as such and retained (broadly) their original shape.
3. Time, artifact lithology, artifact class, sediment lithology, and sediment volume regulated abrasion rates. Increases in these factors generally increased the amounts of abrasion of primary metrics and artifact condition, however, this occurred at different rates, probably because of different abrasion pathways.
4. Artifact condition rates did not aggregate linearly with changes to overall artifact shape. The influence of time, artifact lithology and class, sediment lithology, and sediment mass modulated the rates of change to artifact conditions, indicating that abrasion categories are poor indicators of both artifact change and artifact transport distances.

The last two decades have witnessed a resurgent interest in the archaeology of northern Europe for its ability to provide novel information regarding the climatic tolerances of early hominins (Roebroeks 2008; Ashton *et al.* 2010; Stringer 2007; Pettitt and White 2012). However, a limitation to interpreting the spatial and temporal patterns of hominin settlement has been the modification and transport of Palaeolithic sites as a result of powerful fluvial and glacio-fluvial discharge in gravel-bedded rivers during the Pleistocene (Ashton and Hosfield 2010).

To address the problem of modification and transport of sites and artifacts, researchers have attempted to categorize assemblages into preservation categories (i.e., condition) based on their external appearance (Ashton and Hosfield 2010; Ashton *et al.* 1996). A major supposition built into these categories is that visual preservation categories correspond to both the transport time and changes in artifact morphology. A plethora of geomorphological literature suggests that this may not be the case and that the underlying mechanisms of abrasion are highly variable (e.g., Adams 1978; Abbott and Peterson 1978; Krumbein 1941) and can be influenced by a range of sedimentological variables (e.g., Mikoš and Jaeggi 1995; Schumm and Stevens 1973; Šolc *et al.* 2012). Yet, to date, there are only limited studies exploring their effects on Palaeolithic artifact forms.

This study found that abrasion within coarse-grained sediments had a significant impact on the shapes and preservation conditions of archaeological artifacts, manifested largely in changes to the weight, length, width, elongation ratio, and condition (1–4; mint–heavily abraded) of the artifacts. The results of these experiments also indicated that (1) time, (2) volume of sediment, (3) artifact lithology, (4) sediment lithology, and (5) artifact class all played significant roles in the way artifacts characteristically abraded. It furthermore showed that the impact of these factors was highly variable; most notably, that artifact condition as well as primary metrics varied according to sedimentary parameters, but not necessarily in similar ways. These experiments present the first systematic statistical analysis of artifact morphology and condition of experimentally tumbled Palaeolithic artifacts.

7.4. Fine-grained experiments (flume experiments)

This section discusses the results of the fine-grained flume abrasion experiments from Chapter 6 (refer to Section 3.4 for methodology), which focused on understanding how (1) abrasion time, (2) sediment size (3), and artifact mobility affected the microscopic surface texture, ridge widths, and edge damage of artifacts. In the course of the discussion, possible mechanisms or pathways will be presented and the data will be compared to the earlier studies (detailed in Section 2.5 and summarized in Figure 218). The strengths and limitations of these experiments will also be discussed in addition to their implications for Palaeolithic artifact sites and findspots.

7.4.1. Time

The surface condition, ridge wear, and micro-flaking that were assessed in the flume experiments indicate that artifact abrasion is sensitive to short time scales but also to small changes in sediment size and modes of artifact entrainment. These findings are largely in line with previous abrasion studies (Burroni *et al.* 2002; Chambers 2003; 2004; Fernandes *et al.* 2007; Hosfield 1999; Hosfield and Chambers 2004; Shackley 1974; Tringham *et al.* 1974).

Surface condition

The correlation of increased wear with increased flume time that this experiment found is in agreement with the work of Burroni *et al.* (2002) who found that the degree of abrading of lithic surfaces was strongly linked to abrasion time. However, Burroni *et al.* (2002), who abraded artifacts in sand-sized sediments, found that the surface was demarcated by 'micro-cracks' and then striations as sediment size increased. This was not the case for this study, which found that increased sediment size only resulted in increased abrasion levels. These findings, therefore, were more consistent with the studies of Fernandes *et al.* (2007) and Lenoble (2005) who found that archaeological materials were typified by a removal of all asperities when abraded in a flume. It is not entirely clear why Burroni *et al.* (2002) found such distinctive marks on lithic surfaces, as the experimental parameters were broadly similar to this experiment and those of Fernandes *et al.* (2007) and Lenoble (2005). One reason may be the choice of lithic materials, though this is only speculation because Burroni *et al.* (2002) do not specify their composition. Another reason may be the use of differing magnification levels. Burroni *et al.* (2002) used an optic microscope at 200x while this study, Fernandes *et al.* (2007), and Lenoble (2005) all used SEM between 500–1000x magnifications.

Figure 218: Comparison of this volume's fine-grained experiments and previous experiments. *Experiments are also listed in Figure 217 as they are relevant to both short and fine-grained issues.

Reference	Aim(s)	*n*	Equipment	Duration	Sediment	Velocity	Controls	Dependent variable(s)	Independent variable(s)	Result(s)	Limitation(s)
Tringham *et al.* 1974*	• Distinguish fluvial edge damage from usewear	5	Agitated plastic container	1 hour	Water, sand, stones	N/A	• Artifact lithology (European chalk flint)	• Edge damage	• Artifact lithology	• Randomly oriented micro-flaking along artifact perimeter	• No consideration of artifact morphology
Shackley 1975*	• Distinguish artifact abrasion groupings • Assess abrasion for regional archaeological patterns	6	Tumbling barrel (large plastic barrel)	100 hours (10 intermediate time-steps)	Sand to clay	25.5rpm	• Rotational speed • Lithology	• Ridge widths	• Time • Handaxe shape • Sediment size (sand to clay)	• Ridge abrasion related to time • Ridge abrasion related to abrasive type	• Small sample size • Only small sediments tested • Variability of ridge widths across artifacts
Burroni *et al.* 2002	• Understand how time, artifact lithology, and substrate wetness influence wear and fracturing of lithic surfaces	*Uns.*	Tumbling barrel (800ml, diameter=8cm)	32 hours (6 intermediate steps)	*Uns.* (likely fine-grained)	40rpm	• Rotational speed • Artifact class (handaxe)	• Ridge width • Weight • Edge damage • Surface condition	• Time • Artifact lithology • Grit size • Water volume	• Ridges wear increases in stages • Ridge wear increases when the sediment is wet • Ridge/edge wear increases with chert grain • Edge damage takes time to form	• Use of tumbling barrel
Chambers 2003; 2004*	• Assess artifact transport duration and ridge development	*Uns.*	12m glass walled flume	250m (16 intermediate steps) 1000m (31 intermediate steps	*Uns.*	*Uns.*		• Ridge width • Micro-flaking	• Artifact lithology (flint, greensand chert) • Bedload transport (rolling/saltation/ sliding)	• Raw material affects ridge abrasion rates • Morphology affects transport • Transportation mode affects damage	• Small sample size • Limited distance examined
Lenoble 2005	• Observe effects of abrasion on artifacts in slope processes	20	1 liter rotating glass bottles	24 hours (every 6 hours)	Sand	30rpm		• Micro-surface condition (1–4) • Ridge wear		• Artifacts experience ridge wear/ micro-abrasion in fine-grained conditions	• Time too long to reflect common slope processes • Limited application

Figure 218: *Continued.*

Reference	Aim(s)	*n*	Equipment	Duration	Sediment	Velocity	Controls	Dependent variable(s)	Independent variable(s)	Result(s)	Limitation(s)
Fernandes *et al.* 2007	• Establish surface characteristics of taphonomic history	50	Mio-Pliocene alluvial deposits from Naussac (Lozère)	N/A	Mio-Pliocene alluvial deposits from Naussac (Lozère)	N/A	• N/A	• Surface features (i.e., cracking, pitting)	• Taphonomic history	• Impact stigma on flat surfaces • Flaky appearance • X-shaped cracks	• Only locally applicable • Difficult to extract from other taphonomic processes
Current Study	• Assess role of time, sediment size, and transport mode on micro-abrasion, ridge width, and edge damage	27 (x9)	Miniflume (Amos *et al.* 2000) Lab Carrousel (Sutherland *et al.* 1998)	168 hours (3 intermediate steps)	Quartzite sands from southern UK	Saltation	• Sediment transport Mode	• Ridge width • Surface condition • Edge micro-flaking	• Time • Sediment size • Artifact transport mode	• Time and sediment size affect wear • Time, sediment size and transport mode affect ridge width • Sediment size affects edge damage	• Limited applicability • Limited transport modes examined

Ridge wear

The ridges on artifacts abraded in fine-grained sediments significantly widen over time, as has been shown in all previous studies; the results of this study were no different. What differed were measurement widths. The values achieved here (average ridge widths: 56 and 130µm; up to 168 hours abrasion) were more in line with those of Shackley (1975: 46; 70–150µm) and Burroni *et al.* (2002; approximately 42–85µm after 32 hours) than with those of Chambers (2004), who reported abraded values as high as 2800µm after 50m of transport in a flume, and Hosfield (1999: 114) who reported ridge widths as large as 1567µm after 50 hours. Hosfield (1999: 114) suggests that such discrepancies may be related to differences in microscopic setups and recording methods, which is likely, given that Chambers (2004) recorded comparatively larger initial ridge values than this study in the earliest stages of the experiments when they were presumably mint (*c.* 20µm). In the case of Hosfield (1999: 114), it is likely that these enhanced abrasion levels may be attributed to either the large grain size (>95mm in long-axis) of the sediment supply or the use of a metal cement mixer and the impact of the paddles.

In contrast to Shackley (1975: 46) and Burroni *et al.* (2002), who noted distinct pitting and/or stress cracks in the faces of the ridges, this study observed overwhelmingly smooth ridges throughout the experiment. This pitting may relate to the use of tumbling barrels v. flumes or to Burroni *et al.*'s use of varying unspecified artifact lithologies.

It is safe to conclude that ridge widths are a useful marker of fluvial abrasion, as ridges are extremely sensitive to taphonomic conditions. It must be noted, however, that this study and those outlined above have shown that numerous factors are involved in ridge abrasion. Particularly problematic is that artifact ridges can be abraded *in situ*. Great caution should be taken, therefore, when making inferences from them about artifact transport and depositional histories.

Edge damage

The development of edge damage (micro-flaking) has previously been related to saltating transport of artifacts (Chambers 2003). This study shows that it can also occur when the artifact is stationary, implying that even when found *in situ*, in fluvial contexts, edge damage cannot be automatically regarded as anthropogenic (Hosfield and Chambers 2003). Micro-abrasion and potentially larger scale wear indices such as incipient cones of percussion or breakages may, therefore, be more reliable indicators of an energetic environment than active artifact transport *per se*, and may relate more to time spent in fluvial environments than to distances traveled.

7.4.2. Mobility

The results of these experiments have a series of important implications for decoding the transport history of artifacts in fluvial settings. First, artifacts do not need to be mobile to develop wear or edge damage as is sometimes suggested (Shackley 1974). It is sufficient for sediments to be transported over stationary artifacts. In the case of ridge wear, active artifact transport resulted in reduced ridge width modification over the duration of the experiment, perhaps due to a reduction in the number of sediment impacts. Therefore, 'trapped' artifacts (Hosfield and Chambers 2005c) may still undergo further modification.

7.4.3. Sediment size

Sediment size has been shown in this study to have a varying effect on microscopic features. In the case of edge damage, there appears to be a critical size threshold below which edge damage does not develop, which coincides with the granule/pebble size classification boundary (Wentworth 1922). As solid transmitted stress (T) is a function of sediment grain size (Thompson and Amos 2004), the implication is that there is a critical value for T that will induce edge damage. For ridge width abrasion, sediment size is not significant within the range assessed, possibly because while T increases with sediment size, the number of impacts are reduced (Thompson and Amos 2004). This, along with the small surface area of the ridges, reduces the likely number of impacts as grain size increases. This study showed that ridge widths abraded up to 168 hours corresponded to those found in flints experimentally abraded in tumbler experiments for 100 hours using natural fluvial sandy-gravel sediments (Shackley 1975; 0-300 mm).

There is abundant documentation that many fluvially deposited lithic artifacts have similar edge damage (continuous bifacial micro-flaking) to those that abraded in this experiment (Locht *et al.* 2010; Schoville 2010; Singer *et al.* 1973). This experiment suggests that the effects of sandy-gravel fluvial abrasion share some similarities with certain hominin tool resharpening.

7.4.4. Significance and application to Pleistocene contexts

The results of these experiments have a number of implications for Pleistocene archaeological assemblages:

- First, that artifacts in fine sand-sized sediments can be transported significant distances without showing appreciable damage to their macroscopic morphology.
- Second, that lithic artifacts occuring within medium gravel or larger fluvial settings will experience edge damage. However, this may not relate to actual movement and may have occurred *in situ.*
- Third, that microscopic ridge widths and surface conditions of artifacts can provide valuable proof of artifact interaction with fluvial sediments, especially at the local scale.

7.4.5. Limitations and future work

Because of the novelty of the new micro-abrasion approach, there are still unexplored variables and limitations to the method. This experiment used saltating grains, though it is important to note that saltation, the transport mode in flumes, is not likely to have been the permanent transport mode within Pleistocene fluvial settings, as sediment transport mode must have changed according to the flow conditions (Bagnold 1986). However, using saltation allowed the maximum amount of possible wear to be developed over the timescale of the experiments (Thompson and Amos 2004; Thompson *et al.* 2011).

While it is difficult or even impossible to convert hours in the flume to time or distance traveled in the field, it is likely that the experiments represent long timescales in the natural environment, where variations in flow parameters and stationary periods must be taken into account. However, the use of annular flumes to assess wear insures that sediment transport modes are accurately modeled, including fully developed bottom boundary layers and bedforms, a key factor in reproducing natural transport scenarios (Chambers 2003; Thompson *et al.* 2011). Tumbler experiments (Grosman *et al.* 2011; 2013; Shackley 1974) cannot replicate natural modes of transport and previous experiments are likely to be more representative of extremely energetic flood or storm events (see Section 7.3).

There are uncontrolled variables in the above experiments such as sediment lithology (i.e., softer/harder materials), different artifact classes, and sediment entrainment pathways that may contribute to the process of abrasion. In addition, more precise computerized methods for the quantification of surface abrasion condition and edge damage would assess the effects of microscopic abrasion more accurately either through image recognition (Álvarez *et al.* 2012; Grace *et al.* 1985) or other more high tech alternatives (Evans and Macdonald 2011; Faulks *et al.* 2011; Stemp *et al.* 2010). The application of the present findings is therefore limited. Further work is required to identify the impacts of patina formation and other post-depositional processes that affect artifact surface integrity (Burroni *et al.* 2002; Howard 2002; 1999b; Rottländer 1975).

7.4.6. Section summary

This study demonstrated that although lithic artifacts may appear macroscopically 'fresh,' they might still have been transported from their original contexts. This study's results also outline a possible area of study for lithic micro-abrasion specialists regarding potential overlap between fluvial abrasion and anthropogenic wear patterns (Levi Sala 1986).

The experimentally abraded flint artifacts from the flume experiments described in Chapter 6 showed unequivocal signs of microscopic abrasion when analyzed with an

SEM. These modifications were manifest in surface condition, ridge widths, and edge micro-flaking. These findings provide a potential new pathway for archaeologists to identify fluvial post-depositional modification on lithic artifacts, particularly for those that appear to be macroscopically 'fresh', such as at the open-air sites of Happisburgh 1 and 3 or Elveden, UK (Ashton *et al.* 2005; Parfitt *et al.* 2010). This study presents the first laboratory test results that prove a positive association between fluvial transport mechanisms and microscopic evidence of surface micro-abrasion, edge damage, and ridge widening.

7.5. Archaeological sites

This study has so far focused on experimental research concerning Palaeolithic site formation processes. It has also put forth the implications of the experimental results for archaeological analysis. This section will apply the results of the experimental findings to archaeological sites in northern Europe, specifically from the Lower and Middle Palaeolithic.

When applying experimental data to the archaeological record, it must be pointed out that observations made in modern day fluvial systems and in laboratory settings are inherently complex and these complications increase when the data obtained from the observations are applied to archaeological sites. Archaeological sites commonly delete large amounts of evidence in numbers, diversity, and the variety of materials they contain.

The following analyses represent an attempt to apply the findings of this study to specific archaeological sites and to assess their potential value in disentangling the major forces at work in fluvial site formation. I hope to show not only how experimental findings can be successfully applied to archaeological data, but also to instigate a greater appreciation for the importance and complexity of fluvial processes when it comes to site formation.

7.5.1. Caours (Area 1): An interglacial butchery site?

This section discusses the implications of the findings of this book's experiments for the interpretation of the site of Caours, Picardy, France.

Site background

Caours (50°07'53"N/01°52'59"E) is an open-air archaeological site located in the Middle Somme alluvial terrace system (Figure 219; France), imbedded within a tufa sequence overlaying the lowest terrace (+5m) of the Scardon river (Antoine *et al.* 2000; 2003; 2007; 2010).

The site is characterized by tufa sands and large scale cross-bedded fluvial facies. Excavations at Area 1 exposed a series of sediment facies (Figure 220): glacial stages dominated by braided river systems as well as sand and gravel deposition (i.e., Level 6); and interglacial stages corresponding to meandering river systems with overbank silt deposition (i.e., Level 5), capped by marshy soil formations and cemented by a travertine source (i.e., Level 2) (Antoine *et al.* 2006; Limondin-Lozouet *et al.* 2006; Locht *et al.* 2010).

Synthesis and dating

Litho- and bio-stratigraphy, palaeomagnetism, and U/Th dating of the Caours sequence all point to the site as being from the Eemian (Ipswichian, MIS 5e) interglacial in Northern Europe (Limondin-Lozouet and Antoine 2001; Locht *et al.* 2010; M. Sier *pers comm.*).

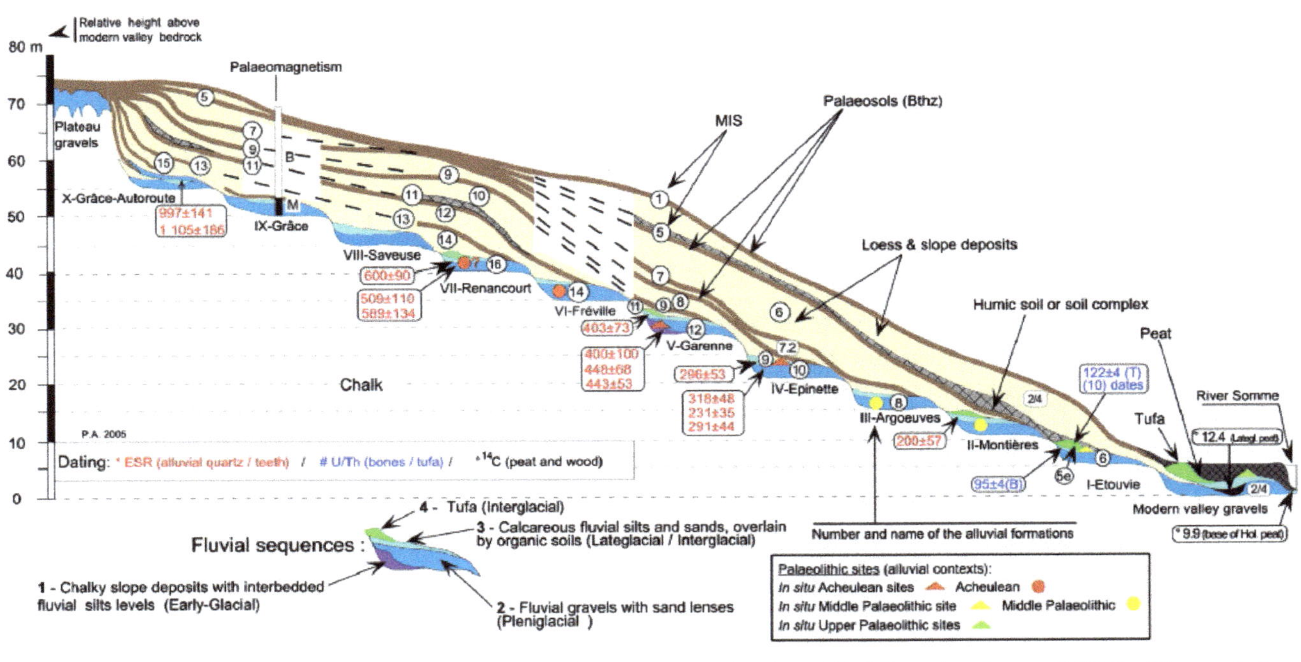

Figure 219: Idealized terraces of the Middle Somme basin (Reproduced and modified from Antoine *et al.* 2007 with permission from Elsevier).

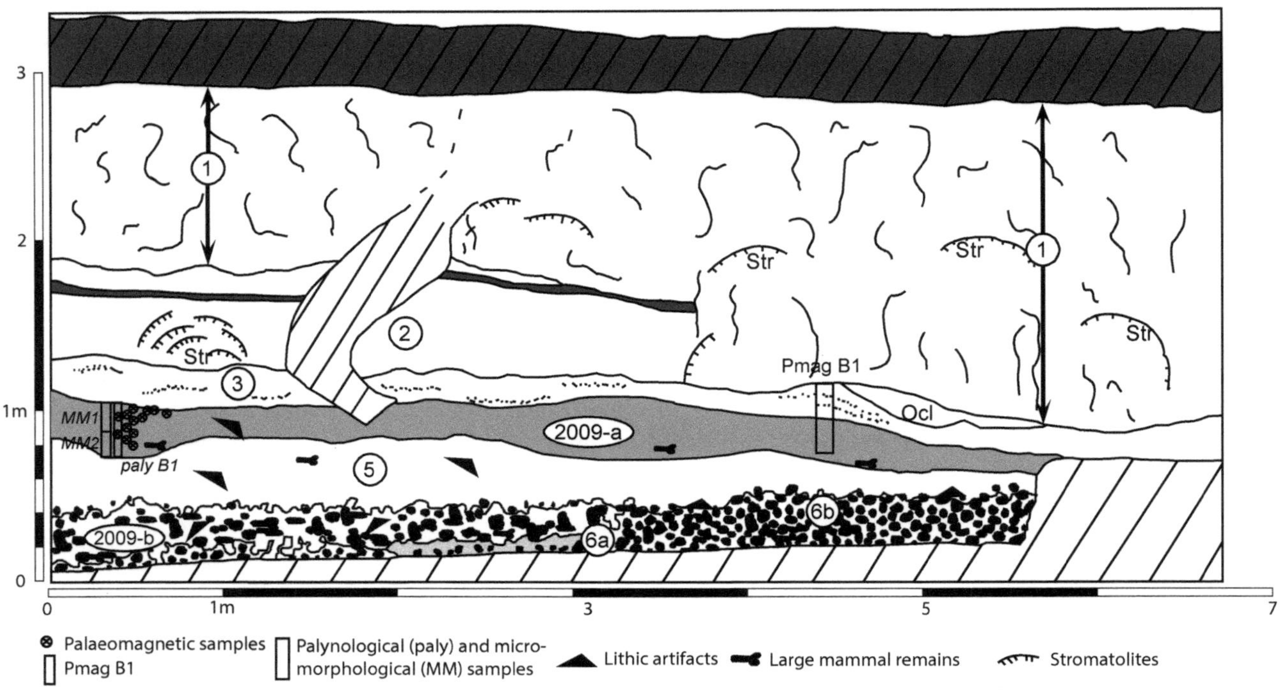

Figure 220: Eastern profile of Area 1 from 2009 (Redrawn after Locht *et al.* 2010, courtesy Jean-Luc Locht).

Archaeology

The artifacts from Caours are strongly Middle Palaeolithic in nature; the majority are unmodified flakes with a large number of 'splinters'. Unipolar, bipolar, and discoidal cores were also found. A large percentage of cortical flakes suggest that the raw materials were locally sourced and that knapping took place on site. A number of Levallois flakes and pseudo-Levallois points are also found throughout the site, though there is a virtual absence of Levallois cores. The stone artifacts are manufactured from local grayish-black flint nodules (>15cm).

The main archaeological finds in Area 1 are situated within the fluvial facies and correspond to four different archaeological levels (from top to bottom):

- Level A (2009a): A silty laminated tufa with sandy lenses and driftwood casts.
- Level B (Level 5): A set of unconsolidated beige textured sandy loam tufas with re-worked flint pebbles (10–20cm in diameter).
- Level B/C (base of Level 5): Found on top of coarse, rounded to sub-rounded chalk and flint river gravels of variant size. This may not represent a discrete unit.
- Level C (base of Level 5): Found on top of coarse, rounded to sub-rounded chalk and flint river gravels of variant size.

Current site interpretation

Multiple lines of reasoning have led researchers to suggest that the main archaeological levels (B/C and C) represent an *in situ* Neandertal hunting site (Locht *et al.* 2010):

1. Artifacts are covered in a fine-grained loamy matrix, apparently deposited in a low-energy environment.
2. Artifacts occur in a dense and relatively thin horizontal layer, implying that they were discarded on a stable surface.
3. Concentrations of artifacts are in close association with cutmarked/modified bones suggesting that the artifact positions are the results of hominin activity consistent with butchery.
4. The stone tool edges and ridges are fresh and well preserved. The large percentage of cortical flakes and a number of refits suggest that the nodules were reduced on site.

Data collected

To test the *in situ* hunting site hypothesis, a total of 642 artifacts from the 2009 Area 1 Levels B/C and C were analyzed. The following data was collected (Figure 221):

1. Artifacts were categorized into basic artifact types: cores, flakes, and débitage (splinters). Flakes were defined as any artifact with a bulb of percussion and a platform. If these were not present, it was classified as débitage. In addition, when specific tool types were recognized, they were classified according to standard archaeological categories (i.e., denticulate, Levallois flake; i.e., Bordes 1961; 1988)
2. Maximum length, width, and platform thickness measurements were taken to the nearest 100th millimeter (Andrefsky 2005: 98). Artifacts were also weighed to the nearest gram.
3. Artifact edge damage was noted as 'flaked' or 'micro-flaked' or both, with flakes being larger than 1mm in

Figure 221: Data collected on artifacts from Caours.

Artifact type	Primary metrics				Edge damage		Ridge damage				Surface condition				Cortex
	Maximum length	Maximum width	Platform width	Weight	Flaked	Micro-flaked	Ridge Abrasion	Rounding	Pitting	Battering	Patination	Concussion marks	Striations	Polish	
core	measured (0.01mm)	measured (0.01mm)	measured (0.01mm)	measured (0.1g)	bifacially continuous	bifacially continuous	yes	yes	yes	yes	black	yes	yes	yes	0% to 100% (by 10%)
flake					bifacially discontinuous	bifacially discontinuous	no	no	no	no	black/white	no	no	no	
débitage					unifacially continuous	unifacially continuous					black/gray				
					unifacially discontinuous	unifacially discontinuous					black/brown				
					no	no					brown/gray				
											gray				
											gray/white				
											white				

diameter. If present, flaking or micro-flaking was classified as either bifacially continuous, bifacially discontinuous, unifacially continuous, or unifacially discontinuous.

4. Ridge conditions were also recorded. If ridges were abraded, it was noted if they were rounded, pitted, or battered.
5. Surface conditions of the artifacts were also recorded. Artifact patination was classified as either black (unpatinated), black/white (a mottled white patina), black/gray (a mottled black and gray patina) black/brown (a mottled brown patina), brown/gray (a mottled brown and gray patina), gray, gray/white (a mottled gray and white patina), and white. In addition, artifacts were noted if they had either concussion marks, striations (Ashton *et al.* 1996; Moir 1914), or surface polish (Howard 2002).
6. Cortical values were also estimated for each artifact to the nearest 10% (i.e., an unmodified cobble=100%; a primary flake (*éclat entame*)=50%; a tertiary flake=0%).
7. Finally, 10 small (<3cm) artifacts were taken to the UK for SEM analysis.

Results

There is little doubt that most of the artifacts found in Caours Area 1, Level 3 are of hominin manufacture as evidenced by the number of heavily worked cores and the strong flake component in the assemblage. Over 75% of the assemblage exhibits prominent bulbs of percussion or at least one flake scar (Figure 222). Débitage comprised less than 22% of the entire assemblage. In addition, some of the artifacts were distinct tools including denticulates and Levallois points (Figure 223).

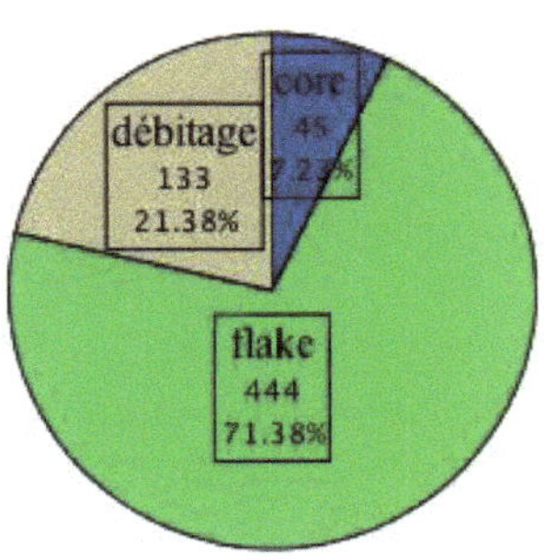

Figure 222: Distribution of artifact types in levels B/C and C of Caours.

It was unclear to the excavators whether the artifacts from Level B/C formed a discrete unit separate from Level C, since B/C occurred at the transition of Level 5 and Level 6. To test if the archaeological materials were metrically different, an independent t-test was performed between the two assemblages to see if there was any statistical differences between their primary metrics (Figure 224). As there was no statistical difference between and because Level B/C was comprised of only 43 flakes, the two levels were combined for this analysis.

The primary metrics of the lithics from Caours showed little variability. Flakes and débitage ranged from 11mm to 271mm in maximum length with an average of 51mm (Figure 232). The low standard deviation (SD) indicates that most flakes were clustered around this value, with low numbers of flakes in the smaller and larger size ranges. The mean maximum width of artifacts was only slightly smaller than the mean maximum length and had a tighter SD. This reflects an overall trend of short, squat flakes at Caours. Weight was the best indicator of assemblage variety, because it was the only measurement carried across artifact types. Artifacts ranged from less than 1g to upwards of 1kg. The mean value however, indicates that most artifacts found are low in that range and the SD demonstrates that the upper range mostly represents outliers.

Many of the artifact edges (42%) at Caours were preserved in a fresh condition with no macroscopic damage (Figure 225; Figure 226a). These edges are smooth and still sharp to the touch. However 58% of artifacts at Caours

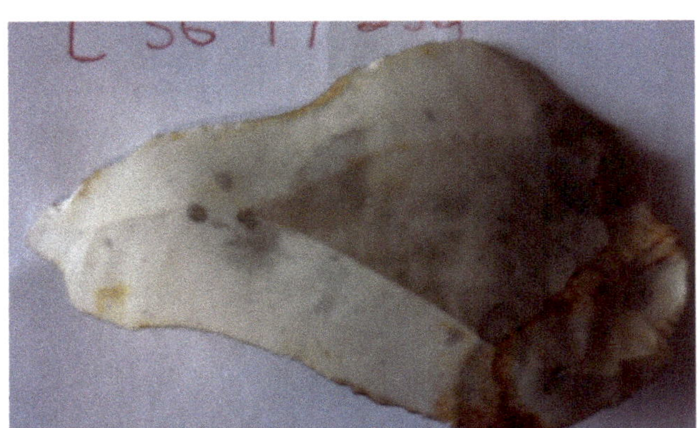

Figure 223: An example of a Levallois point (Artifact L36-11).

Figure 224: Artifact dimensions from Caours. No differences were found between levels B/C and C at $p<0.05$ (see Appendix 15).

		n	Min	Max	Mean±SD
Maximum length (mm)	*B/C*	32	22.99	271.00	40.98±66.88
	C	422	11.31	115.42	50.73±16.46
	Total	454	11.31	271.00	50.80±16.62
Maximum width (mm)	*B/C*	32	13.86	92.32	35.67±14.31
	C	419	11.84	80.12	35.38±11.75
	Total	451	11.84	92.32	35.40±11.94
Platform thickness	*B/C*	25	2.08	12.47	6.52±2.78
	C	303	1.57	51.42	8.08±5.93
	Total	328	1.57	51.42	11.89±52.01
Weight	*B/C*	43	1.0	271.0	40.98±66.88
	C	599	1.0	2103	35.17±11.76
	Total	642	1.0	2103.0	35.56±106.34

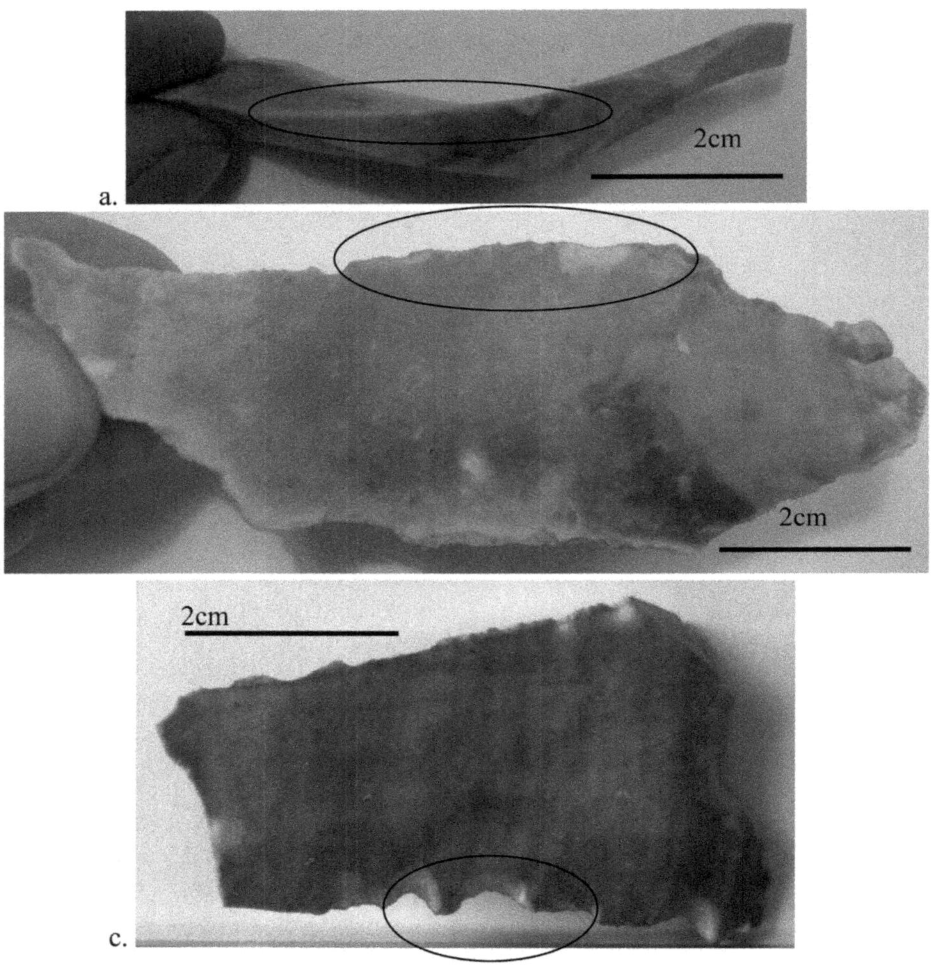

Figure 225: Examples of the different artifact edge damages found at Caours: a. no damage (artifact n34-34); b. edge micro-flaking (artifact h36-11) ; and c. artifact with edge flaking (artifact s39-15).

showed some form of damage in the form of flaking or micro-flaking (Figure 225). For artifacts that were flaked, the majority were continuously bifacially flaked (17%; Figure 226b), followed by discontinously bifacially flaked artifacts (10%). Unifacially flaked artifacts comprised less than 6% of the total assemblage.

Over 30% of the assemblage demonstrated continuous and bifacial micro-flaking (Figure 225b; Figure 226c), followed by discontinuous bifacial retouch (7%). Approximately 6% of the total assemblage showed unifacial retouch.

Most of the ridges (91%; Figure 227) at Caours were preserved in fresh conditions and were still sharp. However, the rest of the artifacts (9%) showed some damage to the ridges, mostly characterized by pitting and battering (Figure 228). In at least three cases, this was observed on cores and was often localized, suggesting that some of the

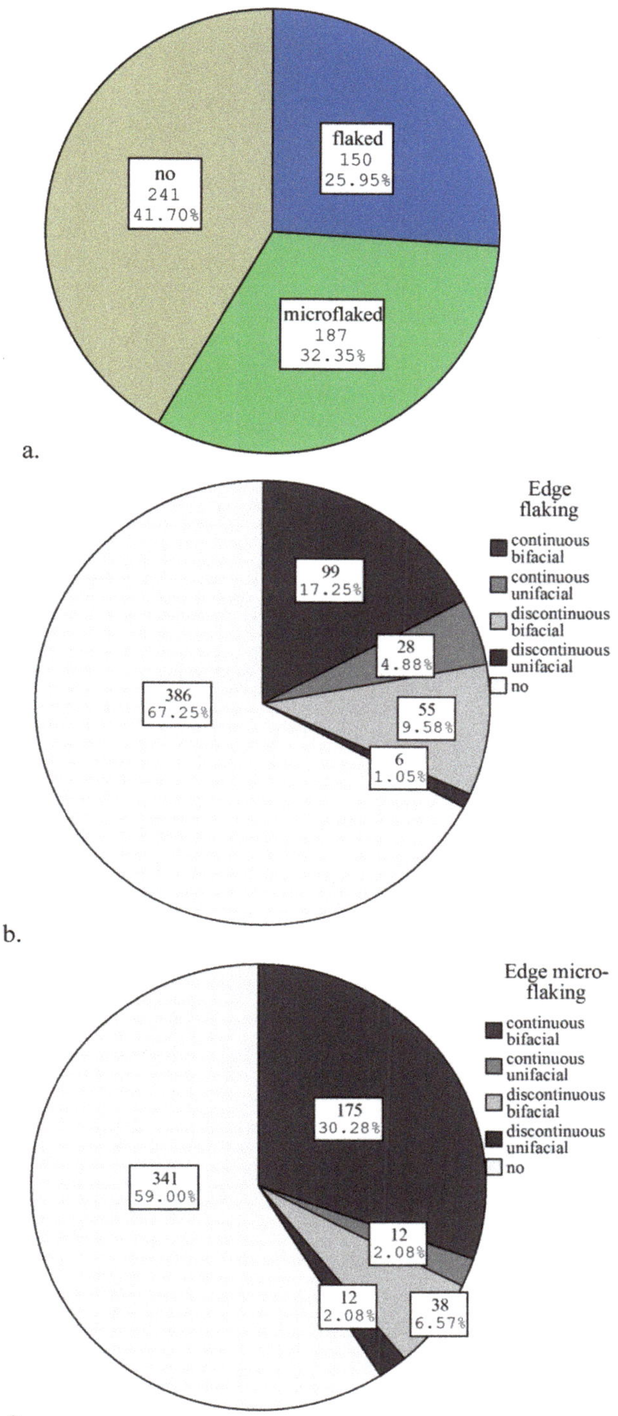

Figure 226: Distribution of artifact edge damage at Caours. a. artifacts showing no damage, flaking or micro-flaking; b. distribution of flaking type; c. distribution of micro-flaking type.

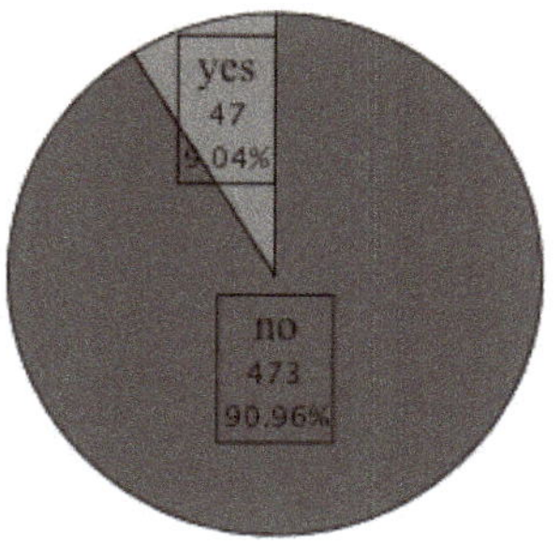

Figure 227: Artifacts at Caours exhibiting ridge abrasion.

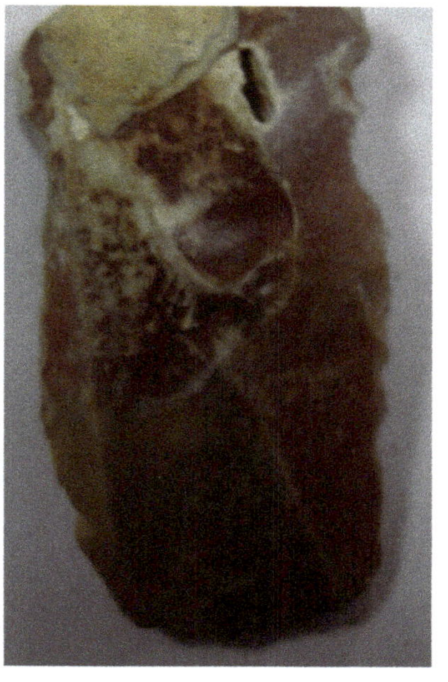

Figure 228: An artifact with ridge damage (artifact d36-10).

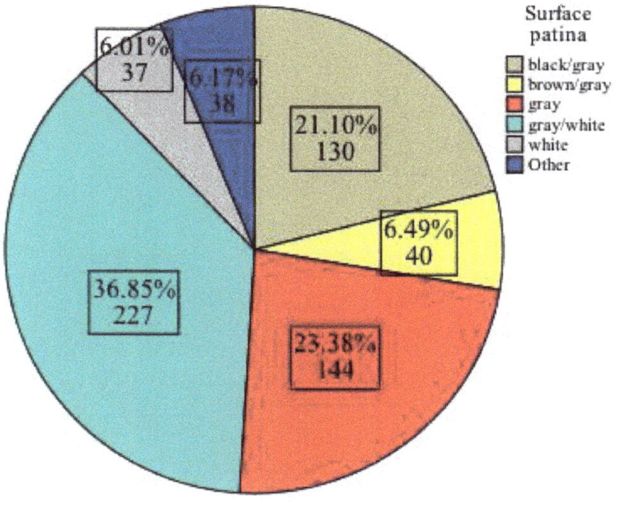

Figure 229: Distribution of the patinas found at Caours.

pitting and ridge damage was the result of the nonexhausted and exhausted cores being used as hammerstones in the absence of suitable alternatives (Locht *et al.* 2010).

More than 95% of the artifacts at Caours were patinated with distinct patinas, despite being from the same local raw materials (Figure 230). 37% of these artifacts were patinated *gray/white* followed by *gray* which formed 23% of the assemblage and *black/gray* which compromised 21% of the assemblage (Figure 229). *Brown* or other patinas compromised less than 20% of the total assemblage.

Artifact surfaces were surprisingly well preserved. Only 10 (2%) artifacts showed striations and 19 (3%) artifacts showed surface polish. 4% (n=20) of the artifacts showed concussion marks. Where surfaces showed concussion

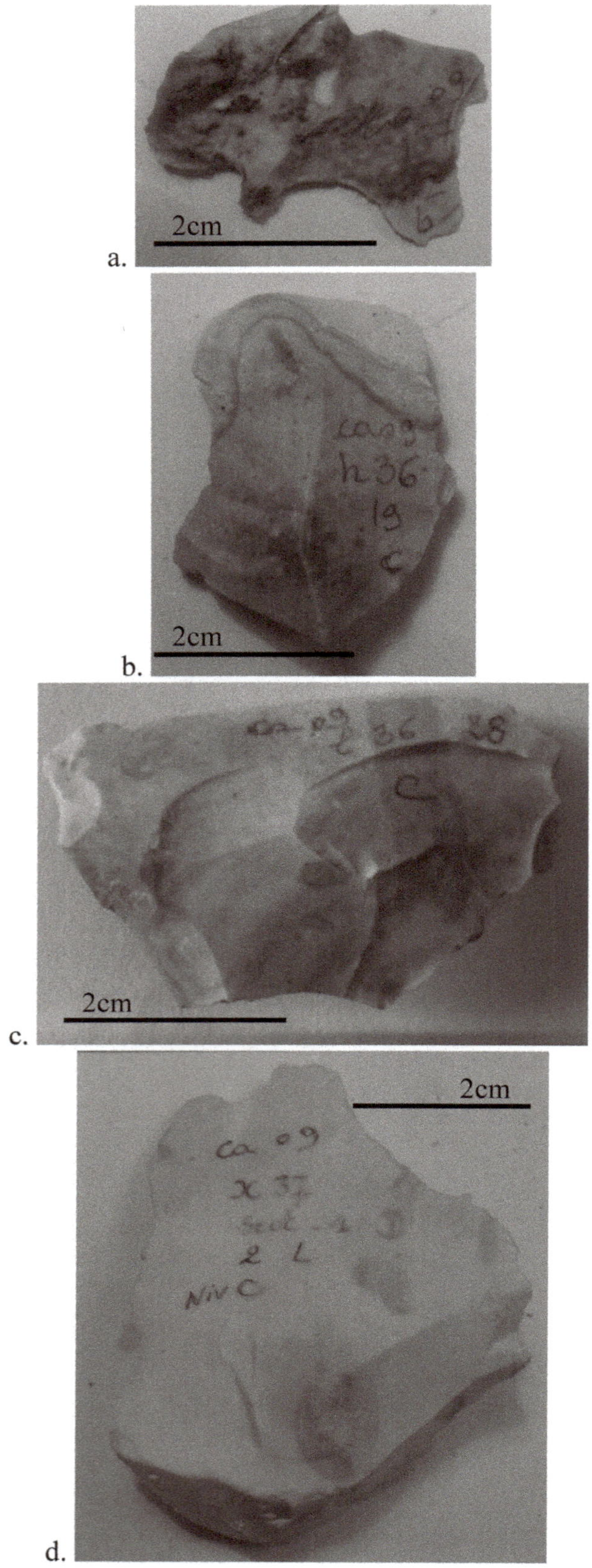

Figure 230: Examples of the major patinas types found at Caours. a. black/gray patina (artifact h36-8); b. brown/gray (artifact h36-19); c. gray (l36-28); d.gray/white (artifact x37).

marks, they were few (n<5) and were generally localized on the same area of the artifact (Figure 231).

Figure 231: An artifact with concussion marks (artifact n39-6).

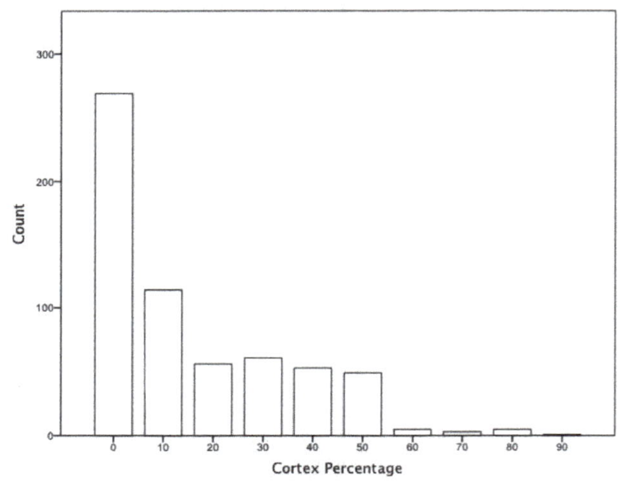

Figure 232: Cortical values of artifacts at Caours.

Artifact cortex values were distributed across the entirety of the percentage ranges, except for 100%, since there were no unknapped hammerstones found on site (Figure 234). The average cortex retained on artifacts was 16%, though most artifacts had no cortex. As cortex values increased, their frequency decreased nonlinearly.

The results of the SEM analysis showed that the artifacts from Caours were largely in a fresh condition (Figure 233). Still, there was some variation in the surface preservation and ridge widening between artifacts, with one micro-flaked artifact showing more rounded edges than the other fresh artifacts (Figure 233; h38-80).

Discussion

The preliminary data analysis from this study provides some indications that Caours may have undergone significant post-depositional transformations. This discussion illustrates why this site appears to have been transported, to what extent, and why its cause may have been fluvial.

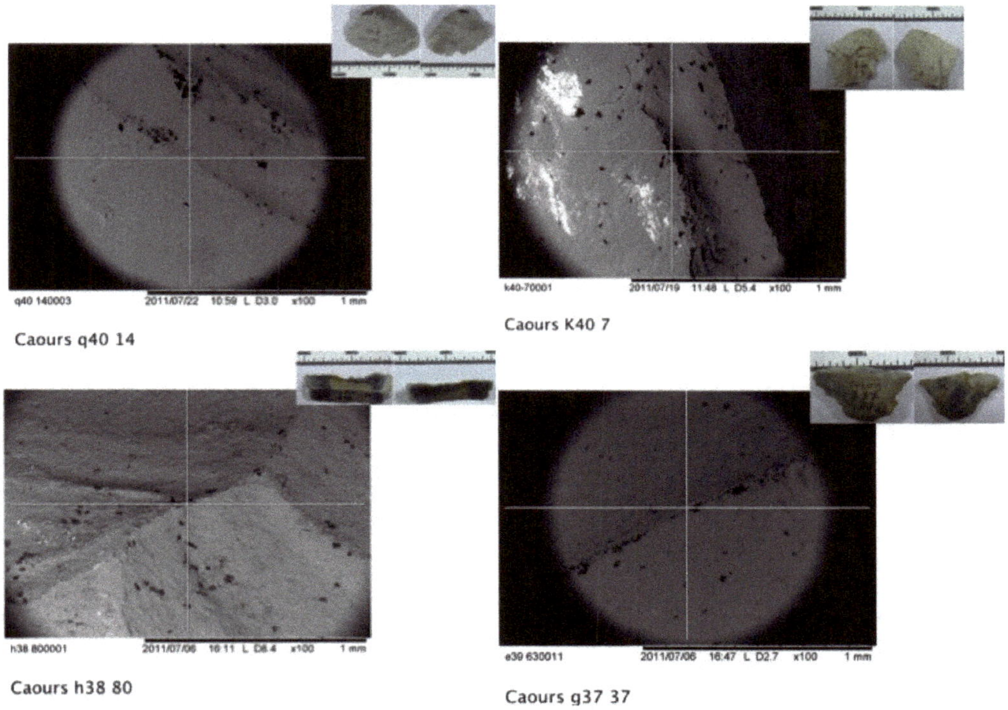

Figure 233: Micrographs from four artifacts from Caours showing low, but variable ridge abrasion and changes to surface texture.

A suspicious marker of transport is the edge damage on stone tools exhibited by over half of the assemblage. Many of the artifacts in Area 1 Level C showed damage in the form of flaking or micro-flaking. Specifically, this damage was largely continuous or near-continuous and bifacial. In most cases, edge damage was not invasive and appeared to be minor, indicating that the artifacts may not have been moved far from their original context. These patterns were consistent with data from both the field and flume experiments.

Edge damage, of course, is not necessarily the result of fluvial transport and can be the result of use wear (Tringham *et al.* 1974) and retouch (Dibble and Holdaway 1993) and it is clear from many of the continuously unifacially flaked tools at Caours that this was sometimes the case. But hominin edge damage is usually localized on artifacts and is related to their specific functions (Schoville 2010).

Cultural processes can modify stone tool edges as well. Trampling, sweeping, and dumping can indiscriminately result in edge damage in a relatively short period (McBrearty *et al.* 1998; Miller *et al.* 2009; Nielsen 1991). Though this can result in randomly distributed edge fracturing, it is more likely to produce a significant amount of notching and snapping that is mostly unifacial in nature (See Section 7.2.8). These characteristics are not observed on materials at Caours Area 1 Level C. Moreover, in open-air sites, soil processes such as solifluction, gelifluction or shock, may create surface modifications *in situ* that appear similar to fluvial transport as a result of sedimentary pressures and grinding (Burroni *et al.* 2002; Stapert 1976; Vallin *et al.* 2001). However, this is an unlikely reason for the damage of the artifacts at Caours, which do not exhibit any of the striations and cracks that often accompany these processes.

Another clue to the disturbance of this site is the distribution of the artifacts. Experimental evidence has shown that small artifacts greatly outnumber larger flakes by several orders of magnitude under normal conditions of manufacture and maintenance (Amick and Mauldin 1989; Bertran *et al.* 2012; Schick 1987; Figure 234d). Figure 234a illustrates that for the total assemblage, using width as a proxy, small sized artifacts were outnumbered by larger size categories. (Note that the absence of very small artifacts in the lower size ranges of this graph is because the sediments were not wet-sieved to recover smaller artifacts (i.e., <5mm)). Their absence strongly suggests a significant winnowing out of the smaller artifacts due to post-depositional processes. However, when damaged artifacts (i.e. flaked or micro-flaked) were separated from each other, it was clear that undamaged artifacts (Figure 234b) more closely approximated experimental scatters and that damaged artifacts (Figure 234c) became more normally distributed.

What these distributions suggest is that at least part of the damaged component of the Caours assemblage has been fluvially sorted (Dibble *et al.* 1997; Shea 1999) and that Level C represents a palimpsest of at least two different archaeological assemblages with the damaged components being washed short distances into the site.

Ridge abrasion is seen in a small percentage of this archaeological assemblage. In this book, macroscopic ridge abrasion was only observed within the tumbling barrel experiments. Applying this finding to Caours suggests that

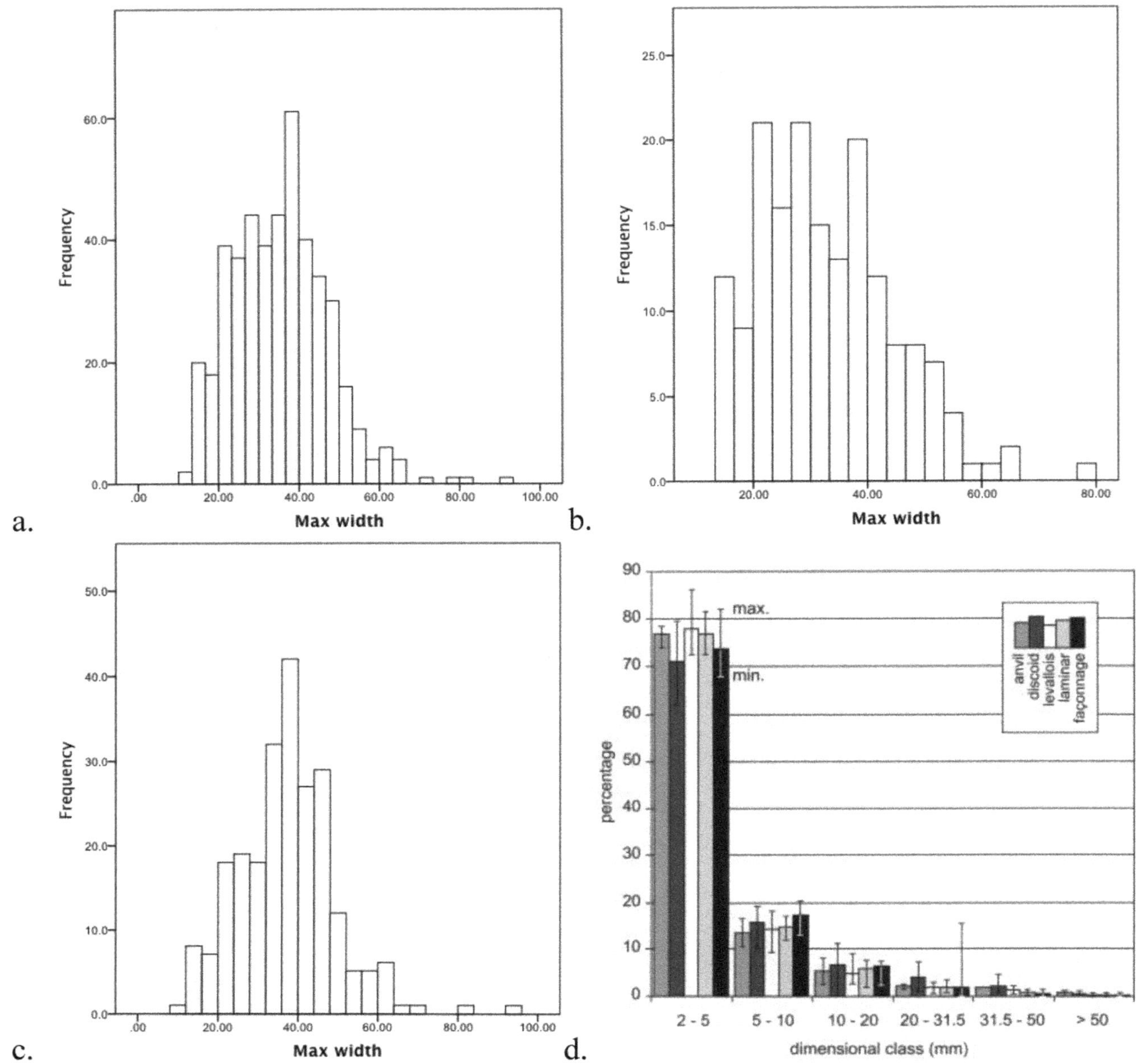

Figure 234: Comparison of artifact widths from Caours to those of experimentally derived widths. a. All artifacts; b. all undamaged; c. all damaged; d. experimentally derived lithic reductions (Redrawn after Bertran *et al.* 2012; Fig. 3).

the majority of the artifacts here were only associated with fine-grained fluvial sediments, indicating that, the artifacts were deposited upon the gravels rather than having been transported along with them. This is further verified by the fact that the surface conditions of artifacts were extremely well preserved, and the appearance of the few and concentrated concussion marks on lithics suggests that they were the result of hominin knapping rather than fluvial abrasion.

Archaeologists have suggested that patinas can be useful in identifying post-depositional processes on archaeological artifacts (Glauberman and Thorson 2012; Goodwin 1960). Most of the patinas were mottled between two different colors, suggesting that the differences between *white* and *gray* patinas for instance, may be varying degrees of the same type of 'porcelain patina' (Glauberman and Thorson 2012). The *brown* patina and its variants may however represent a 'color patina', which is likely the result of a different chemical interaction. This patination at Caours suggest multiple post-depositional processes affecting the lithic artifacts.

A key argument used by the excavators in favor of an *in situ* site was the large amount of cortical flakes found within the assemblage. It was argued that the large cortical retention pattern suggests knapping of complete or semi-complete materials on site, followed by immediate discard. There are no indications from experimental work that natural sorting processes affect the distribution of cortical flakes, so it is impossible to assert that a disturbed site would yield lower percentages. Cortex amounts are already high in discoidal reduction sequences and increase exponentially as spheroid nodules reduce (Figure 235). It is therefore likely that, if

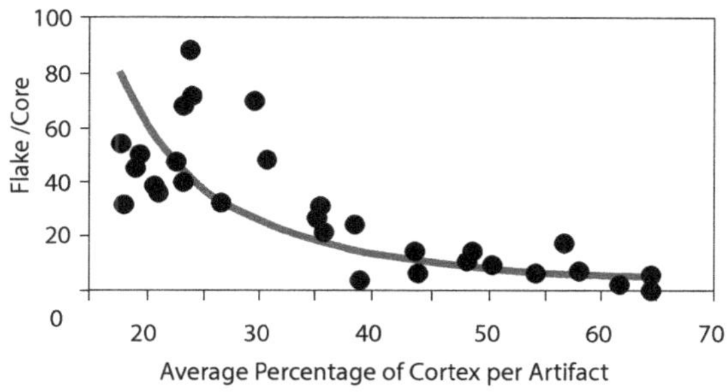

Figure 235: The relationship between the ratio of flakes to core and average cortex for the experimental material (Redrawn after Dibble *et al.* 2005).

there is a high cortical component, it is simply a result of small raw materials and reduction techniques and the assemblage may still be disturbed.

The micrographs of the artifacts at Caours showed very little abrasion and ridge damage confirming the fact that, if artifacts were transported or subjected to moving flow, this was likely not for a long time (more than 24 h; see Chapter 6). However, one artifact did show some variation in ridge width, suggesting that at least some of the assemblage had been subjected to variable flow.

Conclusions and limitations

The taphonomic study of Level C at Caours was aimed at testing if the level represented an *in situ* deposit. Virtually all the data collected suggest that though undoubtably a well-preserved archaeological site, at least some of the material from Level C was disturbed through fluvial action, likely to different degrees. This was in broad accordance with the geological interpretation of the artifacts and damage marks noted by the excavators on the faunal assemblages (Locht *et al.* 2010).

Understanding that a site is disturbed is not necessarily an end goal in itself. It is of greater use for Palaeolithic archaeologists to ascertain the *degree* to which artifact assemblages have been modified (Nash and Petraglia 1987; Petraglia and Potts 1994). One goal for future work is, therefore, to tease out the various assemblages at Caours and to identify how they came into their current position. In the past, archaeologists have attempted to decode palimpsest assemblages by focusing on the relationship of artifact dimensions to nonanthropogenic sediments (Butzer 1982; Goldberg *et al.* 1993; Malinsky-Buller *et al.* 2011). A main problem with these studies is that they rely mainly on weight comparisons which are less applicable to archaeological assemblages due to their unique distribution of small to large artifacts (Bertran *et al.* 2012).

Finally, this study confirms that artifact distributions are a very powerful and useful way to identify fluvial post-depositional processes. It also serves as a reminder that archaeological sites are all palimpsests and can be the result of various occupations and taphonomic disturbances over that time. Therefore, when aiming to explore site taphonomy, archaeologist must use size distributions in conjunction with the entire available suite of taphonomic tools such as abrasion, edge damage, and conditions.

7.5.2. Happisburgh 3

The experimental results from this book are relevant to the interpretation of flake-dominated assemblages in fine-grained fluvial deposits. Happisburgh 3 is an ideal assemblage to test these assumptions on because of the character of its artifacts and its sediments.

Site background

Happisburgh 3 (52°49.36'N, 1°31.58'E) is an open-air Lower Palaeolithic site eroding out of the foreshore of coastal cliffs in northeastern Norfolk, UK (Ashton and Lewis 2012). Site 3 is important for the understanding of the timing of the first presence of hominins in northern Europe after their dispersal out of Africa, as the sediments are dated to the end of the Matuyama Chron between 990,000 and 780,000 years ago (Parfitt *et al.* 2010). Additionally, Happisburgh Site 3 provides an important insight into hominin behavior, adaptation and survival because of the inferred cool conditions (Parfitt *et al.* 2010). Broadly, the site would have presented hominins dispersing from southern Eurasia with a range of new challenges.

Artifacts at Happisburgh Site 3 were found within the Hill House Member, a constituent of the Cromer Forest-bed Formation (Gibbard 2012; Parfitt *et al.* 2010). The succession of the Hill House Member points towards the central or upper reaches of a larger river estuary that may have been the confluence of a northern Thames route and a now-extinct Bytham River (Parfitt *et al.* 2010; Westaway 2011).

The archaeological levels of the site have yielded 78 artifacts, including cores, flakes, scrapers, and various other retouched flakes from different sedimentary contexts, which from top to bottom (Parfitt *et al.* 2010; Supplementary Info 1; Figure 236) are:

- Bed C: Laminated sands and silts (*n*=2);
- Bed D: Sand with clay laminas (*n*=2);
- Bed E: Lag gravel (*n*=74).

Vertebrate, invertebrate, and floral remains indicate 'a fully aquatic through marshy to dry land environment, probably brought together by flood-water from a nearby river' (Parfitt *et al.* 2010 Supplementary Info 1; p.23). Floral remains also point to a nearby conifer forest (Parfitt *et al.* 2010 Supplementary Info 1). Additionally, they indicate a climate with a likely southern Scandinavian analogue. Both the small and large mammal remains indicate some degree of rounding of bones due to fluvial entrainment. Some bones are in perfect condition, while others are heavily rounded, suggesting that they were transported to the site with the rest of the sedimentary particles (Parfitt *et al.* 2010 Supplementary Info 1).

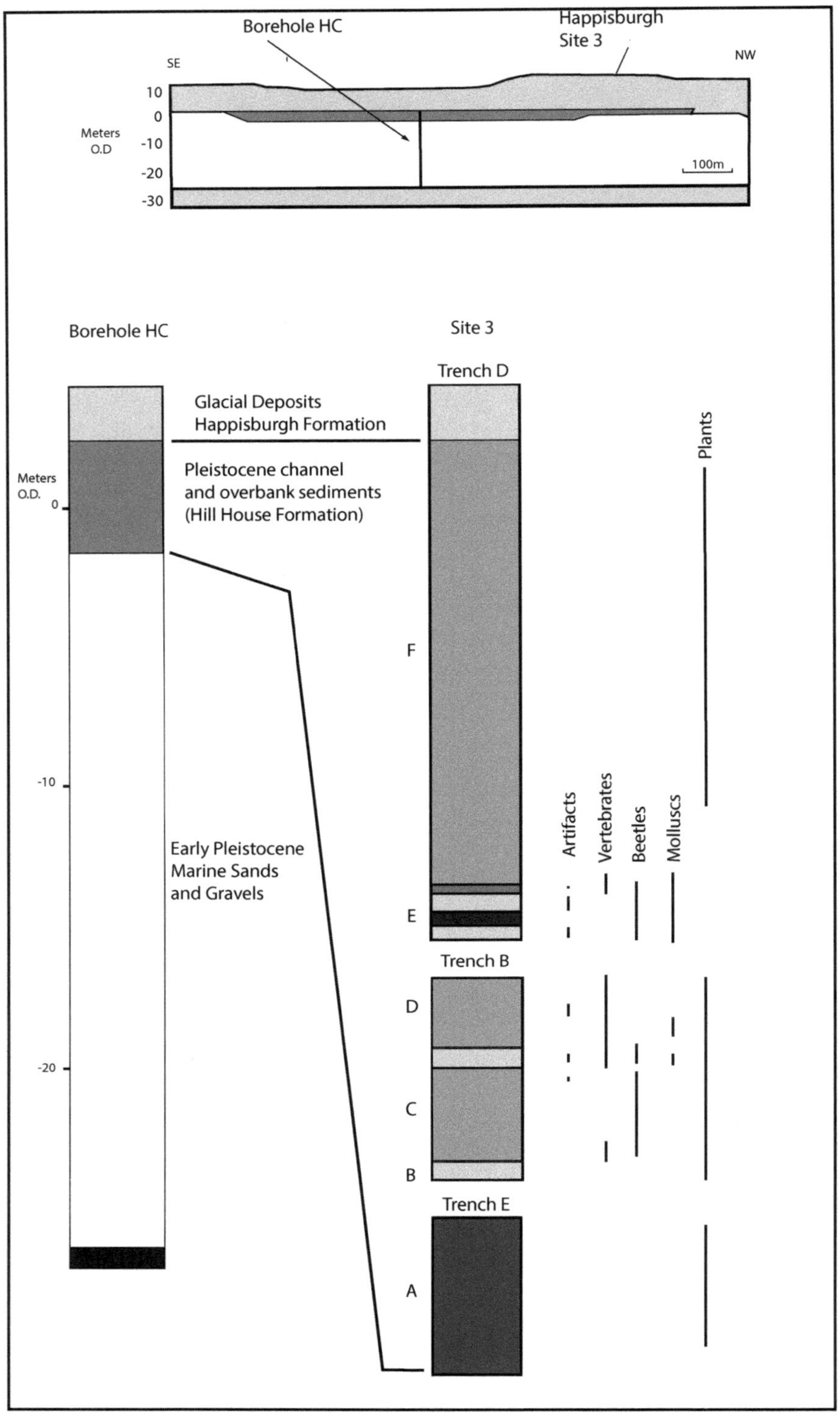

Figure 236: Stratigraphic context of the Happisburgh 3 artifacts and biological remains. Redrawn and reprinted by permission from Macmillan Publishers Ltd: *Nature* (Parfitt *et al.* 2010; fig 3), copyright (2010).

Synthesis and dating

Based on biostratigraphic and palaeomagnetic records, the interpretation of the sedimentary sequence at Happisburgh 3 suggests an age of MIS 25 or 21 (Parfitt *et al.* 2010; Preece and Parfitt 2012).

Archaeology

The archaeology of Happisburgh 3 has previously been summarized as follows (Parfitt *et al.* 2010):

Figure 237: Summary of flint artifacts from HSB3 (Parfitt *et al.* 2010 Supplementary Info Table 6).

Cortex		Butts		Condition	
Cortical	2	*Cortical*	17	*Fresh*	53
>50%	27	*Natural*	3	*Slightly abraded*	23
<50%	31	*Plain*	46	*Abraded*	2
Non-cortical	14	*Dihedral*	3	*Rolled*	0

- Artifacts were made from local flint using medium sized nodules (>25cm).
- Artifacts are consistent with a classic Mode 1 assemblage, predominated by large flakes (up to 145 mm) with sharp cutting edges and opposing cortex.

Current site interpretation

The context of the Happisburgh 3 archaeology has previously interpreted as follows (Parfitt *et al.* 2010, supplementary info 1):

- The artifacts in the lag gravel matrix (Bed E) are assumed to have experienced limited fluvial transport as a result of their fresh conditions (Figure 237) and are therefore interpreted as contemporary with their surroundings.
- The unusual size-range, together with the high proportion of flake tools, suggests that knapping was undertaken elsewhere and artifacts were curated into the area.
- Additionally, the presence of artifacts at three levels in the succession indicates repeated visits to the site.

Data collected

During November of 2011, I collected data on 121 artifacts from Happisburgh 3 (a higher total of artifacts than in the 2010 paper, reflecting more recent fieldwork).

Artifacts were weighed and their maximum length, maximum width, and maximum thickness were recorded (Andrefsky 2005: 98). Ridges were noted for rounding, pitting, or battering, and surfaces were examined for concussion marks, staining, patina, and polish. Additionally, flakes were examined for edge damage, cones of percussion, striations, polish, and patina (Hosfield and Chambers 2004). If edge damage was observed, it was classified as micro-flaking (<1mm), flaking (>1mm), or both. It was also noted if the damage was continuous/ discontinuous, or unifacial/bifacial. Artifacts as well as small pieces of débitage (>1g) were photographed on both their dorsal and ventral sides.

Descriptive statistics

The analysis of the artifacts from Happisburgh 3 yielded the following results:

1. Artifact lengths and widths were not normally distributed as many of the smaller artifact sizes were missing (Figures 238 to 240).

Figure 238: Lithic artifact assemblage from Happisburgh 3, UK, *n*=121.

	Max length (mm)	Max width (mm)	Max thickness (mm)	Weight (g)
Mean	48.53	34.76	13.65	38.86
SD	24.19	18.80	12.85	68.38

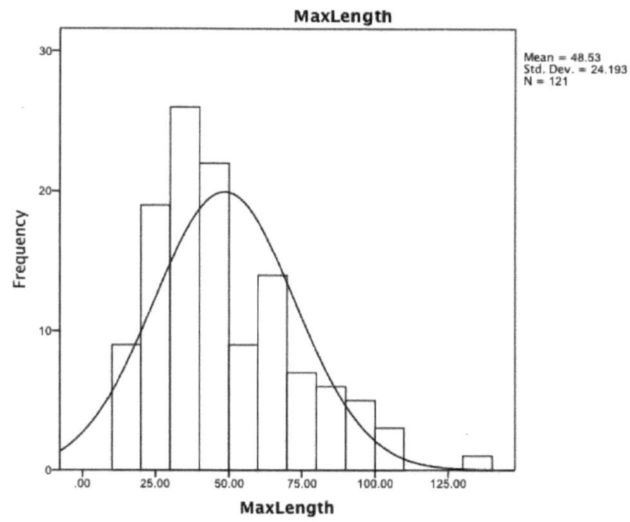

Figure 239: Histogram of the artifact length from Happisburgh 3. K-S(120)=0.123, *p<0*..01.

2. Some of the artifacts from Happisburgh 3 exhibited micro-flaking (79%) and ridge damage (18%; Figure 241; Appendix 16).

Relevant experimental results

The site of Happisburgh 3 presented a perfect opportunity to apply data and conclusions collected in the flume and fluvial experiments, considering:

- That artifact dimensions in gravel contexts are predictors of fluvial transport (field experiment);
- That artifact presence in a fluvial system can produce continuous bifacial edge damage (field and flume experiments);
- That dispersal patterns and size sorting can be an indicator of in-channel transport.

Application of experimental findings

These observations suggest that the current interpretation of assemblage formation at Happisburgh 3 may not be correct.

1. The assumption that the larger size distribution bias is due to curation is unfounded. Rather, the complete lack of smaller artifacts may suggest that the assemblage has been transported along the lag gravel and that smaller artifacts have been winnowed away.
2. The appearance of continuous edge damage, even along the striking platform, and a heavy degree of polish suggests that they have been present in the river system while particles were being entrained (Figure 241).

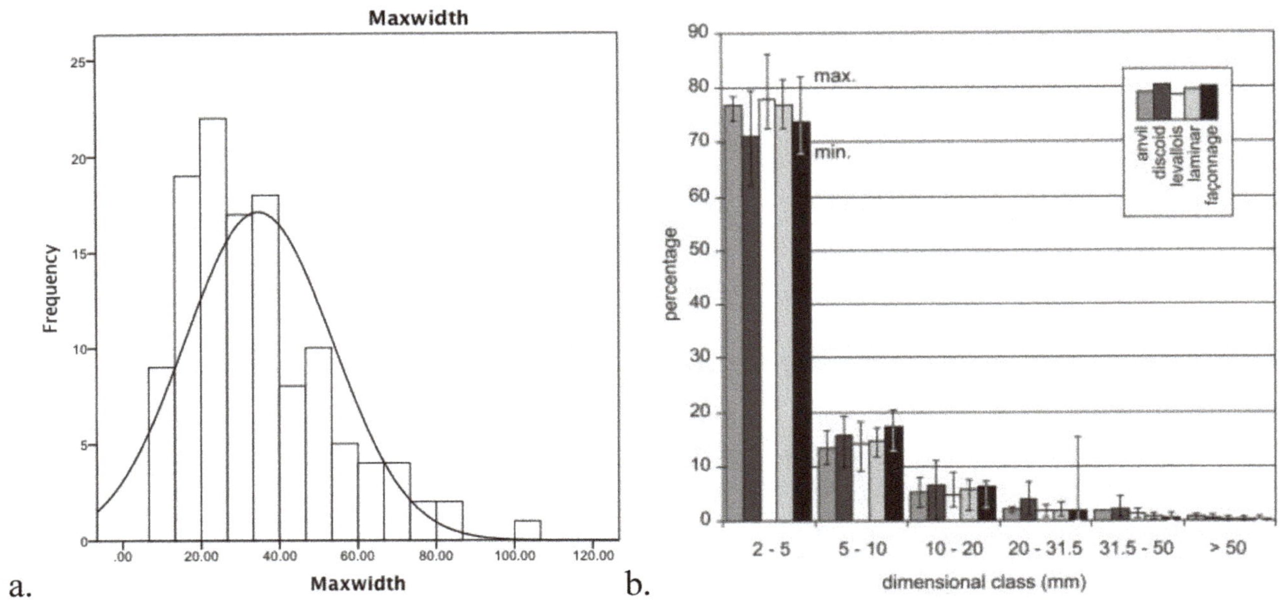

Figure 240: a. Histogram of artifact width from Happisburgh 3. K-S(120)=0.191, *p<0.*.01; b. experimentally derived lithic reductions (Redrawn after Bertran *et al.* 2012; Fig. 3).

Figure 241: Artifact from the Happisburgh 3 excavations exhibiting gloss and bifacial continous edge damage (HSBC-2006-1010).

3. The appearance of artifacts within the lag gravel indicates that they may have been transported within a high-energy fluvial system that would result in edge damage and ridge abrasion.

Conclusions

Significant edge damage, polish, and winnowing suggest that the artifacts from Happisburgh III may not represent an immediate discard environment or the selective introduction of curated artifacts. These findings are provisional and additional weight would be lent to them with finer examination of their micro-surfaces. Given their mostly unpatinated faces, microscopic recording of ridge widths and surface conditions via SEM may be able to give greater indications of their taphonomic histories.

The evidence of lithic redistribution highlighted here may imply that care should be taken in the site interpretation because sorting may have significantly affected the techno-typological balance of the assemblage and the implied associations between the archaeology and the reconstructed palaeoenvironment and landscape. If this is indeed the case, it may also suggest that re-working may have been at the scale to alter some of the other aspects of biostratigraphic record as well.

7.5.3. Pakefield: Continuous or ephemeral occupation?

As noted above, the experimental results of the flume experiment have implications for the re-interpretation of Palaeolithic archaeological assemblages from fine-grained contexts. Another site of interest, in this regards, is Pakefield, UK.

Site background

Pakefield (52°25.9'N, 1°43.8'W) is an open-air site situated on the Suffolk coast of East Anglia, UK. The site is located within the Cromer Forest-bed Formation and associated deposits between Pakefield and Kessingland (Parfitt *et al.* 2005). Already a favorite fossil findspot in Victorian times, it became internationally known in 2005 when artifacts were recovered here that refuted the 'Short Chronology,' as they demonstrated that hominins had been present in northern latitudes for longer than previously believed (Roebroeks 2005; 2006; Roebroeks and Van Kolfschoten 1994).

The combination of vertebrate, invertebrate, and floral remains found in association with the Pakefield artifacts suggest a climate with warm summer and winter temperatures and a strictly seasonal precipitation regimen resembling a warm, seasonally dry Mediterranean climate, leading to the site being dubbed 'Costa del Cromer' (Roebroeks 2005).

Synthesis and dating

The debate about the dating of Pakefield's sequence forms a part of a wider debate regarding the dating of the early middle Pleistocene in East Anglia and has been challenged or modified by researchers (see Lee *et al.* 2006; Gibbard *et al.* 2008; Westaway 2009 for alternatives and dissent). While these debates are of great importance, the focus of this discussion is the taphonomy of the site. To make my argument, I adhere to the original site interpretation that dates it minimally to MIS17 and maximally to MIS 19 and at least to the beginning of the Bruhnes Chron (approximately 700kya; Cande and Kent 1992). If this is subsequently overturned, the timing of my interpretations will need to be corrected.

Geology/stratigraphy

A series of targeted excavations within four different contexts of an interglacial infill of an estuarine river, overlain by a sequence of marine sands and glacio-fluvial sands, recovered 32 worked flints including a core, a crudely retouched flake, and débitage (Figure 242; Parfitt *et al.* 2005; Parfitt 2008; Pettitt and White 2012: 31). Based on the mineral composition of the fluvial sediments, the suggested source is the ancient Bytham River that previously drained the English Midlands.

The excavations revealed four different archaeological contexts (from bottom to top; Parfitt *et al.* 2005).

1. The upper levels of the estuarine silts that had associated marine and brackish-water ostracods, foraminifera, and marine mammals (1 artifact; PaAiv).
2. The 'Rootlet bed,' a series of overbank sediments with well developed soil features (2 artifacts; PaB).
3. The 'Unio-bed,' a sandy lag gravel at the base of the laminated silts, which fill the channel, cut into the overbank sediments (28 artifacts; PaCi).
4. Laminated silts at the edge of the channel (1 artifact; PaCiii).

Archaeology

In the interpretation of the origins of the recovered lithic industry, Parfitt *et al.* (2005) drew the following observations:

- The artifacts were all in 'very sharp condition', on good quality black flint, implying that they had been minimally transported.
- Artifact cortices were water-worn; suggesting the raw material might have been collected in the local vicinity.

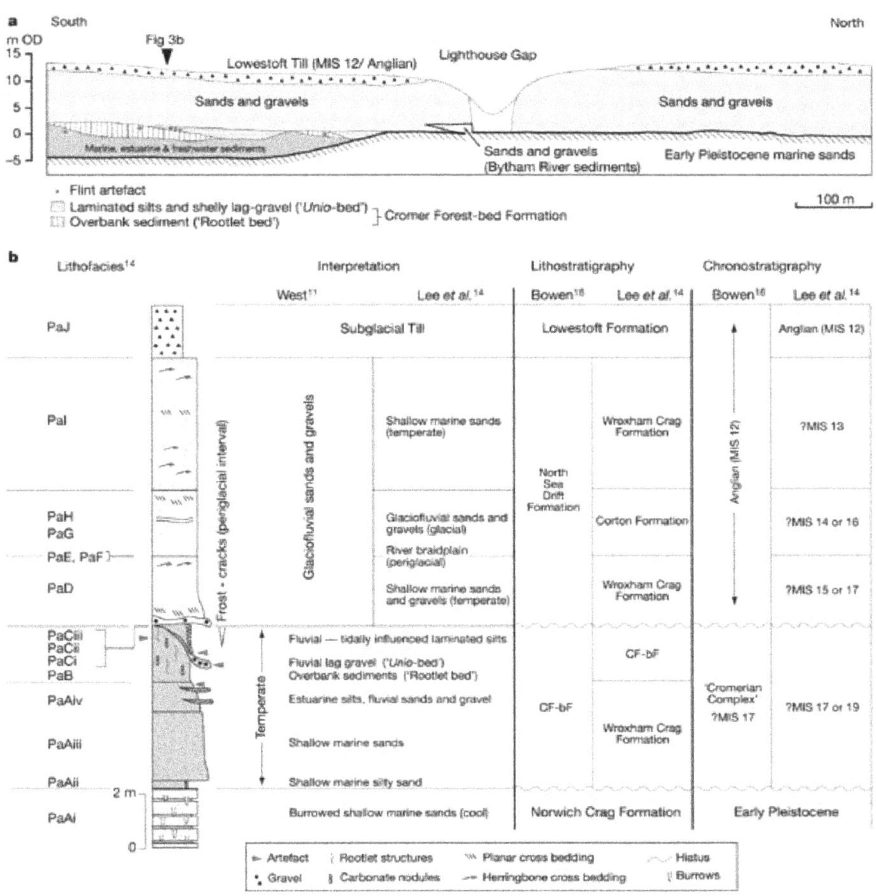

Figure 242: Stratigraphy of Pakefield. Reprinted by permission from Macmillan Publishers Ltd: *Nature* (Parfitt *et al.* 2005; fig 3), copyright (2005).

- Artifacts were provisionally consistent with Mode 1 technology, because they were struck with hard hammers and lack formal tool categories.
- The presence of micro-débitage indicates that flintknapping was undertaken at the site (Parfitt 2008).
- The occurrence of artifacts in four different archaeological contexts has been argued to show that hominins were present for more than one phase of occupation (Ashton *et al.* 2005; Pettitt and White 2012: 32).

Data collected

For 2 weeks during November 2011, data was collected on 20 artifacts from Pakefield at the British Museum in London, UK. It was not possible to examine all of the artifacts because a number of them were being studied abroad. All lithic artifacts from the collection were photographed on the dorsal and ventral sides. Artifacts were weighed and their maximum length, width, and thickness were measured. Edges were examined for continuous bifacial micro-flaking (>1mm). Ridges were noted for rounding, pitting or battering, and surfaces were examined for concussion marks, staining, patina, and polish. The results of this data are presented in Figure 243.

Application of the experimental results

In spite of the lines of evidence used to argue that Pakefield is an *in situ* assemblage, Pettitt and White (2012: 32) have suggested that vertical and horizontal displacement cannot be ruled out. The data collected from Pakefield suggests a number of lines of evidence in support of this view.

- The condition of flints from Pakefield may not be as 'fresh' as previously suggested. Four flakes exhibited bifacial continuous flaking (Figure 243; Figure 244) consistent with the findings of the flume study and the fluvial experiments.
- Other artifacts showed battering to the edges and ridge damage consistent with the entrainment of artifacts within coarser grained sediments (Figure 245). These characteristics imply disturbance or at least exposure to fluvial processes.
- In instances where artifacts were 'fresh' (Figure 246), they may still be derived, particularly for those artifacts found within the silts and the overbank sediments (PaCi, PaCiii).
- The flume experiments also demonstrated that it is unlikely for artifacts to sustain edge damage when they are either transported or in the presence of mobile fine gravels or smaller (approximately >8mm).
- The low spatial density of the artifacts on the gravel lag also suggests that they are the product of a transported assemblage rather than a brief occupation. Artifacts in the fluvial field experiment, when recovered within the channel, were widely dispersed across the gravel bar. All of this suggests that variable fluvial transport may have played a role in structuring the Pakefield assemblage.

Conclusions

In conclusion, the evidence indicates that Pakefield may not represent an on-the-spot knapping location, based on varying degrees of edge micro-flaking to artifacts. Unfortunately, the results of the flume study could not be fully applied to the Pakefield assemblage, because of the impossibility of observing the artifacts under the SEM. Pakefield would be an excellent test case to examine the surface micro-abrasion and ridges of the artifacts, considering the unpatinated nature of the assemblage. This book's experiments make it clear, however, that considerable work is still needed to understand the depositional history of the artifacts at Pakefield, though they do suggest that Pettitt and White's (2012: 32) view that some spatial re-arrangement has taken place may be correct.

7.5.4. Happisburgh 1

The experimental results from the flume study also provide an excellent opportunity to interpret the small artifact assemblage in fine-grained deposits from Happisburgh 1 Site. Of particular interest is the condition of the artifacts recovered from the muds in the 2004 excavations.

Site background

Happisburgh Site 1 (NGR: TG 388307) is an open-air site situated beneath the cliffs of northeastern Norfolk, UK. The site was first excavated in 2004 after artifacts, along with cut-marked bones, were recovered from the foreshore (Ashton *et al.* 2008). Excavations were resumed in 2009 and are currently ongoing (Field 2012). The site is indicative of a fluvial environment within the Cromer Forest Bed formation, for which the stratigraphy is relatively simple. There are two fluvial facies: the upper one is an organic mud containing wood and other plant remains. This is underlain by gray silty sand and capped by the Happisburgh Till Formation (Ashton *et al.* 2008).

Considerable debate has centered on the Middle Pleistocene sedimentary succession of East Anglia, because its interpretation has major bearings on the archaeological materials. At present, two major models, exist—the 'new glacial stratigraphy' model and the 'biostratigraphic' model. This paper follows the biostratigraphic model because it offers multiple lines of evidence based on micro-fauna (Preece *et al.* 2009), invertebrates (molluscs, ostracods, beetles; Preece and Parfitt 2008; Preece *et al.* 2009), and direct dating (Parfitt *et al.* 2010; Pawley *et al.* 2008; Penkman 2009 ; Penkman *et al.* 2013). This suggests a younger age (MIS 13; Preece *et al.* 2009), which is supported here.

In contrast to the 2004 excavation, in more recent excavations no lithic material was recovered from the organic mud, but the underlying gray silty sands with clay lenses have yielded flint artifacts. *In situ* flakes have been recovered from the uppermost part of this facies.

Figure 243: Lithic artifact assemblage from Pakefield, UK. All artifacts were smooth, polished, with no striations or staining.

ID	Max length (mm)	Max width (mm)	Max thick (mm)	Weight (g)	Edges							Ridges			Surface Features		
					Micro-flaking	*Unifacial*	*Continuous*	*Flaking*	*Unifacial*	*Continuous*	*Battering*	*Rounding*	*Pitting*	*Battering*	*Stained*	*Patina*	*Concussion Marks*
Pak 1	39.95	23.97	7.70	7.8	yes	yes	yes	yes	yes	yes	yes	no	no	no	no	no	no
Pak 2	42.09	20.10	9.55	5.9	yes	no	yes	no	n/a	n/a	no	no	no	no	no	no	no
Pak 3	38.00	16.54	6.12	3.4	yes	no	yes	no	n/a	n/a	no	no	yes	no	no	no	no
Pak 4	26.84	14.78	4.53	1.5	yes	no	yes	no	n/a	n/a	no	n/a	n/a	n/a	no	no	no
Pak 5	38.16	33.12	10.42	9.8	no	n/a	n/a	no	n/a	n/a	no	no	no	no	no	no	no
Pak 6	62.96	31.30	17.50	27.0	yes	no	yes	no	n/a	n/a	no	yes	yes	yes	no	yes	yes
Pak 7	39.19	30.96	8.30	8.4	yes	no	yes	yes	no	no	no	n/a	n/a	n/a	no	no	no
Pak 8	39.19	34.41	9.23	9.4	yes	no	no	no	n/a	n/a	no	no	no	no	no	no	no
Pak 9	37.52	20.13	12.24	9.8	yes	no	no	no	n/a	n/a	no	no	yes	no	no	no	no
Pak 10	55.68	49.12	30.82	62.4	yes	yes	no	no	n/a	n/a	no	no	no	no	no	no	no
Pak 11	50.33	42.50	25.21	45.1	yes	no	no	no	n/a	n/a	n/a	no	yes	no	yes	no	no
Pak 12	44.39	32.83	9.46	11.3	yes	no	yes	no	n/a	n/a	no	n/a	n/a	n/a	no	no	no
Pak 13	22.71	14.91	7.71	1.1	no	n/a	n/a	no	n/a	n/a	no	no	no	no	no	no	no
Pak 14	24.55	19.79	6.52	2.7	yes	yes	yes	no	n/a	n/a	no	n/a	n/a	n/a	no	no	no
Pak 15	28.17	20.44	6.21	3.9	yes	no	yes	no	n/a	n/a	yes	yes	yes	no	no	yes	no
Pak 18	30.22	20.41	7.28	3.3	yes	no	no	no	n/a	n/a	no	no	no	no	no	no	no
Pak 19	34.22	28.47	11.09	9.3	no	n/a	n/a	no	n/a	n/a	no	n/a	n/a	n/a	no	no	no
Pak 20	33.39	18.16	8.59	5.8	yes	yes	no	no	n/a	n/a	no	n/a	n/a	no	no	no	no
Pak 21	71.97	41.34	19.57	39.2	yes	no	yes	yes	no	no	no	n/a	n/a	n/a	no	no	no
Pak 22	63.92	39.47	16.58	29.9	yes	no	yes	yes	yes	yes	no	no	yes	no	no	no	yes
n	**20**	**20**	**20**	**20**	**20**	**17**	**17**	**20**	**4**	**4**	**19**	**13**	**13**	**13**	**20**	**20**	**20**
***n*=yes**					**17**	**4**	**11**	**4**	**2**	**2**	**2**	**2**	**6**	**1**	**1**	**2**	**2**
% Yes					**85%**	**24%**	**65%**	**20%**	**50%**	**50%**	**11%**	**15%**	**46%**	**8%**	**5%**	**10%**	**10%**
Mean	**40.16**	**27.27**	**12.12**	**15.09**													
SD	**14.02**	**10.06**	**6.93**	**16.50**													

Figure 244: A flake from Pakefield exhibiting heavy bifacial continuous micro-flaking (Pak 4).

Archaeology

The 2004 excavations yielded a small assemblage made up of 38 flakes and 4 retouched pieces described as notches or denticulates. All were 'thinly dispersed' within the organic muds/clays (Ashton *et al.* 2008). The assemblage is an unprepared core technology including flakes, some of which exhibit notches and denticulates, and multi-platformed cores (Field 2012). The artifacts are fresh and some have been refitted. It is therefore probable that this assemblage is in a primary context, but it is difficult to prove this when excavating under such difficult conditions. The lithic assemblage found during the current excavations is presently being compared by the excavation team to previously discovered material at Happisburgh 1.

Three main observations have been made on these artifacts:

1. The 2004 assemblage appears to include knapping waste: pieces less than 1 gram in size (Ashton *et al.* 2008).
2. The 2004 artifacts are fresh in condition (Ashton *et al.* 2008).
3. The current excavation artifacts are fresh and have some refits (Field 2012).

This has led to the current interpretation of the site:

1. The fresh condition of the artifacts suggest that they have been minimally transported and should be regarded as being in primary context, but probably not *in situ* (Ashton *et al.* 2008).
2. Artifacts and micro-débitage from the 2004 excavation suggest that artifact knapping was taking place close to the site (Ashton *et al.* 2008).
3. Flakes in the uppermost part of the silty sands are *in situ* (Field 2012).

Data collected

During November of 2011, I collected data on 34 of the flakes from the 2004 Happisburgh 1 excavations (Figure 247). Four flakes were being analyzed for flake micro-abrasion at the time and were not available.

Figure 245: A flake from Pakefield exhibiting patination, ridge abrasion and edge damage (Pak 14).

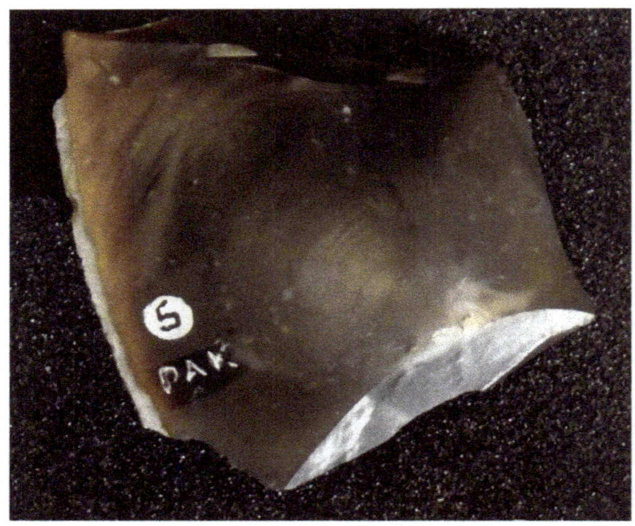

Figure 246: A fresh artifact from Pakfield exhibiting no micro-flaking or ridge abrasion (Pak 5).

1. Artifacts were weighed and their maximum length, maximum width, and maximum thickness were recorded (Andrefsky 2005: 98).
2. Flakes were examined for edge damage and it was noted if there was either flaking (<1mm) or micro-flaking (>1mm) and if it was either continuous/discontinuous unifacial/bifacial.
3. Artifacts' surfaces were noted, specifically, patina, striations (Moir 1914), polish (Howard 2002), staining (Borrazzo 2007), or percussion bulbs (Hosfield and Chambers 2005a).
4. Artifacts as well as 53 small pieces of fractured flint (<1g) were photographed on both their dorsal and ventral sides.

Observations

The collected data showed three key characteristics that deserve highlighting:

Figure 247: Lithic artifact assemblage from the organic clays of Happisburgh 1, UK. All artifacts were smooth, polished, with no striations, staining or concussion marks.

Artifact ID	Max length (mm)	Max width (mm)	Max thickness (mm)	Weight (g)	Edges							Ridges			
					Micro-flaking	Unifacial	Continuous	Flaking	Unifacial	Continuous	Battered	Rounded	Pitting	Battering	Patina
6-8. 10	24.11	19.91	4.73	1.9	yes	no	yes	no	n/a	n/a	no	n/a	n/a	n/a	no
6-8. 11	32.20	24.30	6.80	4.3	no	n/a	n/a	no	n/a	n/a	no	no	no	no	no
6-8. 12	22.53	14.92	4.26	1.2	no	n/a	n/a	no	n/a	n/a	no	no	no	no	no
6-8. 13	19.58	17.46	3.49	1.2	no	n/a	n/a	no	n/a	n/a	no	no	no	no	no
6-8. 17	22.63	17.74	4.74	1.3	no	n/a	n/a	no	n/a	n/a	no	n/a	n/a	n/a	gray
6-8. 18	28.39	23.34	4.54	2.8	yes	no	yes	no	n/a	n/a	no	no	no	no	no
6-8. 19	28.48	21.77	6.86	4.6	no	n/a	n/a	no	n/a	n/a	no	no	no	no	no
6-8. 20	19.76	17.17	6.71	1.5	no	n/a	n/a	no	n/a	n/a	no	no	no	no	no
6-8. 21	28.76	27.22	8.82	5.3	no	n/a	n/a	no	n/a	n/a	no	no	no	no	no
6-8. 22	20.16	15.65	6.20	1.2	no	n/a	n/a	no	n/a	n/a	no	no	no	no	no
6-8. 23	30.41	21.69	7.64	4.9	yes	yes	yes	yes	no	yes	no	no	yes	no	gray
6-8. 27	39.73	15.75	5.86	2.6	yes	no	no	no	n/a	n/a	no	no	no	no	no
6-8. 28	46.67	46.10	11.98	21.4	yes	no	no	no	n/a	n/a	no	no	no	no	no
6-8. 29	22.94	10.93	2.19	0.6	no	n/a	n/a	no	n/a	n/a	no	no	no	no	no
6-8. 30	38.81	14.44	6.43	3.0	yes	no	yes	no	n/a	n/a	no	n/a	n/a	n/a	no
6-8. 32	21.73	16.40	4.15	1.3	no	n/a	n/a	no	n/a	n/a	no	no	no	no	no
6-8. 33	33.60	28.10	7.21	6.2	yes	yes	yes	yes	yes	no	no	no	no	no	gray
6-8. 34	38.26	27.16	7.97	6.5	yes	no	no	no	n/a	n/a	no	n/a	n/a	n/a	no
6-8. 35	37.74	27.02	7.24	8.0	yes	yes	yes	no	n/a	n/a	no	n/a	n/a	n/a	no
6-8. 36	50.85	35.57	16.59	18.6	yes	no	yes	no	n/a	n/a	no	n/a	n/a	n/a	no
6-8. 38	54.00	37.01	16.76	37.7	yes	no	yes	yes	no	yes	yes	yes	yes	yes	brown
6-8. 39	30.44	20.91	4.19	2.2	yes	no	yes	no	n/a	n/a	no	n/a	n/a	n/a	no
6-8. 40	19.72	15.47	4.61	1.3	yes	no	yes	yes	no	no	no	n/a	n/a	n/a	no
6-8. 41	20.60	18.53	3.05	1.5	yes	no	yes	no	n/a	n/a	no	n/a	n/a	n/a	no
6-8. 42	56.77	53.09	14.79	35.9	yes	no	no	no	n/a	n/a	no	no	no	no	no
6-8. 43	85.64	66.14	51.43	264.9	n/a	n/a	n/a	n/a	n/a	n/a	n/a	yes	yes	yes	gray
6-8. 44	91.30	45.06	26.65	82.9	yes	no	no	no	n/a	n/a	no	no	no	no	no
6-8. 45	50.47	39.91	22.04	41.6	no	n/a	n/a	no	n/a	n/a	no	no	yes	no	no
6-8. 46	36.80	22.69	8.68	7.4	no	no	no	no	n/a	n/a	no	n/a	n/a	n/a	no
6-8. 47	21.31	14.25	6.87	1.8	no	n/a	n/a	no	n/a	n/a	no	no	no	no	no
6-8. 49	29.66	14.05	5.07	2.0	yes	no	yes	no	n/a	n/a	no	n/a	n/a	n/a	no
6-8. 5	27.49	20.81	8.23	3.3	no	n/a	n/a	no	n/a	n/a	no	no	no	no	no
6-8. 50	25.43	16.74	5.40	1.9	no	n/a	n/a	no	n/a	n/a	no	n/a	n/a	n/a	no
6-8. 6	39.44	34.97	10.69	15.4	yes	no	no	no	n/a	n/a	yes	n/a	n/a	n/a	no
Mean	35.19	25.36	9.50	17.59											
SD	17.13	12.65	9.17	46.82											
n					33	19	19	33	19	19	33	21	21	21	34
n=yes					18	3	12	4	1	2	2	2	4	2	4
% Yes					55%	16%	63%	12%	5%	11%	6%	10%	19%	10%	12%

1. Most of the 'knapping waste' was abraded and it was the author's view, based on the lack of concoidal ripples and their stark contrast to the knapped artifacts themselves, that it might not have been anthropogenic.
2. All of the artifacts were heavily polished (Figure 248).
3. None of the artifacts exhibited striations, staining or percussion marks. Few of the artifacts exhibited bifacial micro-flaking but three exhibited a brown patina and heavy edge damage, ridge wear, consistent with the results from the coarse-grained tumbling experiments (6-8-25, 6-8-38, 6-8-40; Figure 249).

Relevant experimental results

The experiments presented in this book produced five results that are relevant to the data from Happisburgh 1. Broadly, these are:

1. Artifacts within a relatively low-energy fluvial system can become widely dispersed over short timescales (up to 250m, fluvial experiments; Hosfield and Chambers 2005c).
2. Artifact dimensions influence transport distances (fluvial experiments).
3. Entrainment within a fluvial system can result in a river polish (flume experiments).
4. Entrainment within a fine-grained fluvial system does not necessarily cause micro-flaking (<1mm in in terms of individual scars; flume experiments).
5. Artifacts entrained in coarse-grained sediments acquire edge flaking and ridge damage.

Application of experimental results

Based on these observations, four tentative interpretations of the artifacts from the 2004 excavations can be made:

1. If the débitage is discounted, the artifact density is low compared to *in situ* scatters (approximately 7 artifacts/m^2), suggesting that there may have been some fluvial transport.
2. The presence of gloss on the artifacts probably correlates to surface abrasion and water flow, as witnessed in the flume experiments. This may not confirm that the artifacts have been transported, but it does suggest that they were in the river while it was active (Howard 1999b; 2002).
3. The black unpatinated artifacts do not exhibit major edge damage. Although they were recovered in fine-grained clays, this does not imply that they were *not* transported as in the flume experiments in larger medium sands, entrainment did not result in edge micro-flaking.
4. The three artifacts that are heavily abraded, patinated, and have ridge damage may originate from a different context. This type of damage is consistent with the tumbling barrel experiments and suggests they have been moved from and through different contexts.

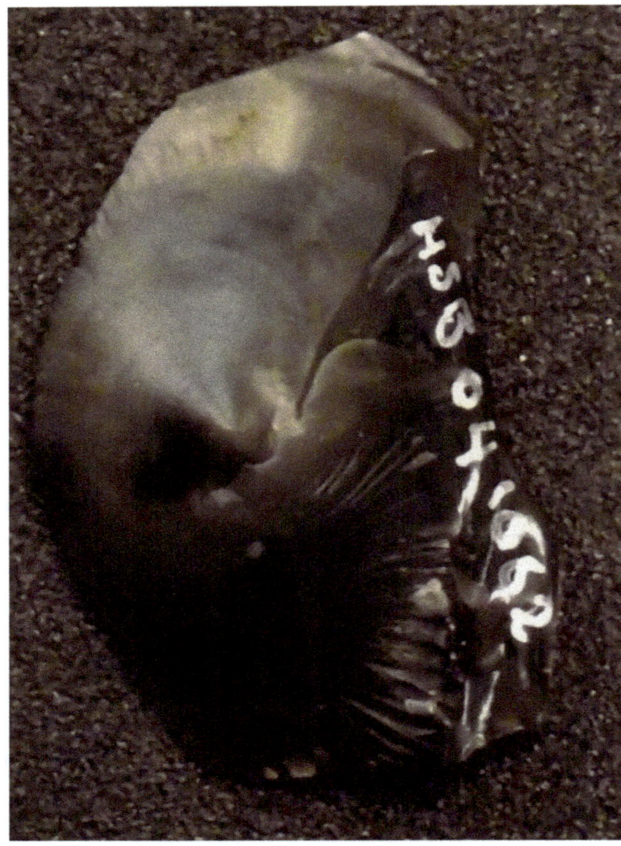

Figure 248: An unabraded, polished flake from Happisburgh 1 (6-8-50).

Figure 249: An artifact from Happisburgh 1 (6-8-38) with heavy ridge and edge battering consistent with the results from the tumbling barrel experiments.

Conclusions

The artifacts from Happisburgh 1 in all likelihood do not represent a single knapping event and assertions that they are *in situ* remain unverified. The similar cortices, artifact condition, and size distribution of artifacts seem to suggest that, in spite of having been in an active fluvial system, they have not been transported far. The greater damage on other artifacts may represent an assemblage from a more distant or older context.

What this study does highlight is the type of information that is necessary to strengthen the conclusions about transport. Examining the artifacts with an SEM would have made it possible to fully assess the surface condition of the artifact as well as the ridge wear.

7.5.5. The Trent Valley: Limited occupation or taphonomic problem?

The results of the tumbling barrel experiments have implications for the interpretation and reinterpretation of Palaeolithic assemblages and findspots in the United Kingdom. Here, the results are applied to the Trent Valley landscape, an area of archaeological importance because it represents a possible northern limit of early hominin occupation (White *et al.* 2004). A key question in recent research has been 'whether this is the genuine northern limit of presapiens human penetration… or whether the boundary is an artifact of preservation' (White *et al.* 2004: 6). This project was well suited to address some of this question as among the raw materials and gravels tested in the tumbling barrel experiments, some were sourced from the Trent Valley.

Landscape background

The Trent is the third longest river in Britain, starting in the West Midlands and flowing eastward approximately 210km to the North Sea (Howard *et al.* 2007). The Trent Valley system has recently been the subject of a major geoarchaeological re-investigation (Bridgland *et al.* 2006). Compared to southern British rivers, the Trent Valley has been an underexplored feature of the British landscape. In spite of collections and remote sensing initiatives (Howard *et al.* 2008), it has yielded relatively few unambiguous Palaeolithic artifacts (Bridgland *et al.* 2006).

Most of the artifacts (pointed and ovate handaxes, together with a range of retouched/unretouched flakes and occasional cores) have been recorded from terrace sediments correlated with MIS8 and MIS4 (White *et al.* 2010), but much of this material is described as heavily rolled and abraded, suggesting significant reworking from older deposits (Howard *et al.* 2007). Recently, two cores and a flake, all of quartzite, have been discovered in a pit at East Leake, where they were associated with terrace/outwash gravel of possible MIS 12 age. Their heavily rolled condition further indicates that they may pre-date MIS 12.

What is the typical interpretation of the archaeological landscape?

Bridgland *et al.* (2006: 16) hypothesize that the paucity of Palaeolithic artifacts in the Trent Valley are not due to a lack of hominin presence, but rather reflect:

- Poor methods of identifying artifacts made on quartzite materials and;
- The susceptibility of local cryptocrystalline raw materials to weathering and abrasion.

These conclusions are based on three suppositions:

- Quartzite artifacts abrade differently from flint artifacts.
- The lithology of the Trent Valley gravels abrades artifacts similarly to that of southern rivers.
- Fluvial abrasion and weathering of quartzite artifacts affects their condition differentially to flint, presumably by chemically altering the outer cortex of the stone.

Application of the experimental findings

This book's tumbling barrel experiments produced results relevant to the assumptions and conclusions of the Trent Valley Project. This study's experimental results found that quartzite had a lower rate of abrasion development when compared to flint. This is most likely related to the larger grain-size of the cryptocrystalline structures in quartzite found by Burroni *et al.* (2002).

The results of this experiment led to two main conclusions:

1. The supposition for artifact lithology differentially affecting abrasion rates was not met in the data charting changes to morphology. The artifact lithologies explored here had no effect on the changes to primary metrics over time.
2. Artifact lithology did affect the changes to artifact condition, which agrees with Bridgland *et al.* (2006), who suggest that lithological hardness plays a role in the development of artifact condition. This was verified when comparing patinated flint to flint, because the abrasions of the softer neocortices of the patinated flint made it abraded rapidly. Bridgland *et al.*'s prediction however was not met when comparing quartzite to flint; quartzite abraded more slowly in spite of being a softer material.

Based upon these conclusions, the following provisional statements are proposed about the interpretation of the Trent Valley project:

1. That the paucity of lithic findspots and artifacts within the Trent Valley may indeed be related to the variant lithology (especially when compared to southern rivers) and the difficulty of identifying lithic findspots and artifacts. However, this may not be primarily related to increased abrasion rates, but instead to the difficult nature of identifying artifacts due to the coarse-grained raw material. Similar issues have been raised concerning the identification of Lower Palaeolithic artifacts in Asia (e.g., Dennell 2004: 227; Chen Shen 2010).
2. That, because of the lower abrasion rates in quartzite artifacts, the heavily abraded artifacts such as those found at East Leake may be indicative of major inter-terrace transport.
3. That the planiform condition of abraded quartzite artifacts (at least for flakes) is largely comparable to those of flint artifacts. This suggests that their changes in dimensions/proportions are comparable thereby rendering inter-assemblage comparisons valid.

Conclusions

What does this all mean for the general lack of artifacts in the Trent Valley? The occasional presence of Palaeolithic artifacts there suggests that post-Anglian hominins were present. The lithic tumbling barrel experiments performed as part of this book support the notion put forth by Bridgland *et al.* (2006: 16) that the paucity of data mainly relates to the difficulties in identifying quartzite artifacts.

Furthermore, while the artifacts at East Leake may indicate a Pre-Anglian deposit, I do not see their abraded condition as lending strong support to this argument. The significance of these finds remains to be fully assessed and support to this assumption would be likely found by discovering artifacts within pre-Anglian deposits.

7.6. Chapter summary

This chapter discussed the results of Chapters 4–6 that examined the taphonomic results of an actualistic field experiment, a tumbling barrel experiment, and a flume experiment on lithic artifacts. They highlight that archaeologists have largely oversimplified the effects of fluvial processes on lithic artifacts by demonstrating that aspects of channel morphology as well as sediment characteristics and transport mode affect the spatial distribution and macro/microscopic morphology of artifacts.

The results of the experiments were then applied to various sites in northern Europe. The results did not strictly agree with current interpretations of the sites, suggesting that the impact of fluvial taphonomy has been largely underplayed in their analyses. What emerges from these brief reviews is a repeated scenario—claims of limited spatial disturbance made for lithic assemblages in fine-grained sediments, are challenged by reanalyses. These case studies highlight the lesson that there is no simple way of understanding lithic taphonomic history. However, through rigorous testing, and the collecting of the right kinds of data during excavation, archaeologists can more adequately address the fluvial taphonomic history of Palaeolithic sites (Dibble et al. 1997).

Chapter 8

Conclusion

Palaeolithic archaeological sites are commonly found within Pleistocene fluvial deposits (Mishra *et al.* 2007; Wymer 1995; 1999). The implications of this association are that archaeological assemblages may have been modified either spatially or morphologically.

Though techniques and models have been previously developed to assess fluvial modifications to sites and artifacts, they largely underestimate the complex nature of fluvial entrainment known from geomorphological literature. Most of these techniques and models were developed in modern African savannahs, where hydrological regimens differ from what is known of Pleistocene Europe. This book has demonstrated that the existing techniques and models are no longer tenable, or, at least, are not broadly applicable in Europe; models of lithic fluvial transport must incorporate depositional contexts, and macroscopic and microscopic investigations of artifacts to assess spatial and morphological fidelity.

To demonstrate this, specific research aims were formulated and tested throughout this study. These were:

1. How do fluvial erosion, transport, and depositional processes in a gravel-bedded meandering river affect the spatial composition and morphological appearance of Palaeolithic assemblages?
2. How does coarse-grained fluvial abrasion affect the size/shape and physical appearance of Palaeolithic artifacts?
3. How does fluvial abrasion in fine-grained sediments affect microscopic morphology of Palaeolithic artifacts?

These questions were targeted in an attempt to increase our interpretative toolkits for northern European Palaeolithic sites in fluvial settings, and answered through the analysis and discussion of three experimental regimens. The main conclusions reached in this book are summarized as follows:

- Fluvial erosion, transport, and deposition of artifacts are dependent on various factors including discard location, burial, and primary metrics.
- The conditions under which artifacts are entrained in coarse-grained sediments affects their macroscopic morphology in terms of micro-flaking with artifact class and lithology as well as sediment size and lithology, playing key roles.
- Fine-grained fluvial transportion of Palaeolithic artifacts may not yield appreciable macroscopic damage, but may be successfully identified microscopically by evaluating surface texture and ridge widths.

Finally, these results were applied to a number of open-air Northern European Palaeolithic sites to help better understand their depositional history.

8.1. Future research

There is necessarily still further work required to understand the impact of fluvial processes as modifiers of lithic assemblages. The findings of this book point to a number of areas where future research could be profitably directed:

General points:

1. To understand the impact of other taphonomic agents including soil processes such as solifluction, gelifluction, or shock. Additionally, the overlap between hominin stone tool resharpening and fluvial micro-flaking/flaking is another area requiring clarification. Understanding these relationships are key to any discussion of hominin behavioral activity and paramount to any discussions of spatial and temporal patterning.
2. To understand the role of weathering (the development of various patinas and glosses) in the preservation or loss of artifact surfaces.
3. To incorporate faunal remains. An explicit methodological approach is lacking to test whether contemporary links exist between re-worked stone tools and hominin-modified faunal remains from the same site (Vaquero 2008). The occurrence of both indicates a hominin presence at a site, but this does not necessarily indicate a direct association. An understanding of past hominin behaviour requires careful identification and analysis of all the potential sources of faunal assemblage accumulation and disturbance.
4. Understanding the role of artifact abrasion in 3 dimensions (3D). Artifact abrasion is an inherently reductive process, and involves the loss of both mass and volume. Developing methods of assessing stone tool morphology and production are increasingly relying on understanding 3D recording methods to underpin robust conclusions (Crompton 2007; Grosman *et al.* 2013; De Groote *et al.* 2010; Lin *et al.* 2010). To better understand the role of abrasion on artifact morphology, future work should focus on experimental modeling using multivariate analyses (i.e., Fourrier's analyses) of 3D landmarks.
5. Continued work into artifact abrasion, particularly in finer-grained sediments, should increasingly look to advancements in microwear studies. Factors such as artifact lithology, and different entrainment modes, require further examination before they are successfully

applied to future archaeological datasets. Furthermore, the successful identification of micro-abrasion would greatly benefit from enhanced microwear studies (i.e., double-blind verification tests) and advancements in quantification (i.e., Evans and Donahue 2005; Faulks *et al.* 2011; Keeley 1980; Kimball *et al.* 1995). These will ultimately provide more accurate assesements for our understanding of archaeological artifacts and site taphonomy.

6. Finally, this study has demonstrated the need for continued and more detailed work on the taphonomic agents unique to northern Europe. As demonstrated throughout this book, fluvial sites remain one of the better understood aspects of lithic taphonomy, even though their effects are still largely unknown and their regional applications are grossly underdeveloped. This study highlights the need for the increasing study of other taphonomic processes in northern European contexts (i.e., bioturbation, diagenesis, gelifluction) as well as region specific taphonomic processes (i.e., fissure systems, sediment dissolution).

8.2. Concluding remarks

Reconstructing the original structure of an archaeological site is invariably both a Herculean and a Sisyphean task. Though it is impossible to 'recreate' an accurate narrative of any archaeological site, with each exhaustive and rigorous test of archaeological material, we may approach better and more accurate interpretive frameworks of hominin prehistory.

While the results of this book represent only a small step in this direction, they feed into a growing body of European literature that improves previous interpretations of site formation and more explicitly, the roles hominins have played in lithic scatter accumulations.

The book also outlines a number of robust conclusions:

First, that it is impossible to understimate the effects of taphonomic processes. No matter how *in situ* an archaeological assemblage may seem, straightforward interpretations of archaeological sites are invariably erroneous.

Second, that taphonomic investigations of archaeological sites must focus on emperical methods of improving the scientific rigor of archaeological epistemologies. Experimental techniques remain a crucial component in improving the inferential frameworks of the archaeological record.

Third, this study re-emphasizes that taphonomic site investigation *must,* in addition to the archaeological material, incorporate local depositional environments and contexts as critical components in the analyses of archaeological artifacts. This re-emphasizes assertions by Dibble *et al.* (1997) that taphonomic analyses are *ideally*, incorporated into the excavation strategy. I would add that targeted experimental work should also form a critical part of this analysis.

Fourth this study outlines an important new method for distinguishing the various causes of artifact wear.

Finally, it is worth reemphasizing that before behavioral interpretations of archaeological sites are made, researchers must actively engage in careful analyses of their site's depositional contexts and post-depositional processes. Without this, our limited understanding of a sparse and incomplete archaeological record will become fraught with errors in spite of an earnest willingness for it to progress.

Bibliography

Abbott, P. L. and Peterson, G. L. 1978. Effects of abrasion durability on conglomerate clast populations; examples from Cretaceous and Eocene conglomerates of the San Diego area, California. *Journal of Sedimentary Research* 48(1): 31–42.

Ackerman, R. E. 1964. Lichens and the patination of chert in Alaska. *American Antiquity* 29(3): 386–387.

Adams, J. 1978. Data for New Zealand pebble abrasion studies. *New Zealand Journal of Science* 21: 607–610.

Allan, J. C., Hart, R. and Tranquili, J. V. 2006. The use of Passive Integrated Transponder (PIT) tags to trace cobble transport in a mixed sand-and-gravel beach on the high-energy Oregon coast, USA. *Marine Geology* 232(1-2): 63–86.

Allen, J. 1985. *Principles of Sedimentology*. London, Allen and Unwin.

Allen, L. G. and Gibbard, P. L. 1993. Pleistocene evolution of the Solent River of southern England. *Quaternary Science Reviews* 12(7): 503–528.

Álvarez, M. *et al.* 2012. Quantifying use-wear traces through RIMAPS and variogram analyses. *Archaeological and Anthropological Sciences* 4(2): 91–101.

Amick, D. S. and Mauldin, R. P. (eds.) 1989. *Experiments in Lithic Technology*. Oxford, British Archaeological Reports.

Amos, C. L. *et al.* 2000. Corrasion of a remoulded cohesive bed by saltating littorinid shells. *Continental Shelf Research* 20(10-11): 1291–1315.

Amos, C. L., Li, M. Z. and Sutherland, T. F. 1998. The contribution of ballistic momentum flux to the erosion of cohesive beds by flowing water. *Journal of Coastal Research* 14(2): 564–569.

Andrefsky, W. 2005. *Lithics: Macroscopic Approaches to Analysis*. 2nd ed. Cambridge; New York, Cambridge University Press.

Anon, SEARCH British Geological Survey Rock Name Database, Available at: http://www.bgs.ac.uk/Lexicon/ [Accessed May 15, 2012].

Antoine, P. *et al.* 2010. Chronostratigraphy and palaeoenvironments of Acheulean occupations in Northern France (Somme, Seine and Yonne valleys). *Quaternary International* 223-224: 456–461.

Antoine, P. *et al.* 2006. Le tuf de Caours (Somme, France): Mise en évidence d'une séquence eemienne et d'un site paléolithique associé. *Quaternaire* 17: 281–319.

Antoine, P. *et al.* 2003. Paléoenvironnements pléistocènes et peuplements paléolithiques dans le bassin de la Somme (nord de la France). *Bulletin de la Société Préhistorique Française* 100(1): 5–28.

Antoine, P. *et al.* 2007. Pleistocene fluvial terraces from northern France (Seine, Yonne, Somme): synthesis, and new results from interglacial deposits. *Quaternary Science Reviews* 26: 2701–2723.

Antoine, P., Lautridou, J. P. and Laurent, M. 2000. Long-term fluvial archives in NW France: response of the Seine and Somme rivers to tectonic movements, climatic variations and sea-level changes. *Geomorphology* 33(3-4): 183–207.

Ascher, R. 1961. Analogy in archaeological interpretation. *Southwestern Journal of Anthropology* 17(4): 317–325.

Ashmore, P. 2003. Braided channels. In M. J. Church and G. Middleton (eds.), *Encyclopedia of Sediments and Sedimentary Rocks: Encyclopedia of Earth Sciences Series*: 85–88. Springer.

Ashton, N. *et al.* 1994. Contemporaneity of Clactonian and Acheulian flint industries at Barnham, Suffolk. *Antiquity* 68: 585–585.

Ashton, N. *et al.* 2005. Excavations at the Lower Palaeolithic site at Elveden, Suffolk. *Proceedings of the Prehistoric Society* 71: 1–61.

Ashton, N. *et al.* 2008. New evidence for complex climate change in MIS 11 from Hoxne, Suffolk, UK. *Quaternary Science Reviews* 27: 652–668.

Ashton, N. *et al.* 2006. Riparian landscapes and human habitat preferences during the Hoxnian (MIS 11) Interglacial. *Journal of Quaternary Science* 21: 497–505.

Ashton, N. and Hosfield, R. 2010. Mapping the human record in the British Early Palaeolithic: evidence from the Solent River system. *Journal of Quaternary Science* 25(5): 737–753.

Ashton, N. and Lewis, S. 2002. Deserted Britain: declining populations in the British late Middle Pleistocene. *Antiquity* 76: 388–396.

Ashton, N., Lewis, S. and Stringer, C. (eds.) 2010. *The Ancient Human Occupation of Britain*. Amsterdam, Elsevier.

Ashton, N. and Lewis, S. G. 2012. The environmental contexts of early human occupation of northwest Europe: The British Lower Palaeolithic record. *Quaternary International* 271: 50–64.

Ashton, N., Lewis, S. G. and Parfitt, S. (eds.) 1996. *Excavations at the Lower Palaeolithic site at East Farm, Barnham, Suffolk 1989-94*. London, British Museum Press.

Ashton, N. M. 1998. The taphonomy of the flint assemblages. In N. Ashton, S. G. Lewis and S. A. Parfitt (eds.), *Excavations at the Lower Palaeolithic Site at East Farm, Barnham, Suffolk*: 183–204. London, British Museum Occasional Paper No. 125.

Aslan, A. and Behrensmeyer, A. K. 1996. Taphonomy and time resolution of bone assemblages in a contemporary fluvial system: the East Fork River, Wyoming. *Palaios* 11(5): 411–421.

Attal, M. and Lavé, J. 2009. Pebble abrasion during fluvial transport: experimental results and implications for the evolution of the sediment load along rivers. *Journal of Geophysical Research* 114(F4): 1–22.

Bagnold, R. 1980. An empirical correlation of bedload transport rates in flumes and natural rivers. *Proceedings of the Royal Society of London. Series A, Mathematical and Physical Sciences* 372: 453–473.

Bagnold, R. 1986. Transport of solids by natural water flow: evidence for a worldwide correlation. *Proceedings of the Royal Society of London. Series A, Mathematical and Physical Sciences* 405: 369–374.

Bahain, J. -J. *et al.* 2007. ESR chronology of the Somme River terrace system and first human settlements in northern France. *Quaternary Geochronology* 2(1-4): 356–362.

Basell, L. S. *et al.* 2011. The geoarchaeology of Paleolithic rivers of southwest Britain. *Geological Society of America Special Papers* 476: 23–36.

Bayer, R. G. 2004. *Mechanical Wear Fundamentals and Testing*. New York, M. Dekker.

Behrensmeyer, A. K. 1991. Terrestrial vertebrate accumulations. In P. Allison and D. E. G. Briggs (eds.), *Taphonomy: Releasing the Data Locked in the Fossil Record*: 291–335. New York, Plenum Publishers.

Behrensmeyer, A. K. 1990. Transport/hydrodynamics of bones. In D. E. G. Briggs and P. R. Crowthers (eds.), *Palaeobiology: A Synthesis*: 232–235. Oxford, Blackwell Scientific Publications.

Behrensmeyer, A. K., Kidwell, S. M. and Gastaldo, R. A. 2000. Taphonomy and paleobiology. *Paleobiology* 26(4): 103–147.

Belshaw, G. 1960. Soil rebound and precise levelling. *Cartography* 3(4): 193–195.

Benito-Calvo, A. *et al.* 2011. Trampling experiments at Cova Gran de Santa Linya, Pre-Pyrenees, Spain: their relevance for archaeological fabrics of the Upper–Middle Paleolithic assemblages. *Journal of Archaeological Science* 38(12): 3652–3661.

Benito-Calvo, A. and de la Torre, I. 2011. Analysis of orientation patterns in Olduvai Bed I assemblages using GIS techniques: implications for site formation processes. *Journal of Human Evolution* 61(1): 50–60.

Bernatchez, J. A. 2010. Taphonomic implications of orientation of plotted finds from Pinnacle Point 13B (Mossel Bay, Western Cape Province, South Africa). *Journal of Human Evolution* 59(3-4): 274–288.

Bertran, P. *et al.* 2012. Particle size distribution of lithic assemblages and taphonomy of Palaeolithic sites. *Journal of Archaeological Science* 39(10): 3148–3166.

Bertran, P. *et al.* 2010. The impact of periglacial processes on Palaeolithic sites: the case of sorted patterned grounds. *Quaternary International* 214(1-2): 17–29.

Bertran, P. and Texier, J. -P. 1995. Fabric analysis: application to Paleolithic sites. *Journal of Archaeological Science* 22(4): 521–535.

Best, J. and Bristow, C. 1993. *Braided Rivers*. London: Geological Society Special Publications.

Bietti, A. 1996. Image processing in microwear studies on flint artifacts. *Archaeologia e Calcolatori* VII: 387–396.

Binford, L. 2002. *In Pursuit of the Past: Decoding the Archaeological Record.* 2nd ed. Berkeley, University of California Press.

Binford, L. R. 1981. Behavioral archaeology and the "Pompeii premise". *Journal of Anthropological Research* 37(3): 195–208.

Binford, L. R. 1977. Olorgesailie deserves more than the usual book review. *Journal of Anthropological Research* 33(4): 493–502.

Binford, L. R. 1979. Organization and formation processes: looking at curated technologies. *Journal of Anthropological Research* 35(3): 255–273.

Bingham, E. C. 1916. An investigation of the laws of plastic flow. *U.S. Bureau of Standards Bulletin* 13: 309–353.

Black, K. S. *et al.* 2007. The use of particle tracking in sediment transport studies: a review. *Geological Society, London, Special Publications* 274(1): 73–91.

Boaz, N. T. and Behrensmeyer, A. K. 1976. Hominid taphonomy: transport of human skeletal parts in an artificial fluviatile environment. *American Journal of Physical Anthropology* 45(1): 53–60.

Boggs, S. 1995. *Principles of Sedimentology and Stratigraphy*. New Jersey, Prentice Hall.

Boismier, W. *et al.* 2003. A Middle Palaeolithic site at Lynford Quarry, Mundford, Norfolk: interim statement. *Proceedings of the Prehistoric Society* 69: 314–324.

Bonney, T. G. 1888. II.—Observations on the rounding of pebbles by alpine rivers, with a note on their bearing upon the origin of the Bunter Conglomerate1. *Geological Magazine (Decade III)* 5(02): 54–61.

Boot, P. 1987. Trampling damage on stone artefacts-some experimental results. *Australian Archaeology* 24: 10–15.

Bordes, F. 1961. *Typologie du Paléolithique Ancien et Moyen*. Bordeaux, Delmas.

Bordes, F. 1988. *Typologie du Paléolithique Ancien et Moyen*. Presses du CNRS.

Borrazzo, K. 2007. Aporte de la tafonomía lítica al estudio de distribuciones artefactuales en ambientes lacustres: el caso del sistema lacustre al sur del lago Argentino. *Comechingonia Virtual* 3: 132–153.

Borrazzo, K. B. 2006. Tafonomía lítica en dunas: una propuesta para el análisis de los artefactos líticos. *Intersecciones en Antropología* 7: 247–261.

Boule, M. 1905. L'origine des éolithes. *L'Anthropologie* XVI: 257–267.

Le Bouteiller, C. *et al*. 2011. A new framework for modeling sediment fining during transport with fragmentation and abrasion. *Journal of Geophysical Research: Earth Surface* 116(F3): 1–15.

Bowers, P. M., Bonnichsen, R. and Hoch, D. M. 1983. Flake dispersal experiments: noncultural transformation of the archaeological record. *American Antiquity* 48(3): 553–572.

Bradley, W. C. 1970. Effect of weathering on abrasion of granitic gravel, Colorado River (Texas). *Geological Society of America Bulletin* 81(1): 61–80.

Brain, C. K. 1983. *The Hunters or the Hunted?: An Introduction to African Cave Taphonomy*. Chicago, University of Chicago Press.

Braun, D. R. *et al*. 2006. Archaeological inference and Oldowan behavior. *Journal of Human Evolution* 51: 106–108.

Bray, M. J. *et al*. 1996. Field measurements of shingle transport using electronic tracers. In *Proceedings 31st MAFF Conference of River and Coastal Engineers*: 10.4.1–10.4.13. London: Ministry of Agriculture, Fisheries and Food.

Bridge, J. S. 2003. Sediment transport by unidirectional water flows. In M. J. Church and G. Middleton (eds.), *Encyclopedia of Sediments and Sedimentary Rocks: Encyclopedia of Earth Sciences Series*: 609–618. Springer.

Bridgland, D. 1994. *Quaternary of the Thames*. London, Chapman and Hall.

Bridgland, D. 2000. River terrace systems in north-west Europe: an archive of environmental change, uplift and early human occupation. *Quaternary Science Reviews* 19: 1293–1303.

Bridgland, D *et al*. 2006. The Palaeolithic occupation of Europe as revealed by evidence from the rivers: data from IGCP 449. *Journal of Quaternary Science* 21(5): 437–455.

Bridgland, D. R. and Westaway, R. 2008. Climatically controlled river terrace staircases: a worldwide Quaternary phenomenon. *Geomorphology* 98: 285–315.

Bridgland, D. R. *et al*. 2001. Middle Pleistocene interglacial deposits at Barling, Essex, England: evidence for a longer chronology for the Thames terrace sequence. *Journal of Quaternary Science* 16(8): 813–840.

Bridgland, D. R. 1996. Quaternary river terrace deposits as a framework for the Lower Palaeolithic record. In C. S. Gamble and A. J. Lawson (eds.), *The English Palaeolithic Reviewed*: 24–39. Old Sarum, Trust for Wessex Archaeology.

Bridgland, D. R. 2010. The record from British Quaternary river systems within the context of global fluvial archives. *Journal of Quaternary Science* 25(4): 433–446.

Bridgland, D. R. 1985. Uniclinal shifting: a speculative reappraisal based on terrace distribution in the London Basin. *Quaternary Newsletter* 47: 26–33.

Bridgland, D. R. and Harding, P. 1993. Middle Pleistocene Thames terrace deposits at Globe Pit, Little Thurrock, and their contained Clactonian industry. *Proceedings of the Geologists' Association* 104(4): 263–283.

Bridgland, D. R., Howard, A. J. and White, M. J. 2006. *The Trent Valley: archaeology and landscape of the ice age*. Durham: University of Durham.

Bridgland, D. R. and Maddy, D. 2002. Global correlation of long Quaternary fluvial sequences: a review of baseline knowledge and possible methods and criteria for establishing a database. *Netherlands Journal of Geosciences* 81(3/4): 265–282.

Bringmans, P. M. M. A. 2006. *Multiple Middle Palaeolithic Occupations in a Loess-Soil Sequence at Veldwezelt-Hezerwater Limburg, Belgium*. Unpublished PhD thesis, Katholieke Universiteit Leuven.

Brink, J. 1977. Frost-heave and archaeological interpretation. *Western Canadian Journal of Anthropology* 7(3): 61–73.

Bromage, T. G. 1984. Interpretation of scanning electron microscopic images of abraded forming bone surfaces. *American Journal of Physical Anthropology* 64(2): 161–178.

Brook, M. S. and Lukas, S. 2012. A revised approach to discriminating sediment transport histories in glacigenic sediments in a temperate alpine environment: a case study from Fox Glacier, New Zealand. *Earth Surface Processes and Landforms* 37(8): 895–900.

Brown, A. *et al*. 2001. Late Holocene channel changes of the Middle Trent: channel response to a thousand-year flood record. *Geomorphology* 39: 69–82.

Brown, A. G *et al.* 2009. Towards a budget approach to Pleistocene terraces: preliminary studies using the River Exe in South West England, UK. *Proceedings of the Geologists' Association* 120(4): 275–281.

Brown, F. H. and Gathogo, P. N. 2002. Stratigraphic relation between Lokalalei 1A and Lokalalei 2C, Pliocene archaeological sites in West Turkana, Kenya. *Journal of Archaeological Science* 29(7): 699–702.

Brown, K. S *et al.* 2009. Fire as an engineering tool of early modern humans. *Science* 325(5942): 859–862.

Brunnacker, K. *et al.* 1982. Correlation of the Quaternary terrace sequence in the Lower Rhine valley and northern Alpine foothills of central Europe. *Quaternary Research* 18(2): 152–173.

Bryan, N. M., Anderson, V. J. and Fugal, R. A. 2011. Disturbance to surface lithic components of archaeological sites by drill seeding. *Rangeland Ecology and Management* 64(2): 171–177.

Bunn, H. *et al.* 1980. FxJj50: an Early Pleistocene site in Northern Kenya. *World Archaeology* 12: 109–136.

Burroni, D. *et al.* 2002. The surface alteration features of flint artefacts as a record of environmental processes. *Journal of Archaeological Science* 29(11): 1277–1287.

Butzer, K. W. 1982. *Archaeology as Human Ecology: Method and Theory for a Contextual Approach.* Cambridge, Cambridge University Press.

Cahen, D. and Moeyersons, J. 1977. Subsurface movements of stone artefacts and their implications for the prehistory of Central Africa. *Nature* 266(5605): 812–815.

Cande, S. C. and Kent, D. V. 1992. A new geomagnetic polarity time scale for the Late Cretaceous and Cenozoic. *Journal of Geophysical Research: Solid Earth* 97(B10): 13917–13951.

Cardillo, M. 2010. Some applications of geometric morphometrics to archaeology. In A. M. T. Elewa (ed.), *Morphometrics for Nonmorphometricians*: 325–341. Berlin; Heidelberg, Springer.

Chambers, J. 2003. Like a rolling stone? The identification of fluvial transportation damage signatures on secondary context bifaces. *Lithics: the Newsletter of the Lithic Studies Society* 24: 66–77.

Chambers, J. 2004. River gravels and handaxes: new experiments in site formation, stone tool transportation and transformation. In M. Fansa (ed.), *Experimentelle Archäologie in Europa, Bilanz 2004*: 25–41. Oldenburg, Isensee Verlag.

Charlton, R. 2008. *Fundamentals of Fluvial Geomorphology*. New York, Routledge.

Chattopadhyay, S. and Chattopadhyay, D. 2003. Colloidal properties of sediments. In M. J. Church and G. Middleton (eds.), *Encyclopedia of Sediments and Sedimentary Rocks: Encyclopedia of Earth Sciences Series*: 88. Springer.

Chen, S., Xing, G. and Qi, W. 2011. The earliest hominin occupations in the Nihewan Basin of northern China: recent progress in field investigations. In D. R. Braun and C. J. Norton (eds.), *Asian Paleoanthropology: From Africa to China and Beyond*: 169–180. Springer Netherlands.

Chu, W. 2009. A functional approach to Paleolithic open-air habitation structures. *World Archaeology* 41: 348–362.

Church, M. 1992. Channel morphology and typology. In P. Calow and G. E. Petts (eds.), *The Rivers Handbook*: 126–143. Oxford, Blackwell.

Church, M. 2003. Grain size and shape. In M. J. Church and G. Middleton (eds.), *Encyclopedia of Sediments and Sedimentary Rocks: Encyclopedia of Earth Sciences Series*: 339–344. Springer.

Church, M. and Hassan, M.A. 1992. Size and distance of travel of unconstrained clasts on a streambed. *Water Resources Research* 28: 299–303.

Clark, D. E. and Purdy, B. A. 1979. Electron microprobe analysis of weathered Florida chert. *American Antiquity* 44(3): 517–524.

Clark, G. 1977. *World Prehistory: In New Perspective*. Cambridge, Cambridge University Press.

Cleveringa, P. *et al.* 1988. Outline of river adjustments in small river basins in Belgium and the Netherlands since the Upper Pleniglacial. In G. Lang and C. Schluchter *(eds.), Lake, Mire and River Environments during the Last 15000 Years*: 123–132. Rotterdam, Balkema.

Coles, J. 1966-67. Experimental Archaeology. *The Proceedings of the Society of Antiquaries of Scotland* 99: 1–20.

Crabtree, D. E. 1972. An introduction to flintworking. *Occasional Papers of the Idaho State University Museum* 28: 1–98.

Crompton, S. 2007. 3D lithic analysis. In A. Posluschny, K. Lambers and I. Herzog (eds.), *Layers of Perception. Proceedings of the 35th International Conference on Computer Applications and Quantitative Methods in Archaeology*: 147–152. Bonn, Koll. Vor- u. Frühgesch.

Daubrée, A. 1879. *Études Synthétiques de Géologie Expérimentale*. Paris, Dunod.

Dennell, R. 2004. *Early Hominin Landscapes in Northern Pakistan: Investigations in the Pabbi Hills*. Oxford, British Archaeological Reports.

Dibble, H. L. *et al.* 1997. Testing the reality of a "living floor" with archaeological data. *American Antiquity* 62(4): 629–651.

Dibble, H. L. and Holdaway, S. J. 1993. The Middle Paleolithic industries of Warwasi. In D. Olszewski and H. Dibble (eds.), *The Paleolithic Prehistory of the Zagros-Taurus*: 75–99. Philadelphia, University Museum Press.

Dibble, H.L. *et al.* 2005. The measurement and interpretation of cortex in lithic assemblages. *American Antiquity* 70(3): 545–560.

Domínguez-Rodrigo, M. *et al.* 2012. Autochthony and orientation patterns in Olduvai Bed I: a re-examination of the status of post-depositional biasing of archaeological assemblages from FLK North (FLKN). *Journal of Archaeological Science* 39(7): 2116–2127.

Dornbusch, U. *et al.* 2002. Life expectancy of shingle beaches: measuring in situ abrasion. *Journal of Coastal Research* 36: 249–255.

Durham, P., Lewis, P. H. and Shennan, S. 1995. Artefact matching and retrieval using the generalised hough transform. *Proceedings of Computer Applications in Archaeology*: 25–30.

Easterbrook, D. J. 1999. *Surface Processes and Landforms.* 2nd ed. Upper Saddle River, Prentice Hall.

Efremov, I. A. 1940. Taphonomy: a new branch of paleontology. *Pan-American Geologist* 74(2): 81–93.

Einstein, H. A. 1950. *The Bed-Load Function for Sediment Transportation in Open Channel Flows*. Washington D.C., United States Department of Agriculture, Soil Conservation Service.

Engelund, F. and Fredsøe, J. 1976. A sediment transport model for straight alluvial channels. *Nordic Hydrology* 7: 293–306.

Enloe, J. G. 2012. Middle Palaeolithic cave taphonomy: discerning humans from hyenas at Arcy-sur-Cure, France. *International Journal of Osteoarchaeology* 22(5): 591–602.

Eren, M. I. *et al.* 2010. Experimental examination of animal trampling effects on artifact movement in dry and water saturated substrates: a test case from south India. *Journal of Archaeological Science* 37(12): 3010–3021.

Eren, M. I. *et al.* 2011. Flaked stone taphonomy: a controlled experimental study of the effects of sediment consolidation on flake edge morphology. *Journal of Taphonomy* 9(3): 201–217.

Eren, M. I. and Lycett, S. J. 2012. Why Levallois? A morphometric comparison of experimental "preferential" Levallois flakes versus debitage flakes. *PLoS ONE* 7(1): p.e29273.

Esdale, J. A., Le Blanc, R. J. and Cinq-Mars, J. 2001. Periglacial geoarchaeology at the Dog Creek site, Northern Yukon. *Geoarchaeology* 16(2): 151–176.

Evans, A. A. and Donahue, R. E. 2005. The elemental chemistry of lithic microwear: an experiment. *Journal of Archaeological Science* 32(12): 1733–1740.

Evans, A. A. and Macdonald, D. 2011. Using metrology in early prehistoric stone tool research: further work and a brief instrument comparison. *Scanning* 33(5): 294–303.

Evans, J. 1862. *Flint implements in the drift: being an account of further discoveries on the continent and in England, communicated to the Society of Antiquaries.* London, J. B. Nichols and Sons.

Faulks, N. R. *et al.* 2011. Atomic force microscopy of microwear traces on Mousterian tools from Myshtylagty Lagat (Weasel Cave), Russia. *Scanning* 33(5): 304–315.

Fernandes, P. *et al.* 2007. Origins of prehistoric flints: the neocortex memory revealed by scanning electron microscopy. *Comptes Rendus Palevol* 6(8): 557–568.

Fernández-Jalvo, Y. and Andrews, P. 2003. Experimental effects of water abrasion on bone fragments. *Journal of Taphonomy* 1(3): 147–164.

Field, A. P. 2009. *Discovering Statistics Using SPSS.* 3rd ed. Los Angeles, SAGE Publications.

Field, M. H. 2012. The first British record of Actinidia faveolata C. Reid and E. M. Reid (Actinidiaceae family). *Quaternary International* 271: 65–69.

Fischer, A., Hansen, P. V. and Rasmussen, P. 1984. Macro and micro wear traces on lithic projectile points. *Journal of Danish Archaeology* 3(1): 19–46.

Fisher, N. I. 1995. *Statistical Analysis of Circular Data.* Cambridge, Cambridge University Press.

Flenniken, J. J. and Haggarty, J. C. 1979. Trampling as an agency in the formation of edge damage: an experiment in lithic technology. *Northwest Anthropological Research Notes* 13(2): 208–214.

Florsheim, J. L. 1985. *Fluvial Requirements for Gravel Bar Formation in Northwestern California.* Unpublished MA Thesis, Humboldt State University.

Frings, R. M. 2008. Downstream fining in large sand-bed rivers. *Earth-Science Reviews* 87(1–2): 39–60.

Gamble, C. S. 1999. *The Palaeolithic Societies of Europe.* Cambridge; New York, Cambridge University Press.

Gamble, C. S. and Boismier, W. A. 2012. The Lynford Neanderthals. In W. Boismier, C. S. Gamble and F. Coward (eds.), *Neanderthals Among Mammoths.* 283–296: London, English Heritage.

Gamble, C. S. 1996. Hominid behaviour in the Middle Pleistocene: an English perspective. In C. S. Gamble and A. J. Lawson (eds.), *The English Palaeolithic Reviewed*: 63–71. Old Sarum, Trust for Wessex Archaeology.

Gaudzinski-Windheuser, S. *et al.* 2010. Testing heterogeneity in faunal assemblages from archaeological sites. Tumbling and trampling experiments at the early Middle Pleistocene site of Gesher Benot Ya`aqov (Israel). *Journal of Archaeological Science* 37(12): 3170–3190.

Gibbard, P. 1985. *The Pleistocene History of the Middle Thames Valley*. Cambridge, Cambridge University Press.

Gibbard, P. 2012. The status of the Hill House "Formation" at Happisburgh, Norfolk, England. *Quaternary International* 271: 29–30.

Gibbard, P. L. *et al.* 2008. Comment: Middle Pleistocene sedimentation at Pakefield, Suffolk, England. J. R. Lee, J. Rose, I. Candy and R. W. Barendregt (2006). Journal of Quaternary Science 21: 155–1791. *Journal of Quaternary Science* 23(1): 85–92.

Gibbard, P. L. 1994. *Pleistocene History of the Lower Thames Valley*. Cambridge; New York, Cambridge University Press.

Gibbard, P. L. and Lewin, J. 2002. Climate and related controls on interglacial fluvial sedimentation in lowland Britain. *Sedimentary Geology* 151(3–4): 187–210.

Gibbard, P. L. and Lewin, J. 2009. River incision and terrace formation in the Late Cenozoic of Europe. *Tectonophysics* 474(1–2): 41–55.

Gifford, D. P. and Behrensmeyer, A. K. 1977. Observed formation and burial of a recent human occupation site in Kenya. *Quaternary Research* 8(3): 245–266.

Gifford-Gonzalez, D. *et al.* 1985. The third dimension in site structure: an experiment in trampling and vertical dispersal. *American Antiquity* 50: 803–818.

Glass, G. V., Peckham, P. D. and Sanders, J. R. 1972. Consequences of failure to meet assumptions underlying the fixed effects analyses of variance and covariance. *Review of Educational Research* 42(3): 237–288.

Glauberman, P. and Thorson, R. 2012. Flint Patina as an Aspect of "Flaked Stone Taphonomy": A Case Study from the Loess Terrain of the Netherlands and Belgium. *Journal of Taphonomy* 10(1): 21–43.

Goerke, B. 1983. Experimental pedifacts. In G. R. Sharma and J. D. Clark (eds.), *Palaeoenvironments and Prehistory in the Middle Son Valley*: 247–250. Allahabad, Abinash Prakashan.

Goldberg, P., Nash, D. T. and Petraglia, M. D. 1993. *Formation Processes in Archaeological Context*. Madison, Prehistory Press.

Goldstein, J. 2003. *Scanning Electron Microscopy and X-Ray Microanalysis*. 3rd ed. New York, Kluwer Academic/Plenum Publishers.

Gölz, E. *et al.* 1995. Fluvial abrasion of broken quartzite used as a substitute for natural bed load. In C. V. J. Varma and A. R. G. Rao (eds.), *Management of Sediment: Philosophy, Aims and Techniques, Sixth International Symposium on River Sedimentation*: 387–395. New Delhi, India.

González-Urquijo, J. E. and Ibáñez-Estévez, J. J. 2003. The quantification of use-wear polish using image analysis. First results. *Journal of Archaeological Science* 30(4): 481–489.

Goodwin, A. J. H. 1960. Chemical alteration (patination) of stone. *The South African Archaeological Bulletin* 15(59): 67–76.

Goren-Inbar, N., Grosman, L. and Sharon, G. 2011. The technology and significance of the Acheulian giant cores of Gesher Benot Ya'aqov, Israel. *Journal of Archaeological Science* 38(8): 1901–1917.

Grace, R. 1989. *Interpreting the Function of Stone Tools: The Quantification and Computerisation of Microwear Analysis*. Oxford, British Archaeological Reports.

Grace, R., Graham, I. D. G. and Newcomer, M. H. 1985. The quantification of microwear polishes. *World Archaeology* 17(1): 112–120.

Green, C. 2012. *Ham Yard, Soho Court, City of Westminster, London: Fieldwork Report*. Quest Scientific, University of Reading.

Green, C. *et al.* 2006. Marine Isotope Stage 9 environments of fluvial deposits at Hackney, north London, UK. *Quaternary Science Reviews* 25: 89–113.

De Groote, I., Lockwood, C. A. and Aiello, L. C. 2010. Technical note: a new method for measuring long bone curvature using 3D landmarks and semi-landmarks. *American Journal of Physical Anthropology* 141(4): 658–664.

Grosman, L. *et al.* 2013. 3D modeling: new method for quantifying post-depositional damages. In H. G. Bock, W. Jäger and M. J. Winckler (eds.), *Scientific Computing and Cultural Heritage: Contributions in Mathematical and Computational Sciences*: 11–20. Springer Berlin Heidelberg.

Grosman, L. *et al.* 2011. Studying post depositional damage on Acheulian bifaces using 3-D scanning. *Journal of Human Evolution* 60(4): 398–406.

Hakon Wadell, 1935. Volume, shape, and roundness of quartz particles. *The Journal of Geology* 43(3): 250–280.

Harding, P. *et al.* 1987. The transport and abrasion of flint handaxes in a gravel-bed river. In G. de G. Sieveking and M. H. Newcomer (eds.), *The Human Uses of Flint and Chert*: 115–126. Cambridge, Cambridge University Press.

Hart, J. K. 1999. Glacial sedimentology: a case study from Happisburgh, Norfolk. In A. P. Jones, M. E. Tucker and J. K. Hart (eds.), *The Description and Analysis of Quaternary Stratigraphic Field Sections. Technical Guide*: 209–234. London, Quaternary Research Association.

Harwell, M. R. *et al.* 1992. Summarizing Monte Carlo results in methodological research: the one- and two-factor fixed effects ANOVA cases. *Journal of Educational and Behavioral Statistics* 17(4): 315–339.

Hassan, M. A. and Church, M. 2001. Sensitivity of bed load transport in Harris Creek: seasonal and spatial variation over a cobble-gravel bar. *Water Resources Research* 37: 813–825.

Hatfield, R. G. and Maher, B. A. 2008. Suspended sediment characterization and tracing using a magnetic fingerprinting technique: Bassenthwaite Lake, Cumbria, UK. *The Holocene* 18(1): 105–115.

Haupt, R. *et al.* 2009. Using the iphone as a brunton: techniques and applications. In 2009 Portland GSA Annual Meeting. Portland, Oregon.

Haward, F. N. 1913. The problem of the eoliths. *Proceedings of the Prehistoric Society of East Anglia* 1: 347–360.

Hayakawa, Y. and Oguchi, T. 2005. Evaluation of gravel sphericity and roundness based on surface-area measurement with a laser scanner. *Computers and Geosciences* 31(6): 735–741.

Henry, D. *et al.* 2004. Human behavioral organization in the Middle Paleolithic: were Neanderthals different?. *American Anthropologist* 106(1): 17–31.

Hicken, E. 2003. Meandering channels. In M. J. Church and G. Middleton (eds.), *Encyclopedia of Sediments and Sedimentary Rocks: Encyclopedia of Earth Sciences Series*: 430–434. Springer.

Hilton, M. R. 2003. Quantifying postdepositional redistribution of the archaeological record produced by freeze–thaw and other mechanisms: an experimental approach. *Journal of Archaeological Method and Theory* 10(3): 165–202.

Hiscock, P. 2002. Quantifying the size of artefact assemblages. *Journal of Archaeological Science* 29(3): 251–258.

Hiscock, P. 1985. The need for a taphonomic perspective in stone artefact analysis. *Queensland Archaeological Research* 2: 82–95.

Hjulström, F. 1935. *Studies of the Morphological Activity of Rivers as Illustrated by the River Fyris*. Uppsala, Almqvist and Wiksells.

Ho Ho Committee, 1979. The Ho Ho classification and nomenclature committee report. In B. Hayden, B. Cotterell and J. Kamminga (eds.), *Lithic Use-Wear Analysis: Studies in Archaeology*: 133–137. New York, Academic Press.

Hogg, A. J. 2003. Turbulent diffusion. In M. J. Church and G. Middleton (eds.), *Encyclopedia of Sediments and Sedimentary Rocks: Encyclopedia of Earth Sciences Series*: 226–229. Springer.

Honea, K. H. 1964. The patination of stone artifacts. *The Plains Anthropologist* 9(23): 14–17.

Hosfield, R. 1999. *The Palaeolithic of the Hampshire Basin: A Regional Model of Hominid Behaviour During the Middle Pleistocene*. Oxford, BAR Publishing.

Hosfield, R. and Chambers, J. 2004. Experimental archaeology on the Afon Ystwyth, Wales, UK. *Antiquity* 78 (299): http://antiquity.ac.uk/ProjGall/chambers/index.html.

Hosfield, R. and Chambers, J. 2005a. Flake modifications during fluvial transportation: three cautionary tales. *Lithics: the Newsletter of the Lithic Studies Society* 24: 57–65.

Hosfield, R. and Chambers, J. 2005b. Pleistocene geochronologies for fluvial sedimentary sequences: an archaeologist's perspective. *Journal of Quaternary Science* 20: 285–296.

Hosfield, R. and Chambers, J. 2005c. River gravels and flakes: new experiments in site formation, stone tool transportation and transformation. In M. Fansa (ed.), *Experimentelle Archäologie in Europa, Bilanz 2004*: 57–74. Oldenburg, Isensee Verlag.

Hosfield, R. *et al.* 2000. Interpreting secondary context sites: a role for experimental archaeology. *Lithics: the Newsletter of the Lithic Studies Society* 21: 29–35.

Hosfield, R. 2011. Rolling stones: understanding river-rolled Paleolithic artifact assemblages. *Geological Society of America Special Papers* 476: 37–52.

Hosfield, R. 2001. The Lower Palaeolithic of the Solent: site formation and interpretive frameworks. In F. F. Wenban-Smith & R. Hosfield (eds.), *Palaeolithic Archaeology of the Solent River*: 85–97. Lithic Studies Society Occasional Paper 7. Lithic Studies Society, London.

Hosfield, R. and Chambers, J. 2003. Flake modifications during fluvial transportation: three cautionary tales. *Lithics* 24: 57–65.

Howard, A. J. *et al.* 2007. The British Pleistocene fluvial archive: East Midlands drainage evolution and human occupation in the context of the British and NW European record. *Quaternary Science Reviews* 26: 2724–2737.

Howard, A. J. *et al.* 2008. Archaeological resource modelling in temperate river valleys: a case study from the Trent Valley, UK. *Antiquity* 82(318): 1040–1054.

Howard, C. D. 1999a. Amorphous silica, soil solutions, and archaeological flint gloss. *North American Archaeologist* 20(3): 209–216.

Howard, C. D. 1999b. River patina on flint artifacts: features and genesis. *The Plains Anthropologist* 44(169): 293–295.

Howard, C. D. 2002. The gloss patination of flint artifacts. *The Plains Anthropologist* 47(182): 283–287.

Van Huissteden, J., Gibbard, P. and Briant, R. 2001. Periglacial fluvial systems in northwest Europe during marine isotope stages 4 and 3. *Quaternary International* 79: 75–88.

Hutton, J. 1788. Theory of the earth; or an investigation of the laws observable in the composition, dissolution, and restoration of land upon the globe. *Transactions of the Royal Society of Edinburgh* 1(02): 209–304.

Iovita, R. and McPherron, S. P. 2011. The handaxe reloaded: a morphometric reassessment of Acheulian and Middle Paleolithic handaxes. *Journal of Human Evolution* 61(1): 61–74.

Isaac, G. 1967. Towards the interpretation of occupation debris: some experiments and observations. *Kroeber Anthropological Society Papers* 37: 31–57.

Johnson, D. L. and Hansen, K. L. 1974. The effects of frost-heaving on objects in soils. *The Plains Anthropologist* 19(64): 81–98.

Johnson, D. L., Muhs, D. R. and Barnhardt, M. L. 1977. The effects of frost heaving on objects in soils, II: laboratory experiments. *The Plains Anthropologist* 22(76): 133–147.

Jones, L. S. and Humphrey, N. F. 1997. Weathering-controlled abrasion in a coarse-grained, meandering reach of the Rio Grande: implications for the rock record. *Geological Society of America Bulletin* 109(9): 1080–1088.

Kaufmann, C. A. *et al.* 2011. Fluvial dispersal potential of guanaco bones (Lama guanicoe) under controlled experimental conditions: the influence of age classes to the hydrodynamic behavior. *Journal of Archaeological Science* 38(2): 334–344.

Keeley, L. H. 1980. *Experimental Determination of Stone Tool Uses: A Microwear Analysis*. Chicago, University of Chicago Press.

Kelly, R. L. 2011. Why did Binford's middle-range program outcompete Schiffer's formation process program?. *Journal of Archaeological Method and Theory* 18(4): 284–290.

Kimball, L. R., Kimball, J. F. and Allen, P. E. 1995. Microwear polishes as viewed through the atomic force microscope. *Lithic Technology* 20(1): 6–28.

Kluskens, S. 1995. Archaeological taphonomy of Combe-Capelle Bas from artifact orientation and density analysis. In H. L. Dibble and M. Lenoir (eds.), *The Middle Paleolithic Site of Combe-Capelle Bas (France)*: 199–243. Philadelphia, The University Museum Press.

Knight, D. and Howard, A. J. 2004. *Trent Valley Landscapes: The Archaeology of 500,000 Years of Change*. King's Lynn, Heritage Marketing and Publications.

Kodama, Y. 1994a. Downstream changes in the lithology and grain size of fluvial gravels, the Watarase River, Japan; evidence of the role of abrasion in downstream fining. *Journal of Sedimentary Research* 64(1a): 68–75.

Kodama, Y. 1994b. Experimental study of abrasion and its role in producing downstream fining in gravel-bed rivers. *Journal of Sedimentary Research* 64(1a): 76–85.

Kodama, Y. *et al.* 1992. Effect of abrasion on downstream gravel-size reduction in the Watarase River, Japan: field work and laboratory experiment. *Environmental Research Center Papers, the University of Tsukuba* 15: 1–88.

Komar, P. D. 2003. Grain threshold. In M. J. Church and G. Middleton (eds.), *Encyclopedia of Sediments and Sedimentary Rocks: Encyclopedia of Earth Sciences Series*: 345–347. Springer.

Krumbein, W. C. 1941. The effects of abrasion on the size, shape and roundness of rock fragments. *The Journal of Geology* 49(5): 482–520.

Krumbein, W. C. and Sloss, L. L. 1963. *Stratigraphy and Sedimentation*. San Francisco and London, Freeman.

Kuenen, P. H. 1956. Experimental abrasion of pebbles: 2. Rolling by current, *The Journal of Geology* 64(4): 336–368.

Lamarre, H. 2005. Using passive integrated transponder (PIT) tags to investigate sediment transport in gravel-bed rivers. *Journal of Sedimentary Research* 75(4): 736–741.

Lamarre, H. and Roy, A. G. 2008. A field experiment on the development of sedimentary structures in a gravel-bed river. *Earth Surface Processes and Landforms* 33(7): 1064–1081.

Laronne, J. B. and Duncan, M. J. 1992. Bedload transport paths and gravel bar formation. In P. Billi *et al.* (eds.), *Dynamics of Gravel-Bed Rivers*: 177–202. Chichester, Wiley.

Lauer, T. *et al.* 2010. Fluvial aggradation phases in the Upper Rhine Graben—New insights by quartz OSL dating. *Proceedings of the Geologists' Association* 121(2): 154–161.

Lee, H *et al.* 2000. Investigations of continuous bed load saltating process. *Journal of Hydraulic Engineering* 126: 691–700.

Lee, J. R. *et al.* 2006. Sea-level changes, river activity, soil development and glaciation around the western margins of the southern North Sea Basin during the Early and early Middle Pleistocene: evidence from Pakefield, Suffolk, UK. *Journal of Quaternary Science* 21(2): 155–179.

Lee, M. W. E *et al.* 2000. Coastal shingle tracing: a case study using the Electronic Tracer System (ETS). In I. Foster (ed.), *Tracers in Geomorphology*: Chichester, Wiley and Sons.

Lenoble, A. 2005. *Ruissellement et Formation des Sites Préhistoriques : Référentiel Actualiste et Exemples d'Application au Fossile.* Oxford, British Archaeological Reports.

Lenoble, A. and Bertran, P. 2004. Fabric of Palaeolithic levels: methods and implications for site formation processes. *Journal of Archaeological Science* 31(4): 457–469.

Lenoble, A., Bertran, P. and Lacrampe, F. 2008. Solifluction-induced modifications of archaeological levels: simulation based on experimental data from a modern periglacial slope and application to French Palaeolithic sites. *Journal of Archaeological Science* 35: 99–110.

Leopold, L. B. and Wolman, M. G. 1957. River channel patterns: braided, meandering, and straight. *United States Geological Survey Professional Paper* 282B: 39–85.

Levi Sala, I. 1986. Use wear and post-depositional surface modification: a word of caution. *Journal of Archaeological Science* 13(3): 229–244.

Lewin, J. and Brewer, P. A. 2002. Laboratory simulation of clast abrasion. *Earth Surface Processes and Landforms* 27(2): 145–164.

Lewin, J., Macklin, M. G. and Johnstone, E. 2005. Interpreting alluvial archives: sedimentological factors in the British Holocene fluvial record. *Qüaternary Science Reviews* 24(16–17): 1873–1889.

Lewis, S. 1998. Quaternary stratigraphy and Lower Palaeolithic archaeology of the Lark valley, Suffolk. In N. Ashton, F. Healy and P. Pettitt (eds.), *Stone Age Archaeology: Essays in Honour of John Wymer*: 43–51. Lithic Studies Society Occasional Papers. Oxford, Oxbow.

Lhomme, V. 2007. Tools, space and behaviour in the Lower Palaeolithic: discoveries at Soucy in the Paris basin. *Antiquity* 81: 536–554.

Limondin-Lozouet, N. *et al.* 2006. Le tuf calcaire de la Celle-Sur-Seine (Seine et Marne): Nouvelles données sur un site clé du Stade 11 dans le nord de la France. *Quaternaire* 17(2): 5–29.

Limondin-Lozouet, N. and Antoine, P. 2001. Palaeoenvironmental changes inferred from malacofaunas in the Lateglacial and early Holocene fluvial sequence at Conty, northern France. *Boreas* 30(2): 148–164.

Lin, S. C. H. *et al.* 2010. The application of 3D laser scanning technology to the assessment of ordinal and mechanical cortex quantification in lithic analysis. *Journal of Archaeological Science* 37(4): 694–702.

Lisiecki, L. E. and Raymo, M. E. 2005. A Pliocene-Pleistocene stack of 57 globally distributed benthic δ18O records. *Paleoceanography* 20(1): 1–17.

Lix, L. M., Keselman, J. C. and Keselman, H. J. 1996. Consequences of assumption violations revisited: a quantitative review of alternatives to the one-way analysis of variance f test. *Review of Educational Research* 66(4): 579–619.

Locht, J. -L., Antoine, P. and Limondin-Lozouet, N. 2010. *Caours. Rapport Triennal de Fouille Programmée 2009*. Amiens, L'Institut national de recherches archéologiques préventives.

Lombard, M. and Pargeter, J. 2008. Hunting with Howiesons Poort segments: pilot experimental study and the functional interpretation of archaeological tools. *Journal of Archaeological Science* 35(9): 2523–2531.

Lopinot, N. H. and Ray, J. H. 2007. Trampling experiments in the search for the earliest Americans. *American Antiquity* 72(4): 771–782.

Lowe, J. J. and Walker, M. J. C. 1997. *Reconstructing Quaternary environments*. Harlow, Essex: Addison Wesley Longman.

Luedtke, B. 1992. *An Archaeologist's Guide to Chert and Flint*. Los Angeles, Institute of Archaeology: University of California.

De Lumley, H. 1975. Cultural evolution in France in its paleoecological setting during the Middle Pleistocene. In K. W. Butzer and G. L. Isaac (eds.), *After the Australopithecines: Stratigraphy, Ecology and Culture in the Middle Pleistocene*: 745–808. The Hague, Mouton.

De Lumley, H. 1969. A Paleolithic camp at Nice. *Scientific American* 220: 42–50.

Lyman, R. L. 1994. *Vertebrate Taphonomy*. Cambridge, Cambridge University Press.

Lyman, R. L. 2010. What taphonomy is, what it isn't, and why taphonomists should care about the difference. *Journal of Taphonomy* 8(1).

Machin, A. J., Hosfield, R. T. and Mithen, S. J. 2007. Why are some handaxes symmetrical? Testing the influence of handaxe morphology on butchery effectiveness. *Journal of Archaeological Science* 34(6): 883–893.

Macklin, M. G. 1995. Archaeology and the river environment of Britain: a prospective review. In A. J. Barham and R. I. Macphail (eds.), *Archaeological Sediments and Soils: Analysis, Interpretation and Management*: 205–220. London, Institute of Archaeology.

Macklin, M. G. *et al.* 2006. Past hydrological events reflected in the Holocene fluvial record of Europe. *CATENA* 66(1–2): 145–154.

Maddy, D. and Bridgland, D. R. 2000. Accelerated uplift resulting from Anglian glacioisostatic rebound in the Middle Thames Valley, UK: evidence from the river terrace record. *Quaternary Science Reviews* 19(16): 1581–1588.

Maddy, D., Bridgland, D. R. and Green, C. P. 2000. Crustal uplift in southern England: evidence from the river terrace records. *Geomorphology* 33(3–4): 167–181.

Malinsky-Buller, A., Hovers, E. and Marder, O. 2011. Making time: 'living floors', 'palimpsests' and site formation processes—a perspective from the open-air Lower Paleolithic site of Revadim Quarry, Israel. *Journal of Anthropological Archaeology* 30(2): 89–101.

Mallouf, R. J. 1982. An analysis of plow-damaged chert artifacts: the Broken Creek cache (41HI86), Hill County, Texas. *Journal of Field Archaeology* 9(1): 79–98.

Mania, D., Litt, T. and Weber, T. 1990. *Neumark-Gröbern, Beiträge zur Jagd des Mittelpaläolithischen Menschen.* Berlin, Deutscher Verlag der Wissenschaften.

Marshall, G. D. *et al.* 2002. *Acheulian Biface Database. ADS, York.* Available at: http://ads.ahds.ac.uk/catalogue/specColl/ bifaces/bf_query.cfm.

Martin, R. E. 1999. *Taphonomy: A Process Approach.* Cambridge; New York, Cambridge University Press.

Matthews, E. R. 1983. Measurements of beach pebble attrition in Palliser Bay, southern North Island, New Zealand. *Sedimentology* 30(6): 787–799.

McBrearty, S. *et al.* 1998. Tools underfoot: human trampling as an agent of lithic artifact edge modification. *American Antiquity* 63(1): 108–129.

McNabb, J. 2007. *The British Lower Palaeolithic: Stones in Contention.* Abingdon, Routledge.

McPherron, S. P. and Dibble, H. L. 1999. Stone tool analysis using digitized images: examples from the Lower and Middle Paleolithic. *Lithic Technology* 24(1): 38–52.

Mikoš, M. and Jaeggi, M. N. R. 1995. Experiments on motion of sediment mixtures in a tumbling mill to study fluvial abrasion. *Journal of Hydraulic Research* 33(6): 751–772.

Miller, C. E. *et al.* 2009. Dumping, sweeping and trampling: experimental micromorphological analysis of anthropogenically modified combustion features. *Paléthologie* 2: 25–37.

Mishra, S. *et al.* 2007. Fluvial deposits as an archive of early human activity. *Quaternary Science Reviews* 26(22–24): 2996–3016.

Moeyersons, J. 1978. The behaviour of stones and stone implements, buried in consolidating and creeping Kalahari sands. *Earth Surface Processes and Landforms* 3(2): 115–128.

Moir, J. R. 1914. The striation of flint surfaces. *Man* 14: 177–181.

Morris, P. H. and Williams, D. J. 1999. A worldwide correlation for exponential bed particle size variation in subaerial aqueous flows. *Earth Surface Processes and Landforms* 24(9): 835–847.

Munsell, A. H. 1905. *A Color Notation.* Boston, G. H. Ellis Company.

Nanson, G. and Croke, J. 1992. A genetic classification of floodplains. *Geomorphology* 4: 459–486.

Nash, D. and Petraglia, M. (eds.), 1987. *Natural Formation Processes and the Archaeological Record.* Oxford, British Archaeological Reports.

Van Nest, J. 1985. Patination of Knife River flint artifacts. *The Plains Anthropologist* 30(110): 325–339.

Nichols, G. 2009. *Sedimentology and Stratigraphy.* London, Wiley-Blackwell.

Nielsen, A. E. 1991. Trampling the archaeological record: an experimental study. *American Antiquity* 56(3): 483–503.

Nowak, R. M. 1999. *Walker's Mammals of the World.* Baltimore, Johns Hopkins University Press.

Nowell, A. and d'Errico, F. 2007. The art of taphonomy and the taphonomy of art: Layer IV, Molodova I, Ukraine. *Journal of Archaeological Method and Theory* 14(1): 1–26.

O'Connor, A. 2007. *Finding Time for the Old Stone Age: A History of Palaeolithic Archaeology and Quaternary Geology in Britain, 1860–1960.* Oxford; New York, Oxford University Press.

O'Connor, A. 2003. Geology, archaeology, and "the raging vortex of the 'eolith' controversy". *Proceedings of the Geologists Association* 114(3): 255–262.

Odell, G. H. and Cowan, F. 1987. Estimating tillage effects on artifact distributions, *American Antiquity* 52(3): 456–484.

Paddayya, K. and Petraglia, M. D. 1993. Formation processes of Acheulian localities in the Hunsgi and Baichbal Valleys, peninsular India. In P. Goldberg, D. T. Nash and M. D. Petraglia (eds.), *Formation Processes in Archaeological Context*: Monographs in World Archaeology: 61–82. Madison, Prehistory Press.

Pante, M. C. and Blumenschine, R. J. 2010. Fluvial transport of bovid long bones fragmented by the feeding activities of hominins and carnivores. *Journal of Archaeological Science* 37(4): 846–854.

Parfitt, S. A. *et al.* 2010. Early Pleistocene human occupation at the edge of the boreal zone in northwest Europe. *Nature* 466(7303): 229–233.

Parfitt, S. A. 2008. Pakefield Cliffs: archaeology and palaeoenvironment of the Cromer Forest-bed Formation. In I. Candy, J. R. Lee, and A. M. Harrison (eds.), *The Quaternary of Northern East Anglia*: 130–136. Edinburgh, Quaternary Research Association.

Parfitt, S. A. *et al.* 2005. The earliest record of human activity in northern Europe. *Nature* 438: 1008–12.

Pargeter, J. 2011. Assessing the macrofracture method for identifying Stone Age hunting weaponry. *Journal of Archaeological Science* 38(11): 2882–2888.

Pargeter, J. and Bradfield, J. 2012. The effects of Class I and II sized bovids on macrofracture formation and tool displacement: results of a trampling experiment in a southern African Stone Age context. *Journal of Field Archaeology* 37(3): 238–251.

Pawley, S. M. *et al.* 2008. Age limits on Middle Pleistocene glacial sediments from OSL dating, north Norfolk, UK. *Quaternary Science Reviews* 27(13): 1363–1377.

Penkman, K. 2009. Amino acid geochronology: its impact on our understanding of the Quaternary stratigraphy of the British Isles. *Journal of Quaternary Science* 25: 501–514.

Penkman, K. E. H. *et al.* 2013. An aminostratigraphy for the British Quaternary based on Bithynia opercula. *Quaternary Science Reviews* 61: 111–134.

Petraglia, M. and Nash, D. 1987. The impact of fluvial processes on experimental sites. In D. Nash and M. Petraglia (eds.), *Natural Formation Processes and the Archaeological Record*: 108–130. Oxford, British Archaeological Reports.

Petraglia, M. D. and Potts, R. 1994. Water flow and the formation of Early Pleistocene artifact sites in Olduvai Gorge, Tanzania. *Journal of Anthropological Archaeology* 13(3): 228–254.

Pettitt, P. and White, M. 2012. *The British Palaeolithic: Human Societies at the Edge of the Pleistocene World.* London, Routledge.

Pitts, M. W. and Roberts, M. 2000. *Fairweather Eden: Life Half a Million Years Ago as Revealed by the Excavations at Boxgrove.* New York, Fromm International.

Potter, P. E. and Pettijohn, F. J. 1977. *Paleocurrents and Basin Analysis.* Berlin; New York, Springer-Verlag.

Preece, R. C. *et al.* 2009. Biostratigraphic and aminostratigraphic constraints on the age of the Middle Pleistocene glacial succession in north Norfolk, UK. *Journal of Quaternary Science* 24(6): 557–580.

Preece, R. C. and Parfitt, S. A. 2008. The Cromer Forest-bed Formation: some recent developments relating to early human occupation and lowland glaciation. In I. Candy, J. R. Lee and A. M. Harrison (eds.), *The Quaternary of Northern East Anglia. Field Guide*: 60–83. London, Quaternary Research Association.

Preece, R. C. and Parfitt, S. A. 2012. The Early and early Middle Pleistocene context of human occupation and lowland glaciation in Britain and northern Europe. *Quaternary International* 271: 6–28.

Prestwich, J. 1860. On the occurrence of flint-implements, associated with the remains of animals of extinct species in beds of a late geological period, in France at Amiens and Abbeville, and in England at Hoxne, *Philosophical Transactions of the Royal Society of London* 150: 277–317.

Prestwich, S. J. 1857. *The Ground Beneath Us, Its Geological Phases and Changes: Being Three Lectures on the Geology of Clapham and the Neighbourhood of London Generally.* London, J. Van Voorst.

Ragan, D. M. 2009. *Structural Geology: An Introduction to Geometrical Techniques.* Cambridge; New York, Cambridge University Press.

Rayleigh, Lord, 1942. The ultimate shape of pebbles, natural and artificial. *Proceedings of the Royal Society of London. Series A. Mathematical and Physical Sciences* 181(985): 107–118.

Reid, I. and Frostick, L. E. 1985. Arid zone slopes and their archaeological materials. In A. F. Pitty (ed.), *Themes in Geomorphology*: 141–157. London, Croom Helm.

Van Rijn, L. C. 1984. Sediment transport, part II: suspended load transport. *Journal of Hydraulic Engineering* 110(11): 1613–1641.

Roe, D. A. 1968. British Lower and Middle Palaeolithic handaxe groups. *Proceedings of the Prehistoric Society* 34: 1–82.

Roe, D. A. 1980. Introduction: precise moments in remote time. *World Archaeology* 12: 107–108.

Roe, D. A. 1964. The British Lower and Middle Palaeolithic: some problems, methods of study and preliminary results. *Proceedings of the Prehistoric Society* 30: 245–267.

Roe, D. A. 1981. *The Lower and Middle Paleolithic Periods in Britain.* London, Routledge.

Roe, G. 2006. In defense of Milankovitch. *Geophys. Res. Lett., 33, L24703.*

Roebroeks, W. *et al.* 1997. «Site J»: an early Weichselian (Middle Palaeolithic) flint scatter at Maastricht-Belvedere, The Netherlands. *Paléo* 9(1): 143–172.

Roebroeks, W. 2005. Life on the Costa del Cromer. *Nature* 438(15): 921–922.

Roebroeks, W. 1996. The English Palaeolithic record: absence of evidence, evidence of absence and the first occupation of Europe. In C. S. Gamble and A. J. Lawson (eds.), *The English Palaeolithic Reviewed*: 57–62. Old Sarum, Trust for Wessex Archaeology.

Roebroeks, W. 2006. The human colonisation of Europe: where are we?. *Journal of Quaternary Science* 21(5): 425–435.

Roebroeks, W. 2008. Time for the Middle to Upper Paleolithic transition in Europe. *Journal of Human Evolution* 55(5): 918–26.

Roebroeks, W. and Corbey, R. 1999. Periodisations and double standards in the study of the Palaeolithic. In W. Roebroeks *et al.* (eds.), *Hunters of the Golden Age: The Mid Upper Palaeolithic of Eurasia, 30,000-20,000 bp*: 77–86. Leiden, Analecta Praehistorica Leidensia.

Roebroeks, W. and Van Kolfschoten, T. 1994. The earliest occupation of Europe: a short chronology. *Antiquity* 68(260): 489–503.

Rottländer, R. 1975. The formation of patina on flint. *Archaeometry* 17(1): 106–110.

Saragusti, I. *et al.* 2005. Quantitative analysis of shape attributes based on contours and section profiles in artifact analysis. *Journal of Archaeological Science* 32(6): 841–853.

Saville, A. 1997. Palaeolithic handaxes in Scotland. *Proceedings of the Society of Antiquaries of Scotland* 127: 1–16.

Schick, K. 1987. Experimentally-derived criteria for assessing hydrological disturbances of archaeological sites. In D. Nash and M. Petraglia (eds.), *Natural Formation Processes and the Archaeological Record*: 86–107. Oxford, British Archaeological Reports.

Schick, K. 1991. On making behavioral inferences from early archaeological sites. In D. Clark (ed.), *Cultural Beginnings: Approaches to Understanding Early Hominid Life-Ways in the African Savanna*: 79–107. Bonn, Dr. Rudolf Habelt.

Schick, K. D. 1993. *Making Silent Stones Speak: Human Evolution and the Dawn of Technology*. New York, Simon and Schuster.

Schick, K. D. 1986. *Stone Age Sites in the Making: Experiments in the Formation and Transformation of Archaeological Occurrences*. Oxford, British Archaeological Reports.

Schiffer, M. B. 1972. Archaeological context and systemic context. *American Antiquity* 37(2): 156–165.

Schiffer, M. B. 1987. *Formation Processes of the Archaeological Record*. Albuquerque, University of New Mexico Press.

Schiffer, M. B. 1985. Is there a "Pompeii premise" in archaeology?. *Journal of Anthropological Research* 41(1): 18–41.

Schiffer, M. B. 1983. Toward the identification of formation processes. *American Antiquity* 48(4): 675–706.

Schmidt, K. and Ergenzinger, P. 1992. Bedload entrainment, travel lengths, step lengths, rest periods–studied with passive (iron, magnetic) and active (radio) tracer techniques. *Earth Surface Processes and Landforms* 17: 147–165.

Schneider, J. *et al.* 2010. Studying sediment transport in mountain rivers by mobile and stationary RFID antennas. In A. Dittrich *et al.* (eds.), *River Flow*: 1723–1730. Switzerland, Bundesanstalt für Wasserbau.

Schoville, B. J. 2010. Frequency and distribution of edge damage on Middle Stone Age lithic points, Pinnacle Point 13B, South Africa. *Journal of Human Evolution* 59(3-4): 378–391.

Schreve, D. C. 1998. *Mammalian biostratigraphy of the later Middle Pleistocene in Britain*. Unpublished PhD thesis. University College London.

Schreve, D. C. 2006. The taphonomy of a Middle Devensian (MIS 3) vertebrate assemblage from Lynford, Norfolk, UK, and its implications for Middle Palaeolithic subsistence strategies. *Journal of Quaternary Science* 21(5): 543–556.

Schumm, S. A. and Stevens, M. A. 1973. Abrasion in place: a mechanism for rounding and size reduction of coarse sediments in rivers. *Geology* 1(1): 37–40.

Shackley, M. 1975. *A study of the Mousterian of Acheulian tradition industries of southern Britain*. Unpublished PhD thesis. University of Southampton.

Shackley, M. 1974. Stream abrasion of flint implements. *Nature* 248: 501–502.

Shackley, M. L. 1978. The behaviour of artefacts as sedimentary particles in a fluviatile environment. *Archaeometry* 20(1): 55–61.

Shea, J. J. 1999. Artifact abrasion, fluvial processes, and living floors from the Early Paleolithic site of Ubeidiya (Jordan Valley, Israel). *Geoarchaeology* 14(2): 191–207.

Shea, J. J. and Klenck, J. D. 1993. An experimental investigation of the effects of trampling on the results of lithic microwear analysis. *Journal of Archaeological Science* 20(2): 175–194.

Sheppard, P. J. and Pavlish, L. A. 1992. Weathering of archaeological cherts: a case study from the Solomon Islands. *Geoarchaeology* 7(1): 41–53.

Shipman, P. 1981. Applications of scanning electron microscopy to taphonomic problems. *Annals of the New York Academy of Sciences* 376(1): 357–385.

Shipman, P. and Rose, J. 1983. Early hominid hunting, butchering, and carcass-processing behaviors: approaches to the fossil record. *Journal of Anthropological Archaeology* 2(1): 57–98.

Shott, M. 1998. Status and role of formation theory in contemporary archaeological practice. *Journal of Archaeological Research* 6(4): 299–329.

Shott, M. J. 2008. Lower Paleolithic industries, time, and the meaning of assemblage variation. In S. Holdaway and L. Wandsnider (eds.), *Time and Archaeology, Time Perspectivism Revisited*: 46–60. Salt Lake City, University of Utah Press.

Simon, A. and Collison, A. J. C. 2002. Quantifying the mechanical and hydrologic effects of riparian vegetation on streambank stability. *Earth Surface Processes and Landforms* 27(5): 527–546.

Simons, D. B. and Şentürk, F. 1992. *Sediment Transport Technology: Water and Sediment Dynamics*. Littleton, Water Resources Publications.

Singer, R. *et al.* 1973. Excavation of the Clactonian industry at the golf course, Clacton-on-Sea, Essex. *Proceedings of the Prehistoric Society* 39: 6–74.

Smith, F. A. *et al.* 2003. Body mass of late Quaternary mammals. *Ecology* 84(12): 3403–3403.

Smith, G. M. 2012. Middle Palaeolithic subsistence: the role of hominins at Lynford, Norfolk, UK. *Quaternary International* 252: 68–81.

Šolc, T. *et al.* 2012. Application of an instrumented tracer in an abrasion mill for rock abrasion studies. *Journal of Mechanical Engineering* 58(4): 263–270.

De Sonneville-Bordes, D. and Piveteau, J. 1960. *Le paléolithique supérieur en Périgord.* Bordeaux, Delmas.

Soulsby, R. 1997. *Dynamics of Marine Sands*. London, T. Telford.

Speth, J. D. 2006. Housekeeping, Neandertal-style. In E. Hovers and S. L. Kuhn (eds.), *Transitions Before the Transition: Interdisciplinary Contributions to Archaeology:* 171–188. New York, Springer.

Stachowiak, G. W. and Batchelor, A. W. 2005. *Engineering Tribology* 3rd ed. Oxford, Butterworth-Heinemann.

Stapert, D. 1976. Some natural surface modifications on flint. *Palaeohistoria* 18: 7–41.

Stemp, J., Childs, B. E. and Vionnet, S. 2010. Laser profilometry and length-scale analysis of stone tools: second series experiment results. *Scanning* 32(4): 233–243.

Sternberg, H. 1875. Untersuchungen über längen- und querprofil geschiebeführender Flüsse. *Zeitschrift für Bauwesen* 25: 483–506.

Stout, D. 2011. Stone toolmaking and the evolution of human culture and cognition. *Philosophical Transactions of the Royal Society B: Biological Sciences* 366(1567): 1050–1059.

Stringer, C. 2007. *Homo Britannicus: The Incredible Story of Human Life in Britain.* London, Penguin.

Sundborg, Å. 1956. The river Klarälven: a study of fluvial processes. *Geografiska Annaler* 38(2): 125–237.

Sutherland, T. F. Amos, C. L. and Grant, J. 1998. The effect of buoyant biofilms on the erodibility of sublittoral sediments of a temperate microtidal estuary. *Limnology and Oceanography* 43(2): 225–235.

Texier, J. -P. *et al.* 1998. TRANSIT, an experimental archaeological program in periglacial environment: problem, methodology, first results. *Geoarchaeology* 13: 433–473.

Thieme, H. 1997. Lower Palaeolithic hunting spears from Germany. *Nature* 385(6619): 807–810.

Thompson, C. E. L. *et al.* 2011. The abrasion of modern and archaeological bones by mobile sediments: the importance of transport modes. *Journal of Archaeological Science* 38(4): 784–793.

Thompson, C. E. L. *et al.* 2003. The manifestation of fluid-transmitted bed shear stress in a smooth annular flume—a comparison of methods. *Journal of Coastal Research* 19(4): 1094–1103.

Thompson, C. E. L. and Amos, C. L. 2004. Effect of sand movement on a cohesive substrate. *Journal of Hydraulic Engineering* 130(11): 1123–1125.

Thompson, C. E. L. and Amos, C. L. 2002. The impact of mobile disarticulated shells of Cerastoderma edulis on the abrasion of a cohesive substrate. *Estuaries* 25(2): 204–214.

Tringham, R. *et al.* 1974. Experimentation in the formation of edge damage: a new approach to lithic analysis. *Journal of Field Archaeology* 1(1/2): 171–196.

Tuffreau, A. *et al.* 1995. Le gisement Acheuléen de Cagny-l'Épinette (Somme). *Bulletin de la Société Préhistorique Française* 92(2): 169–192.

Tuffreau, A. *et al.* 1986. Les niveaux Acheuléens de la moyenne terrasse du bassin de la Somme à Cagny-l'Epinette (Somme). *L'Anthropologie* 90(1): 9–27.

Vallin, L. Masson, B. and Caspar, J. -P. 2001. Taphonomy at Hermies, France: a Mousterian knapping site in a loessic context, *Journal of Field Archaeology* 28(3/4): 419–436.

Vandenberghe, J. 1993. Changing fluvial processes under changing periglacial conditions. *Zeitschrift für Geomorphologie* 88: 17–28.

Vandenberghe, J. 1995. Timescales, climate and river development. *Quaternary Science Reviews* 14: 631–638.

Vaquero, M. *et al.* 2001. Neandertal behavior at the Middle Palaeolithic site of Abric Romaní, Capellades, Spain. *Journal of Field Archaeology* 28: 93–114.

Vaquero, M. 2008. The history of stones: behavioural inferences and temporal resolution of an archaeological assemblage from the Middle Palaeolithic. *Journal of Archaeological Science* 35(12): 3178–3185.

Viklander, P. 1998. Laboratory study of stone heave in till exposed to freezing and thawing. *Cold Regions Science and Technology* 27(2): 141–152.

Villa, P. 1982. Conjoinable pieces and site formation processes. *American Antiquity* 47(2): 276–290.

Villa, P. 1977. Sols et niveaux d'habitat du Paléolithique inférieur en Europe et au Proche-Orient. *Quaternaria* 19: 107–134.

Villa, P. 1978. *The Stone Artifact Assemblage From Terra Amata: A Contribution to the Comparative Study of Acheulian Industries in Southwestern Europe.* Unpublished PhD thesis, University of California, Berkeley.

Villa, P. and Courtin, J. 1983. The interpretation of stratified sites: a view from underground. *Journal of Archaeological Science* 10(3): 267–281.

Warren, S. H. 1905. On the origin of "eolithic" flints by natural causes, especially by the foundering of drifts. *The Journal of the Royal Anthropological Institute of Great Britain and Ireland* 35: 337–364.

Warren, S. H. 1914. The experimental investigation of flint fracture and its application to problems of human implements. *The Journal of the Royal Anthropological Institute of Great Britain and Ireland* 44: 412–450.

Wentworth, C. K. 1919. A laboratory and field study of cobble abrasion. *The Journal of Geology* 27(7): 507–521.

Wentworth, C. K. 1922. A scale of grade and class terms for clastic sediments. *The Journal of Geology* 30(5): 377–392.

Westaway, R. 2009. Quaternary vertical crustal motion and drainage evolution in East Anglia and adjoining parts of southern England: chronology of the Ingham River terrace deposits. *Boreas* 38(2): 261–284.

Westaway, R. 2011. The Pleistocene terrace staircase of the River Thame, central-southern England, and its significance for regional stratigraphic correlation, drainage development, and vertical crustal motions. *Proceedings of the Geologists' Association* 122(1): 92–112.

Westaway, R., Maddy, D. and Bridgland, D. 2002. Flow in the lower continental crust as a mechanism for the Quaternary uplift of south-east England: constraints from the Thames terrace record. *Quaternary Science Reviews* 21(4–6): 559–603.

Whalley, W. B. 2003. Surface textures. In M. J. Church and G. Middleton (eds.), *Encyclopedia of Sediments and Sedimentary Rocks: Encyclopedia of Earth Sciences Series*: 712–717. Springer.

Whalley, W. B., Marshall, J. R. and Smith, B. J. 1982. Origin of desert loess from some experimental observations. *Nature* 300(5891): 433–435.

White, M., Bridgland, D. R. and Howard, A. J. 2004. The Lower and Middle Palaeolithic Occupation of the Middle and Lower Trent Catchment and Adjacent Areas, as Recorded in the River Gravels. Project outline for a project funded by the Aggregates Levy Sustainability Fund (ALSF). Universities of Durham and University of Newcastle.

White, M. J. 2012. The lithic assemblage from Lynford Quarry and its bearing on Neanderthal behaviour in Late Pleistocene Britain. In W. Boismier, C. S. Gamble, and F. Coward (eds.), *Neanderthals Among Mammoths: Excavations at Lynford Quarry, Norfolk*: 219–261. Swindon, English Heritage.

White, T. S. *et al.* 2010. Evidence from the Trent terrace archive, Lincolnshire, UK, for lowland glaciation of Britain during the Middle and Late Pleistocene. *Proceedings of the Geologists' Association* 121(2): 141–153.

White, T. S. *et al.* 2009. Palaeolithic and Quaternary research in the Trent Valley (UK): contributions by early collectors. *Proceedings of the Geologists' Association* 120(4): 223–232.

Wiberg, P. and Smith, J. 1985. A theoretical model for saltating grains in water. *Journal of Geophysical Research* 90: 7341–7354.

Wiberg, P. and Smith, J. 1989. Model for calculating bed load transport of sediment. *Journal of Hydraulic Engineering* 115: 101–123.

Wilcock, P. 1997. Entrainment, displacement and transport of tracer gravels. *Earth Surface Processes and Landforms* 22: 1125–1138.

Wilcock, P. 2001. Toward a practical method for estimating sediment-transport rates in gravel-bed rivers. *Earth Surface Processes and Landforms* 26: 1395–1408.

Willis, B. 1993. Ancient river systems in the Himalayan foredeep, Chinji Village area, northern Pakistan. *Sedimentary Geology* 88(1–2): 1–76.

Wooldridge, S. W. 1939. *Structure, Surface and Drainage in Southeast England.* London, G. Philip and Son.

Wymer, J. 1968. *Lower Palaeolithic Archaeology in Britain as Represented by the Thames Valley*. London, John Baker.

Wymer, J. 1988. Palaeolithic archaeology and the British Quaternary sequence. *Quaternary Science Reviews* 7(1): 79–98.

Wymer, J. 1999. *The Lower Palaeolithic Occupation of Britain.* Salisbury, Wessex Archaeology and English Heritage.

Wymer, J. J. 1995. The context of Palaeoliths. In A. J. Schofield (ed.), *Lithics in Context: Suggestions for the Future Direction of Lithic Studies*: 45–51. London, Lithic Studies Society.

Yang, S. and Lim, S. 2003. Total load transport formula for flow in alluvial channels. *Journal of Hydraulic Engineering* 129: 68–72.

Yaroshevich, A. *et al.* 2010. Design and performance of microlith implemented projectiles during the Middle and the Late Epipaleolithic of the Levant: experimental and archaeological evidence. *Journal of Archaeological Science* 37(2): 368–388.

Zeuner, F. E. 1945. *The Pleistocene Period: Its Climate, Chronology and Faunal Successions.* 2nd ed. London, Ray Society.

Appendices

Appendix 1: SPSS v20.0 syntax for MIXED used in Chapter 4.

```
*MAIN EFFECTS & TWO_WAY INTERACTIONS
DATASET ACTIVATE DataSet1.
MIXED Dependent Variable BY Independent Variables
 /CRITERIA=CIN(95) MXITER(100) MXSTEP(10) SCORING(1)
SINGULAR(0.000000000001) HCONVERGE(0,
 ABSOLUTE)    LCONVERGE(0,    ABSOLUTE)    PCONVERGE(0.000001,
ABSOLUTE)
 /FIXED=Independent Variables and Interactions | SSTYPE(3)
```

Appendix 2: Pearson's correlation matrix for independent factors used in the transport distance analysis.

		Time	**Trapped**	**Buried**	**Trampled**	**Edge Damage**	**Distance Traveled**	**High Flow**	**High mAOD**	**Height**	**Dip**	**Strike**	**Scatter ID**	**Max Length**	**Max Thickness**	**Weight**	**Width**
	Pearson Correlation	1	.142**	-.792**	.217**	.019	.034	-.181**	.990**	.070*	.393**	.333**	.000	.000	.000	.000	.000
	Sig. (2-tailed)		.000	.000	.000	.594	.422	.000	.000	.023	.000	.000	1.000	1.000	1.000	1.000	1.000
	N	1820	796	794	796	788	557	785	784	1056	852	1037	1820	1700	1688	1700	1692
Trapped	*Pearson Correlation*	.142**	1	.120**	.629**	.163**	-.225**	-.392**	.094**	.247**	-.235**	.116**	-.172**	.157**	.095*	.143**	.177**
	Sig. (2-tailed)	.000		.001	.000	.000	.000	.000	.009	.000	.000	.001	.000	.000	.011	.000	.000
	N	796	796	794	796	787	555	782	780	783	550	763	796	733	727	734	728
Buried	*Pearson Correlation*	-.792**	.120**	1	.231**	-.073*	.079	-.389**	-.865**	-.244**	-.259**	-.130**	.169**	-.087*	-.087*	-.056	-.085*
	Sig. (2-tailed)	.000	.001		.000	.040	.064	.000	.000	.000	.000	.000	.000	.018	.020	.133	.022
	N	794	794	794	794	785	554	780	778	781	548	761	794	732	726	733	727
Trampled	*Pearson Correlation*	.217**	.629**	.231**	1	.057	-.094*	-.712**	.119**	.386**	.[c]	.135**	-.041	.183**	.151**	.126**	.192**
	Sig. (2-tailed)	.000	.000	.000		.110	.026	.000	.001	.000	.000	.000	.252	.000	.000	.001	.000
	N	796	796	794	796	787	555	782	780	783	550	763	796	733	727	734	728
Edge Damage	*Pearson Correlation*	.019	.163**	-.073*	.057	1	-.369**	.122**	.028	.109**	-.022	.084*	-.153**	.002	-.023	.009	.021
	Sig. (2-tailed)	.594	.000	.040	.110		.000	.001	.441	.002	.603	.020	.000	.959	.536	.808	.572
	N	788	787	785	787	788	551	775	773	776	544	757	788	725	719	726	720
Distance Traveled	*Pearson Correlation*	.034	-.225**	.079	-.094*	-.369**	1	-.306**	.007	-.325**	.119*	-.051	.306**	-.176**	-.096*	-.082	-.148**
	Sig. (2-tailed)	.422	.000	.064	.026	.000		.000	.875	.000	.029	.237	.000	.000	.029	.061	.001
	N	557	555	554	555	551	557	557	554	557	334	543	557	524	518	526	519
High Flow	*Pearson Correlation*	-.181**	-.392**	-.389**	-.712**	.122**	-.306**	1	-.039	-.009	-.178**	-.128**	-.001	-.060	-.038	-.047	-.076*
	Sig. (2-tailed)	.000	.000	.000	.000	.001	.000		.276	.806	.000	.000	.971	.104	.314	.204	.041
	N	785	782	780	782	775	557	785	782	785	552	765	785	724	718	725	719
High mAOD	*Pearson Correlation*	.990**	.094**	-.865**	.119**	.028	.007	-.039	1	.276**	.333**	.215**	-.195**	.139**	.122**	.091*	.152**
	Sig. (2-tailed)	.000	.009	.000	.001	.441	.875	.276		.000	.000	.000	.000	.000	.001	.014	.000
	N	784	780	778	780	773	554	782	784	782	552	762	784	722	716	723	717
Height	*Pearson Correlation*	.070*	.247**	-.244**	.386**	.109**	-.325**	-.009	.276**	1	-.093**	.137**	-.042	.307**	.292**	.149**	.290**
	Sig. (2-tailed)	.023	.000	.000	.000	.002	.000	.806	.000		.007	.000	.177	.000	.000	.000	.000
	N	1056	783	781	783	776	557	785	782	1056	823	1036	1056	994	985	995	987
Dip	*Pearson Correlation*	.393**	-.235**	-.259**	.[c]	-.022	.119*	-.178**	.333**	-.093**	1	.093**	-.028	-.029	-.047	-.044	-.029
	Sig. (2-tailed)	.000	.000	.000	.000	.603	.029	.000	.000	.007		.008	.407	.420	.189	.223	.416
	N	852	550	548	550	544	334	552	552	823	852	817	852	786	779	787	781
Strike	*Pearson Correlation*	.333**	.116**	-.130**	.135**	.084*	-.051	-.128**	.215**	.137**	.093**	1	-.087**	.100**	.043	.082*	.101**
	Sig. (2-tailed)	.000	.001	.000	.000	.020	.237	.000	.000	.000	.008		.005	.002	.182	.010	.002
	N	1037	763	761	763	757	543	765	762	1036	817	1037	1037	976	969	978	971
Scatter ID	*Pearson Correlation*	.000	-.172**	.169**	-.041	-.153**	.306**	-.001	-.195**	-.042	-.028	-.087**	1	-.152**	-.064**	-.101**	-.181**
	Sig. (2-tailed)	1.000	.000	.000	.252	.000	.000	.971	.000	.177	.407	.005		.000	.008	.000	.000
	N	1820	796	794	796	788	557	785	784	1056	852	1037	1820	1700	1688	1700	1692
Max Length	*Pearson Correlation*	.000	.157**	-.087*	.183**	.002	-.176**	-.060	.139**	.307**	-.029	.100**	-.152**	1	.469**	.516**	.627**
	Sig. (2-tailed)	1.000	.000	.018	.000	.959	.000	.104	.000	.000	.420	.002	.000		.000	.000	.000
	N	1700	733	732	733	725	524	724	722	994	786	976	1700	1700	1684	1696	1688
Max Thickness	*Pearson Correlation*	.000	.095*	-.087*	.151**	-.023	-.096*	-.038	.122**	.292**	-.047	.043	-.064**	.469**	1	.599**	.652**
	Sig. (2-tailed)	1.000	.011	.020	.000	.536	.029	.314	.001	.000	.189	.182	.008	.000		.000	.000
	N	1688	727	726	727	719	518	718	716	985	779	969	1688	1684	1688	1684	1688
Weight	*Pearson Correlation*	.000	.143**	-.056	.126**	.009	-.082	-.047	.091*	.149**	-.044	.082*	-.101**	.516**	.599**	1	.754**
	Sig. (2-tailed)	1.000	.000	.133	.001	.808	.061	.204	.014	.000	.223	.010	.000	.000	.000		.000
	N	1700	734	733	734	726	526	725	723	995	787	978	1700	1696	1684	1700	1688
Width	*Pearson Correlation*	.000	.177**	-.085*	.192**	.021	-.148**	-.076*	.152**	.290**	-.029	.101**	-.181**	.627**	.652**	.754**	1
	Sig. (2-tailed)	1.000	.000	.022	.000	.572	.001	.041	.000	.000	.416	.002	.000	.000	.000	.000	
	N	1692	728	727	728	720	519	719	717	987	781	971	1692	1688	1688	1688	1692

**. Correlation is significant at the 0.01 level (2-tailed).

*. Correlation is significant at the 0.05 level (2-tailed).

c. Cannot be computed because at least one of the variables is constant.

Appendix 3: Table displaying Schmidt diagrams, Rose diagrams, and statistics of artifact orientations for all artifact long axes during the course of the experiment. Statistics reported are Rayleigh's, Kuiper's, and Watson's tests. Bootstrapped paired t-tests show differences between each measurement. All results are reported significant at $p<0.05$.

		February	April	September	July
All Artifacts		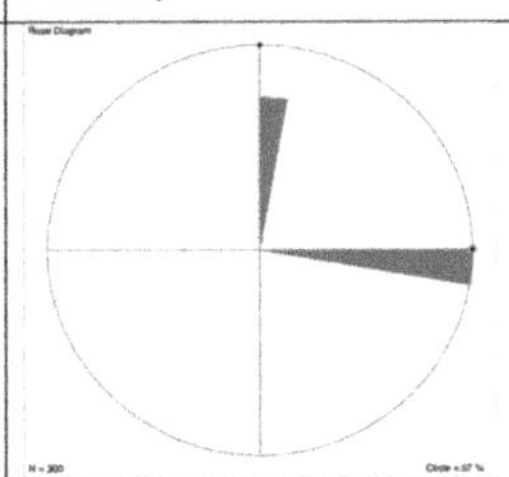	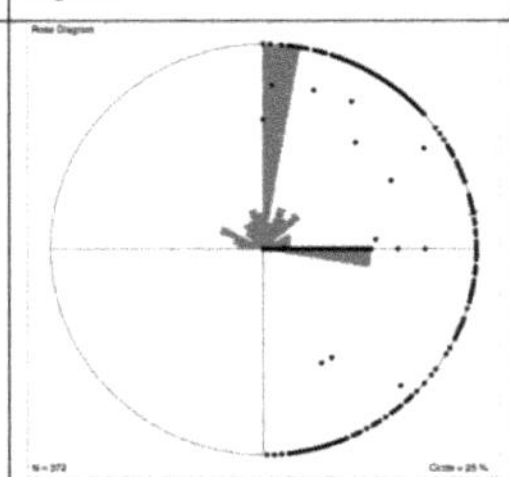	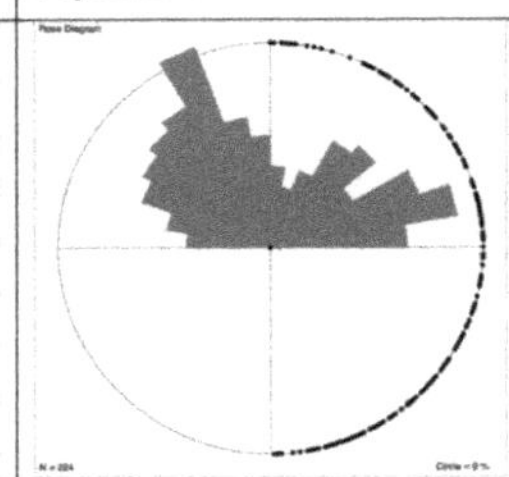	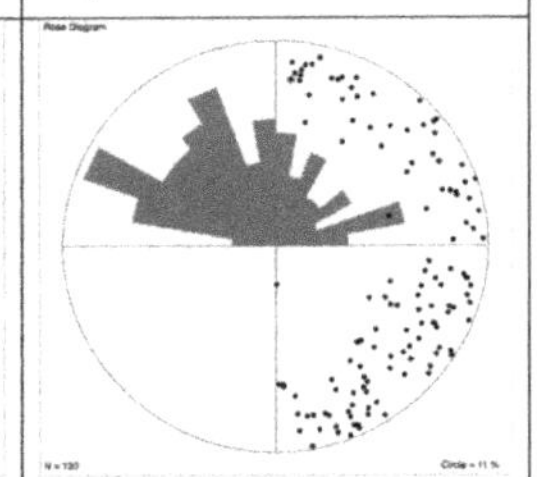
	Mean ± SD	51.644 ± 44.581	88.2920 ± 53.638	97.561± 49.875	100.323 ± 51.890
	t-test	**t(248)=-8.069 *p*<0.001**			
			t(172)=-1.881, *p*=0.062		
				t(102)=0.453, *p*=0.651	
	Ray-leigh		R=0.0631, *p*=0.36	R=0.0846, *p*=0.19	**R=0.1906 *p*=0.007**
	Kuiper		K=1.41, *p*>0.10	K=1.7051, *p*>0.05	K=1.4162, *p*>0.15
	Watson		W=0.0598, *p*>0.05	W=0.0412, *p*>0.10	**W=0.0565, *p*>0.05**

Scatter 1		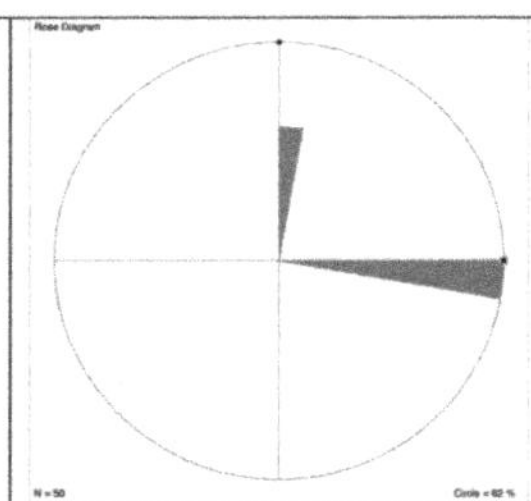	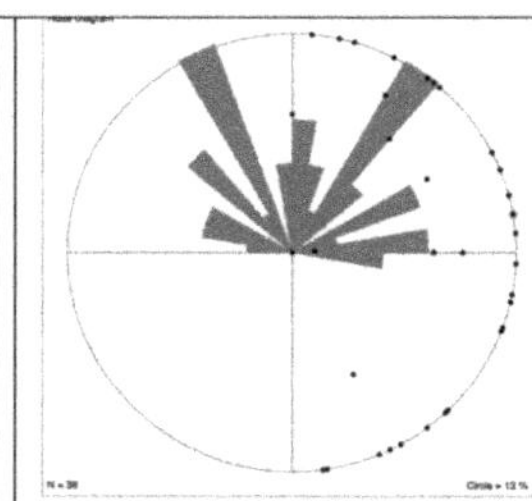	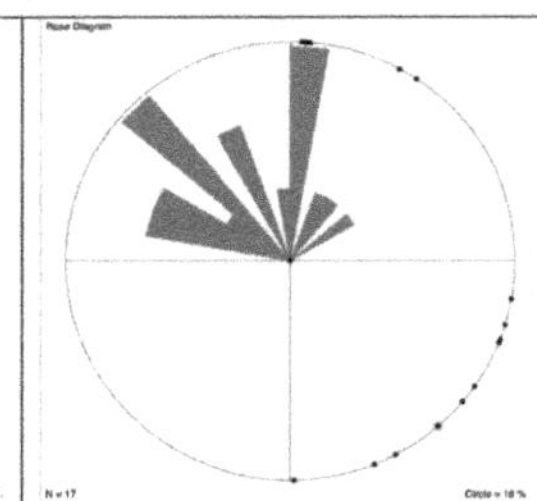	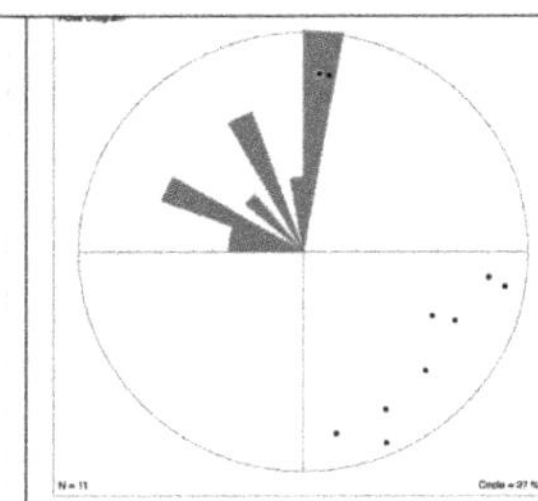
	Mean ± SD	55.8 ± 44.128	84.2105 ± 44.128	93.2353 ± 58.748	96.64 ± 62.522
	t-test	t(37)=-2.157. *p*=0.043			
			t(14)=-1.904, *p*=0.107		
				t(6)=0.659 *p*=0.534	
	Ray-leigh		R=0.0413 *p*=0.93	R=0.3030 *p*=0.21	R=0.4435 *p*=0.11
	Kuiper		K=1.1741 *p*>0.15	K=1.381 *p*>0.15	K=1.0791 *p*>0.15
	Watson		U2=0.0583 *p*>0.10	U2=0.0457 *p*>0.10	U2=0.0491 *p*>0.10

Scatter 2		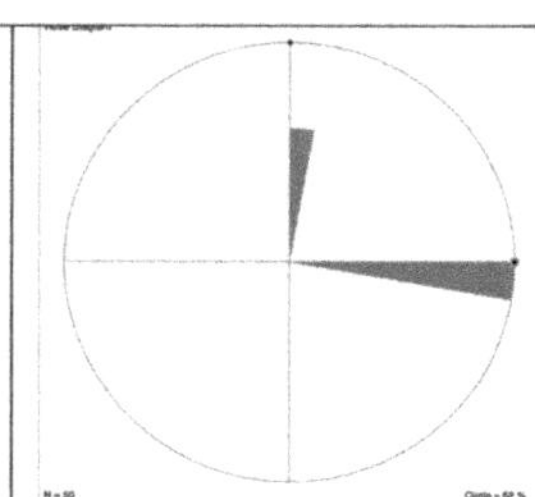	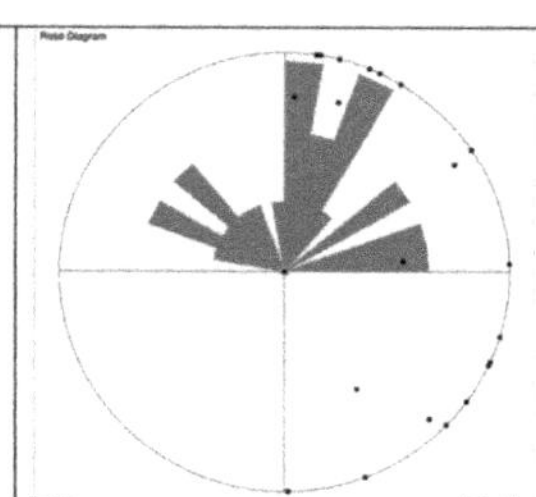	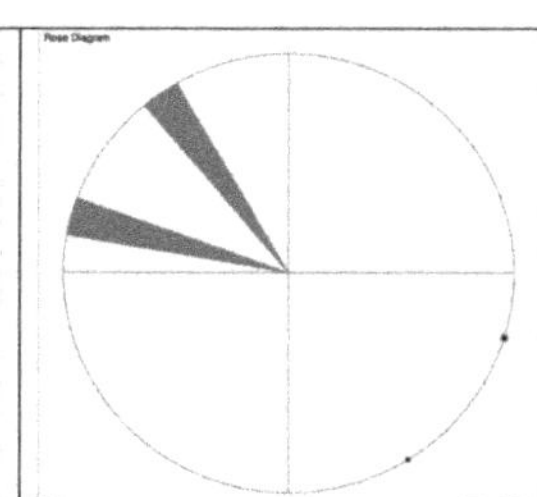	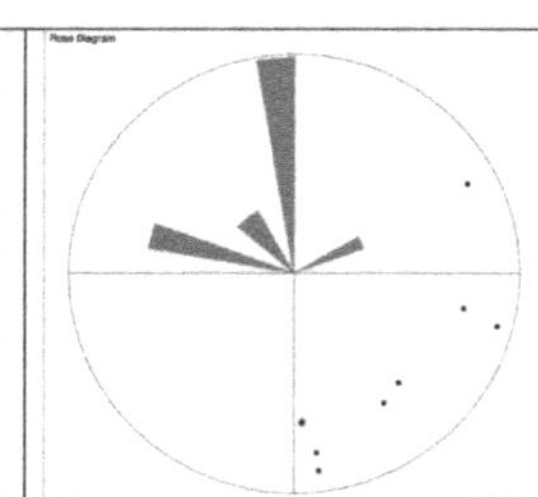
	Mean ± SD	54 ±44.53	75.13 ± 54.981	127.50 ± 28.992	134.37 ± 41.562
	t-test	t(23)=-1.858, *p*=0.076			
	t(df)=t,		Not Enough data		
	p=			t(4)=-.485, *p*=0.653	
	Ray-leigh		R=0.1166 *p*=0.72	Not enough data	R=0.3490 *p*=0.39
	Kuiper		K=0.9984 *p*>0.15	Not enough data	K=1.1279 *p*>0.15
	Watson		U2=0.0469 *p*>0.10	Not enough Data	U2=0.0232 *p*>0.10

Scatter 3					
		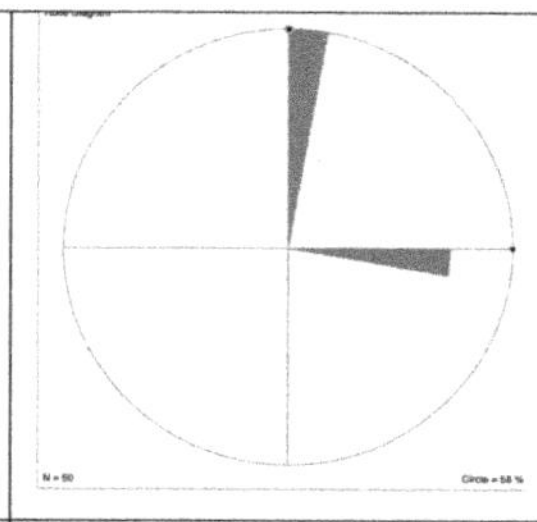	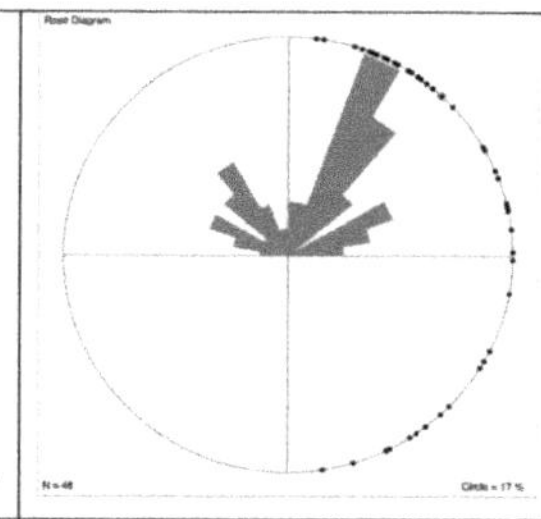	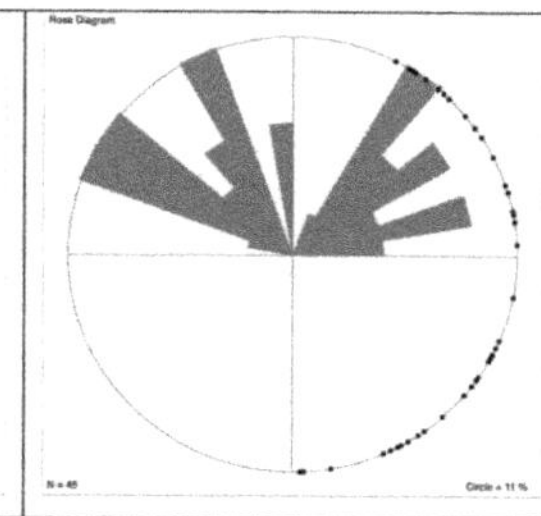	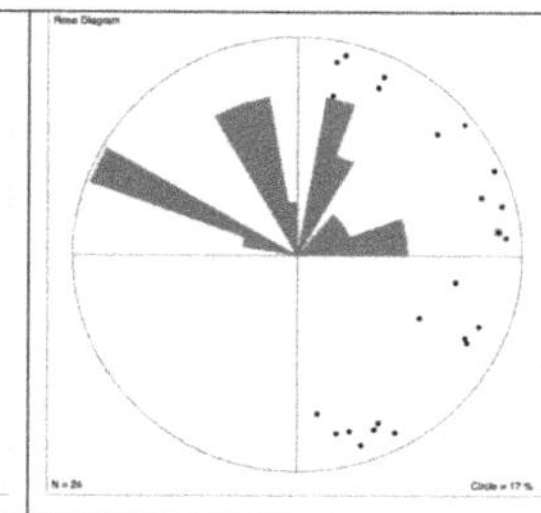
	Mean ± SD	37.8 ± 44.87	75.7917 ±50.279	98.6 ± 46.849	94.29 ±55.085
	t-test t(df)=t, p=	**t(47)=-4.353, *p*=0.001**			
			t(43)=-2.629, *p*=0.019		
				t(20)=-.57, *p*=0.964	
	Ray-leigh		R=0.1743 *p*=0.23	R=0.0927 *p*=0.68	R=0.0668 *p*=0.90
	Kuiper		K=0.9883 *p*>0.15	K=0.814 *p*>0.15	K=1.1589 *p*>0.15
	Watson		U2=0.0212 *p*>0.10	U2=0.025 *p*>0.10	U2=0.0361 *p*>0.10

Scatter 4					
		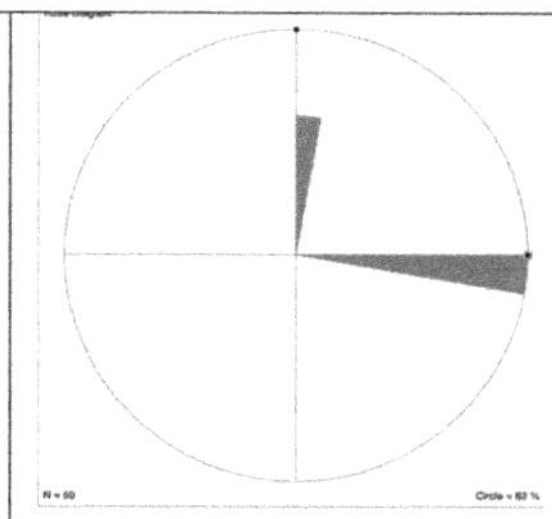	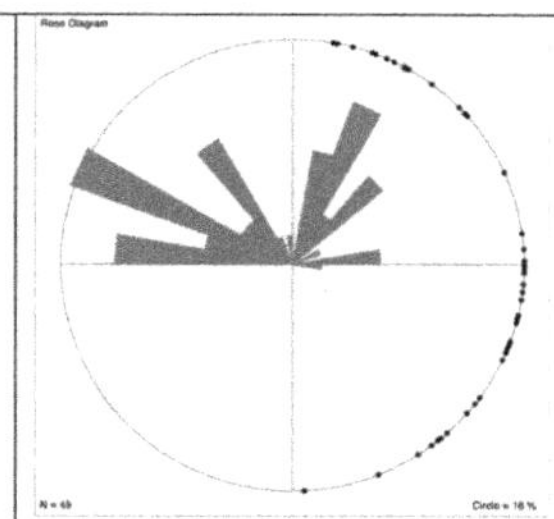	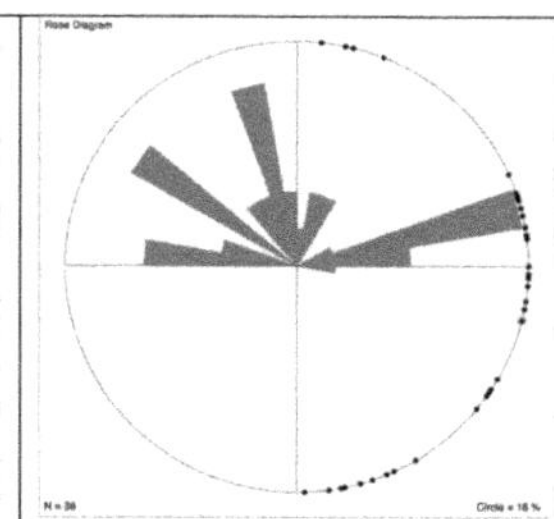	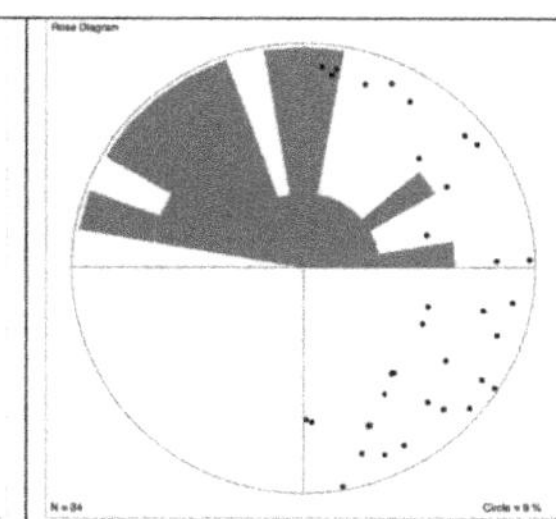
	Mean ± SD	55.8 ± 44.12	86.1429 ± 45.961	102.842 ± 48.94	101.617 ± 53.458
	t-test t(df)=t, p=	**t(48)=-3.081 *p*=0.005**			
			t(37)=-1.676 *p*=0.102		
				t(24)=0.759 *p*=0.435	
	Ray-leigh		R=0.1926 *p*=0.16	R=0.1759 *p*=0.31	R=0.2240 *p*=0.1822
	Kuiper		K=1.4354 *p*>0.15	K=0.954 *p*>0.15	**K=1.9474 *p*<0.05**
	Watson		U2=0.0365 *p*>0.10	U2=0.0255 *p*>0.10	U2=0.0232 *p*>0.10

Scatter 5					
		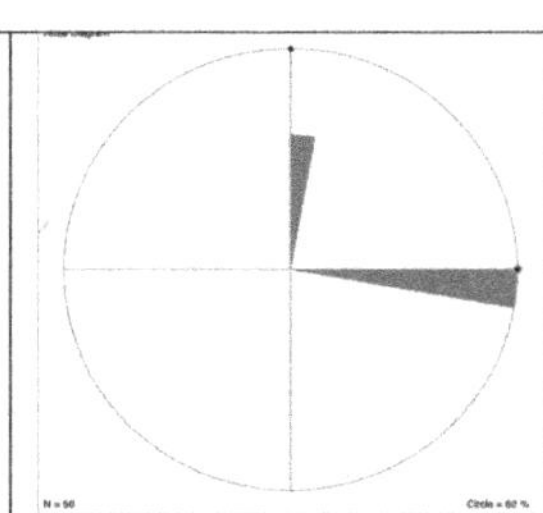	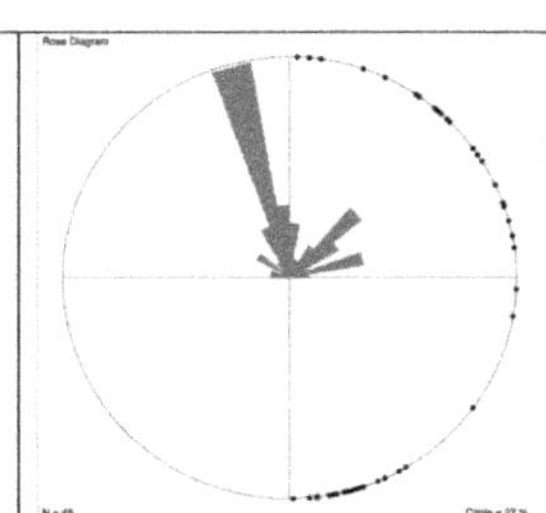	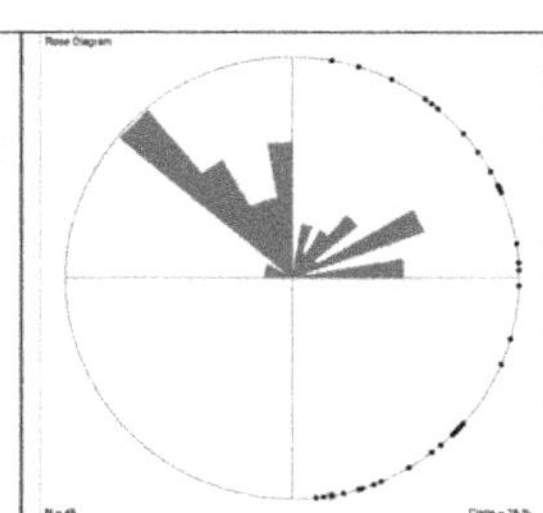	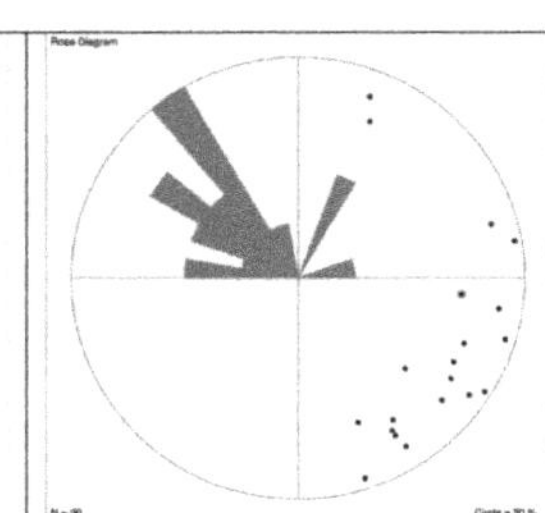
	Mean ± SD	56.47 ± 43.946	104.044 ± 60.149	109.044 ± 50.18	118.79 ± 34.030
	t-test t(df)=t, p=	t(44)=-4.076, *p*=0.001			
			t(40)=-.385, *p*=0.702		
				t(16)=-1.029, *p*=0.303	
	Ray-leigh		**R=0.3035 *p*=0.01**	R=0.1817 *p*=0.22	**R=0.5125 *p*=0.00**
	Kuiper		K=1.0559 *p*>0.15	K=1.2092 *p*>0.15	**K=1.869 *p*<0.05**
	Watson		U2=0.0292 *p*>0.10	U2=0.0342 *p*>0.10	U2=0.0224 *p*>0.10

Scatter 6					
		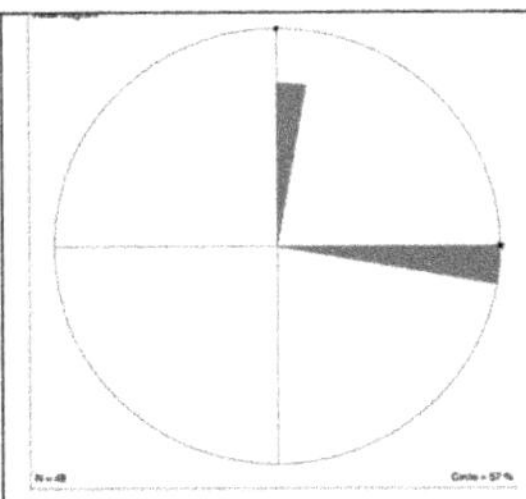	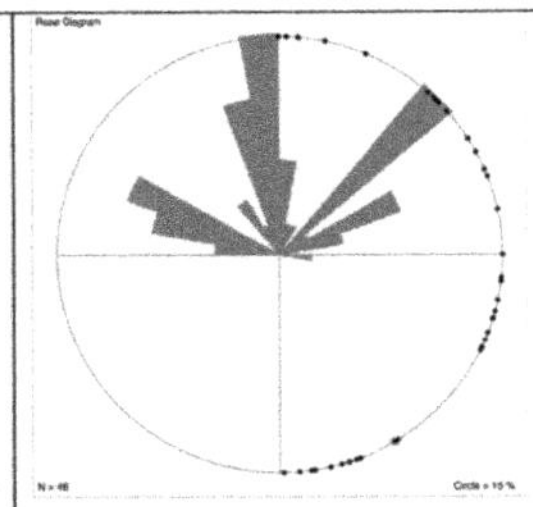	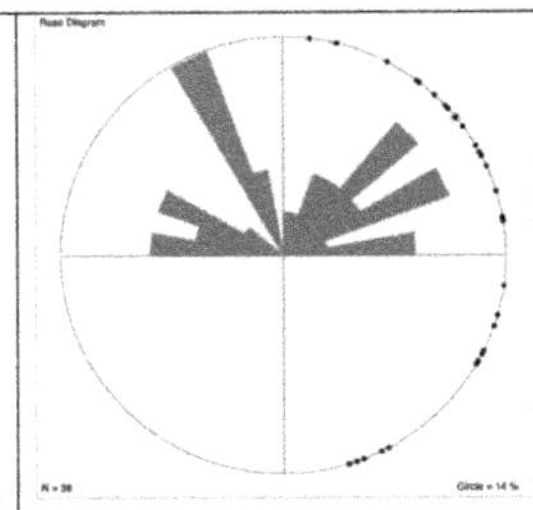	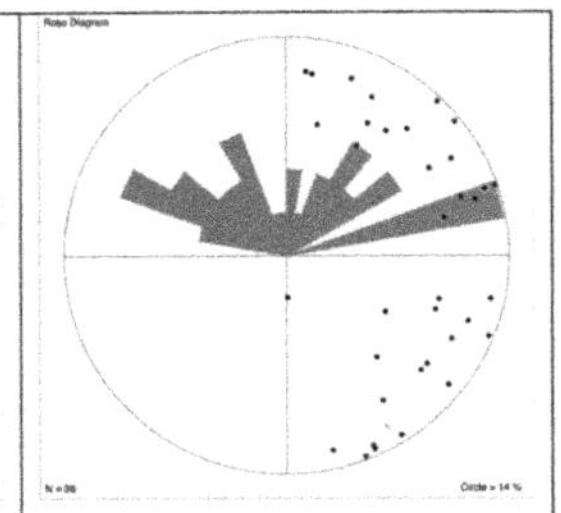
	Mean ± SD	51.42 ± 45	99.913 ± 55.39	85.58 ± 45.958	89.14 ± 51.178
	t-test	t(43)=-3.722, p=0.003			
	t(df)=t,		t(31)=0.824, p=0.412		
	p=			t(28)=-302=, p=0.738	
	Ray-leigh		R=0.0505 p=0.89	R=0.1947 p=0.25	R=0.0147 p=0.99
	Kuiper		K=1.5935 p>0.10	K=1.4998 p>0.15	K=1.5104 p>0.15
	Watson		U2=0.0404 p>0.10	U2=0.0512 p>0.10	U2=0.0574 p>0.05

Scatter 7					
		N/A	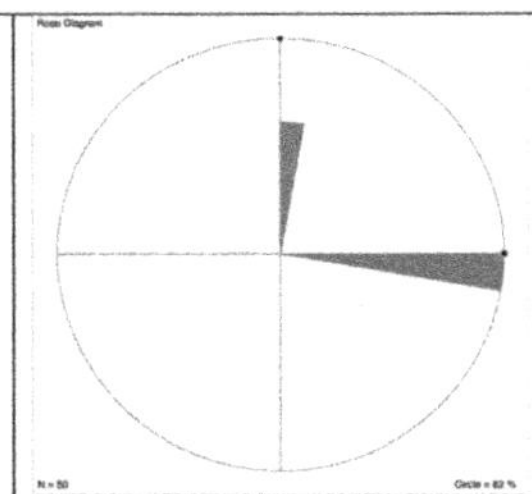	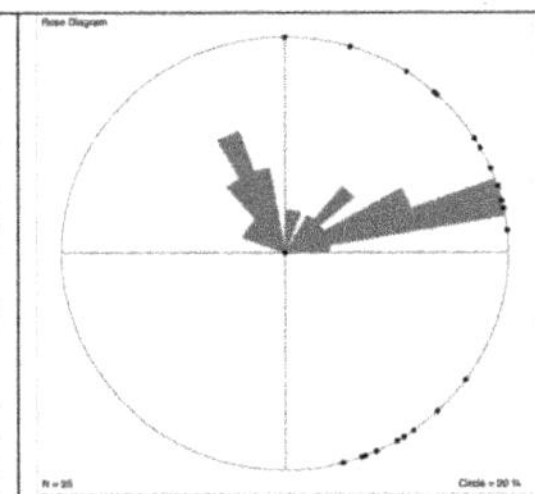	Not recovered
	Mean ± SD	-	55.8 ± 44.128	92.08 ± 49.537	N/A
	t-test	N/A			
			t(24)=-2.714, p=0.017		
				N/A	
	Ray-leigh	-	-	R=0.8032 p=0.80	-
	Kuiper	-	-	K=0.7533 p>0.15	-
	Watson	-	-	U2=0.0181 p>0.10	-

Scatter 8					
		N/A	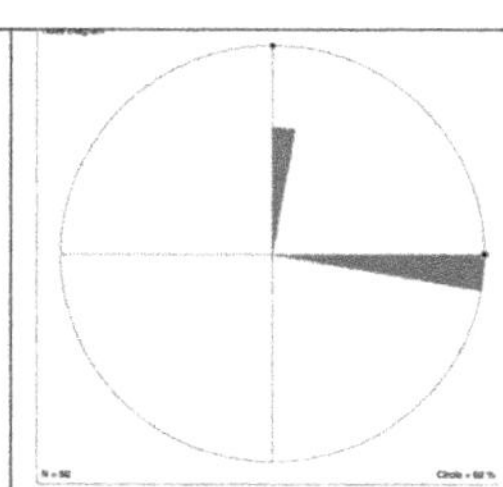	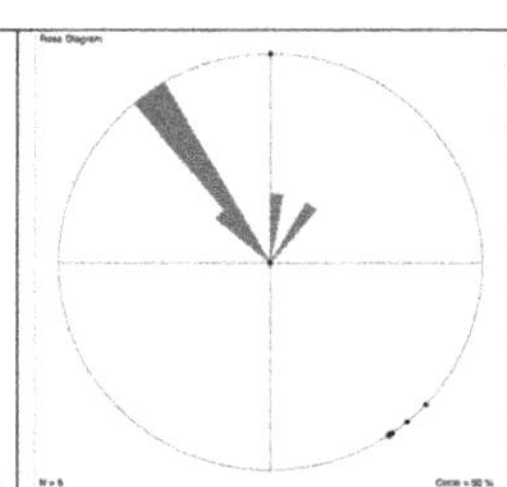	not recovered
	Mean ± SD	N/A	55.8 ± 44.128	85.86 ± 69.684	N/A
	t-test	N/A			
			t(6)=-.393, p=0.679		
				N/A	
	Ray-leigh	-	N/A	R=0.6089 p=0.06	-
	Kuiper	-	-	K=1.2974 p>0.15	-
	Watson	-	-	U2=0.0291 p>0.10	-

Scatter 9					
		N/A	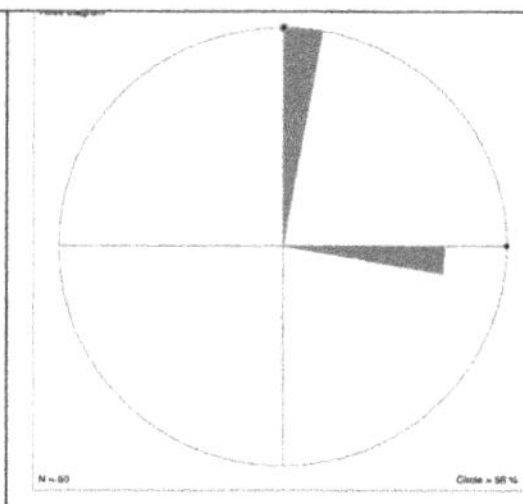	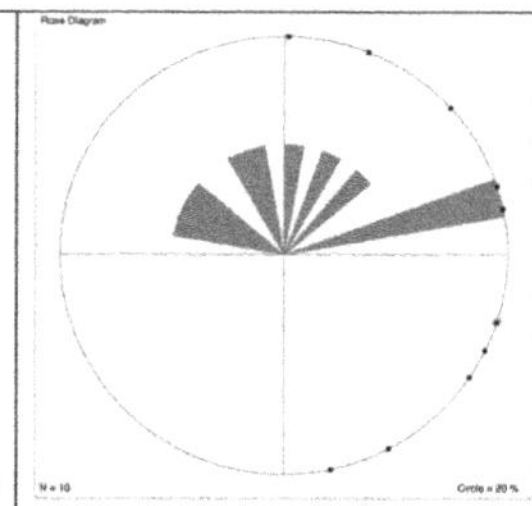	Not recovered
	Mean ± SD	N/A	98.6 ± 46.849	89.85 ± 53.111	N/A
	t-test	N/A			
			t(8)=0.535, p=0.593		
				N/A	
	Ray-leigh	-	N/A	R=0.0741, p=0.93	-
	Kuiper	-	-	K=1.3061 p>0.15	-
	Watson	-	-	U2=0.0652 p>0.05	-

Appendix 4: Summary of replica artifacts tumbling conditions.

Flake group	N	Handaxe names	Sediment lithology	Wentworth classification	φ range	Size range (mm)	Artifact lithology	Sediment mass
zł (Zloty)	10	P, Q, R, Asquare, Dsquare, Jsquare, Msquare, Nsquare, Rsquare, Ssquare	Ham Yard	Unsifted	-6.0–0	63–0	Flint	2kg
^ (Caret)	10		Ham Yard	Unsifted	-6.0–0	63–0	Patinated flint	2kg
₫ (Dong)	10		Ham Yard	Unsifted	-6.0–0	63–0	Quartzite	2kg
% (Percent)	10	U,S,T	Ham Yard	Very coarse gravel	-6.0–5.5	63–45	Flint	2kg
~ (Tilde)	10		Ham Yard	Very coarse gravel	-5.5–5.0	45–31.5	Flint	2kg
# (Hash)	10		Ham Yard	Coarse gravel	-5.0–4.5	31.5–22.4	Flint	2kg
₲ (Guarani)	10		Ham Yard	Coarse gravel	-4.5–4.0	22.4–16	Flint	2kg
฿ (Baht)	10		Ham Yard	Medium gravel	-4.0–3.5	16–11.2	Flint	2kg
₴ (Hryvnia)	10		Ham Yard	Medium gravel	-3.5–3.0	11.2–8	Flint	2kg
₳ (Austral)	10		Kennet	Very coarse gravel/ coarse gravel	-5.5–4.5	45–22.4	Flint	2kg
¢ (Cent)	10	F*,G*,h	Kennet	Very coarse gravel	-5.5–5.0	45–31.5	Flint	2kg
***f* (Florin)**	10	D*,E*,i	Kennet	Very coarse gravel	-5.5–5.0	45–31.5	Flint	1kg
₩ (Won)	10		Kennet	Coarse gravel	-5.0–4.5	31.5–22.4	Flint	2kg
¥ (Yen)	10	L,M,O	Trent	Cobble	-7.0–6.5	128–90	Flint	2kg
₮ (Togrog)	10		Trent	Coarse gravel	-4.5–4.0	22.4–16	Flint	2kg
N/A		f,n,t	Trent	Medium gravel	-3.5–3.0	11.2–8	Flint	2kg
₣ Franc	10	g, C*,H*	None	No sediment	0.0	0.0	Flint	2kg

Appendix 5: Comparison of artifact primary metrics for the groups used in the sediment mass test (t-test).

	t	df	sig. (2-tailed)	Mean Difference	Std. error Difference
Weight	-0.344	12.001	0.737	-3.49	10.142
Length	0.301	17.584	0.767	0.05571	0.18486
Width	-0.263	15.455	0.796	-0.523	1.98505
Thickness	0.414	16.988	0.684	2.328	5.627
Elongation	0.222	14.698	0.827	0.14948	0.67265
Refinement	0.93	17.899	0.365	6.905	7.427

Appendix 6: Table of planned comparisons for time and the interaction of time and sediment mass.

		Weight					**Length**					**Width**				
	Time (minutes)	*Type III Sum of Squares*	*df*	*Mean Square*	*F*	*Sig.*	*Type III Sum of Squares*	*df*	*Mean Square*	*F*	*Sig.*	*Type III Sum of Squares*	*df*	*Mean Square*	*F*	*Sig.*
Time	0 to 1															
	1 to 5	0.033	1	0.033	7.521	0.013	0.075	1	0.075	12.341	0.002	0.003	1	0.003	2.102	0.164
	5 to 15	0.017	1	0.017	2.235	0.152	0.029	1	0.029	4.694	0.044	0.003	1	0.003	2.572	0.126
	15 to 30	0.025	1	0.025	15.114	0.001	0.017	1	0.017	22.225	0	0.009	1	0.009	3.335	0.084
	30 to 60	0.04	1	0.04	11.63	0.003	0.049	1	0.049	14.968	0.001	0.019	1	0.019	7.346	0.014
	60 to 120	0.067	1	0.067	34.695	0	0.044	1	0.044	19.268	0	0.07	1	0.07	12.836	0.002
Time * Sediment mass	0 to 1															
	1 to 5	0.014	1	0.014	3.269	0.087	0.03	1	0.03	5.016	0.038	0	1	0	0.169	0.685
	5 to 15	0.007	1	0.007	0.862	0.365	0.006	1	0.006	0.944	0.344	0	1	0	0.204	0.657
	15to 30	0.001	1	0.001	0.574	0.458	8.00E-07	1	8.00E-07	0.001	0.975	0	1	0	0.084	0.775
	30 to 60	0.014	1	0.014	4.131	0.057	0.016	1	0.016	4.846	0.041	0	1	0	0.047	0.831
	60 to 120	3.65E-05	1	3.65E-05	0.019	0.892	9.80E-06	1	9.80E-06	0.004	0.949	0.007	1	0.007	1.227	0.283

		Elongation					**Refinement**				
	Time	*Type III Sum of Squares*	*df*	*Mean Square*	*F*	*Sig.*	*Type III Sum of Squares*	*df*	*Mean Square*	*F*	*Sig.*
Time (minutes)	0 to 1	0.007	1	0.007	1.121	0.304	0.192	1	0.192	0.622	0.441
	1 to 5	0.162	1	0.162	6.999	0.016	0.05	1	0.05	0.469	0.503
	5 to 15	0.085	1	0.085	1.755	0.202	0.064	1	0.064	1.353	0.261
	15 to 30	0.006	1	0.006	0.691	0.417	0.066	1	0.066	0.204	0.657
	30 to 60	0.026	1	0.026	2.716	0.117	0.906	1	0.906	3.781	0.069
	60 to 120	0.003	1	0.003	0.14	0.713	3.533	1	3.533	11.554	0.003
Time * Sediment mass	0 to 1	0.001	1	0.001	0.105	0.75	0.4	1	0.4	1.297	0.271
	1 to 5	0.115	1	0.115	4.951	0.039	0.246	1	0.246	2.304	0.147
	5 to 15	0.04	1	0.04	0.831	0.374	0.051	1	0.051	1.066	0.316
	15 to 30	0.001	1	0.001	0.158	0.696	0.189	1	0.189	0.581	0.456
	30 to 60	0.053	1	0.053	5.55	0.03	0.09	1	0.09	0.376	0.548
	60 to 120	0.009	1	0.009	0.487	0.494	0	1	0	0.001	0.977

Appendix 7: Bootstrapped one way ANOVA (Brown-Forsythe) illustrating the significance between groups' primary metrics used in the artifact lithology experiment (*df*M=2).

	df_R	F	Sig.
Weight	17.182	2.212	0.14
Length	22.183	2.252	0.129
Width	21.237	4.301	0.270
Thickness	25.866	3.297	0.053
Elongation	22.874	0.914	0.415
Refinement	17.182	2.212	0.14

Appendix 8: Table of planned comparisons for time and the interaction of time and artifact lithology.

		Weight					**Length**					**Width**				
	Time (minutes)	*Type III Sum of Squares*	*df*	*Mean Square*	*F*	*Sig.*	*Type III Sum of Squares*	*df*	*Mean Square*	*F*	*Sig.*	*Type III Sum of Squares*	*df*	*Mean Square*	*F*	*Sig.*
Time	0 to 1															
	1 to 5	0.004	1	0.004	5.615	0.025	0.017	1	0.017	19.156	0	0.008	1	0.008	8.282	0.008
	5 to 15	0.006	1	0.006	9.559	0.005	0.006	1	0.006	9.04	0.006	0.009	1	0.009	5.82	0.023
	15 to 30	0.008	1	0.008	16.966	0	0.004	1	0.004	10.502	0.003	0.003	1	0.003	9.037	0.006
	30 to 60	0.012	1	0.012	14.939	0.001	0.023	1	0.023	12.915	0.001	0.009	1	0.009	16.526	0
	60 to 120	0.03	1	0.03	8.796	0.006	0.026	1	0.026	7.472	0.011	0.004	1	0.004	0.746	0.395
	120to 180	0.022	1	0.022	4.217	0.05	0.024	1	0.024	2.744	0.109	0.061	1	0.061	3.609	0.068

		Elongation					**Refinement**				
	Time (minutes)	*Type III Sum of Squares*	*df*	*Mean Square*	*F*	*Sig.*	*Type III Sum of Squares*	*df*	*Mean Square*	*F*	*Sig.*
Time	0 to 1	0.071	1	0.071	1.973	0.171	0.201	1	0.201	6.684	0.015
	1 to 5	0.001	1	0.001	0.778	0.385	0.029	1	0.029	0.374	0.546
	5 to 15	8.33E-05	1	8.33E-05	0.032	0.86	0.611	1	0.611	6.919	0.014
	15 to 30	0	1	0	0.279	0.602	0.015	1	0.015	9.69	0.004
	30 to 60	0.025	1	0.025	0.699	0.41	0.157	1	0.157	4.119	0.052
	60 to 120	0.058	1	0.058	1.224	0.278	0	1	0	0.001	0.974
	120to 180	0.03	1	0.03	0.875	0.358	2.2	1	2.2	3.261	0.082

Appendix 9: Bootstrapped one way ANOVA illustrating the significance between groups primary metrics used in the sediment size experiment (dfR=72 dfM=7).

Factor	F	Sig.
Weight	0.965	0.463
Length	1.65	0.135
Thickness	0.241	0.973
Elongation	1.08	0.385
Refinement	0.723	0.652

Appendix 10: Table of planned comparisons for time and the interaction of time and sediment size.

		%originalweight					**%originallength**					**%originalwidth**				
	Time (minutes)	*Type III Sum of Squares*	*df*	*Mean Square*	*F*	*Sig.*	*Type III Sum of Squares*	*df*	*Mean Square*	*F*	*Sig.*	*Type III Sum of Squares*	*df*	*Mean Square*	*F*	*Sig.*
Time	0 to 1															
	1 to 5	0.033	1	0.033	6.217	0.015	0.024	1	0.024	23.162	0	0.055	1	0.055	16.998	0
	5 to 15	0.019	1	0.019	47.061	0	0.044	1	0.044	35.009	0	0.031	1	0.031	20.567	0
	15to 30	0.085	1	0.085	11.546	0.001	0.081	1	0.081	16.579	0	0.044	1	0.044	17.349	0
	30 to 60	0.035	1	0.035	24.096	0	0.024	1	0.024	18.552	0	0.021	1	0.021	16.976	0
	60 to 120	0.13	1	0.13	11.29	0.001	0.117	1	0.117	12.42	0.001	0.131	1	0.131	10.77	0.002
	120 to 180	0.053	1	0.053	4.709	0.033	0.044	1	0.044	3.64	0.06	0.037	1	0.037	3.518	0.065
Time ****Phi* **group	0 to 1															
	1 to 5	0.075	7	0.011	1.994	0.068	0.01	7	0.001	1.395	0.221	0.052	7	0.007	2.283	0.037
	5 to 15	0.009	7	0.001	3.226	0.005	0.017	7	0.002	1.912	0.08	0.03	7	0.004	2.818	0.012
	15to 30	0.054	7	0.008	1.04	0.411	0.071	7	0.01	2.056	0.06	0.033	7	0.005	1.867	0.088
	30 to 60	0.014	7	0.002	1.362	0.235	0.01	7	0.001	1.161	0.336	0.006	7	0.001	0.675	0.692
	60 to 120	0.194	7	0.028	2.402	0.029	0.084	7	0.012	1.275	0.275	0.132	7	0.019	1.561	0.161
	120 to 180	0.103	7	0.015	1.308	0.259	0.09	7	0.013	1.064	0.395	0.124	7	0.018	1.656	0.134

		Elongation					**Refinement**				
	Time (minutes)	*Type III Sum of Squares*	*df*	*Mean Square*	*F*	*Sig.*	*Type III Sum of Squares*	*df*	*Mean Square*	*F*	*Sig.*
Time	0 to 1	0.001	1	0.001	0.275	0.601	0.881	1	0.881	5.572	0.021
	1 to 5	0.022	1	0.022	2.654	0.108	0.924	1	0.924	2.511	0.018
	5 to 15	3.35E-05	1	3.35E-05	0.005	0.945	0.905	1	0.905	9.354	0.003
	15to 30	0.018	1	0.018	1.524	0.221	1.214	1	1.214	10.829	0.002
	30 to 60	0.001	1	0.001	0.1	0.753	0.296	1	0.296	14.537	0
	60 to 120	0.001	1	0.001	0.085	0.771	3.409	1	3.409	7.333	0.009
	120 to 180	0.003	1	0.003	0.155	0.695	0.055	1	0.055	0.403	0.527
Time * ***Phi*** **group**	0 to 1	0.023	7	0.003	1.022	0.424	0.394	7	0.056	0.356	0.924
	1 to 5	0.071	7	0.01	1.2	0.314	3.804	7	0.543	1.476	0.19
	5 to 15	0.031	7	0.004	0.642	0.72	1.082	7	0.155	1.597	0.151
	15to 30	0.104	7	0.015	1.253	0.286	1.029	7	0.147	1.312	0.258
	30 to 60	0.027	7	0.004	0.401	0.899	0.12	7	0.017	0.845	0.554
	60 to 120	0.032	7	0.005	0.312	0.946	4.226	7	0.604	1.299	0.264
	120 to 180	0.161	7	0.023	1.307	0.26	1.338	7	0.191	1.408	0.216

Appendix 11: Bootstrapped one way ANOVA (Brown-Forsythe) illustrating the significance between groups' primary metrics used in the sediment lithology experiment (*df*M=3).

	F	df	Sig.
Weight	1.511	25.931	0.235
Length	1.256	29.898	0.155
Width	1.994	33.223	0.134
Thickness	0.288	29.119	0.834
Elongation	0.047	32.18	0.986
Refinement	0.487	27.677	0.694

Appendix 12: Table of planned comparisons for time and the interaction of time and sediment lithology.

		Weight					**Length**					**Width**				
	Time (minutes)	*Type III Sum of Squares*	*df*	*Mean Square*	*F*	*Sig.*	*Type III Sum of Squares*	*df*	*Mean Square*	*F*	*Sig.*	*Type III Sum of Squares*	*df*	*Mean Square*	*F*	*Sig.*
Time	0 to 1															
	1 to 5	0.009	1	0.009	8.327	0.007	0.012	1	0.012	24.115	0	0.027	1	0.027	10.054	0.003
	5 to 15	0.014	1	0.014	59.41	0	0.028	1	0.028	16.718	0	0.026	1	0.026	12.078	0.001
	15 to 30	0.044	1	0.044	6.736	0.014	0.062	1	0.062	9.405	0.004	0.023	1	0.023	8.992	0.005
	30 to 60	0.077	1	0.077	3.535	0.068	0.055	1	0.055	2.36	0.133	0.06	1	0.06	2.667	0.111
	60 to 120	0.014	1	0.014	29.229	0	0.009	1	0.009	11.189	0.002	0.01	1	0.01	23.703	0
	120 to 180	0.005	1	0.005	9.6	0.004	0.007	1	0.007	5.341	0.027	0.005	1	0.005	22.912	0
Time* Sediment lithology	0 to 1															
	1 to 5	0.004	3	0.001	1.235	0.311	0.003	3	0.001	2.042	0.125	0.006	3	0.002	0.682	0.569
	5 to 15	0.005	3	0.002	7.592	0	0.007	3	0.002	1.316	0.284	0.022	3	0.007	3.377	0.029
	15 to 30	0.044	3	0.015	2.224	0.102	0.047	3	0.016	2.398	0.084	0.013	3	0.004	1.644	0.196
	30 to 60	0.099	3	0.033	1.51	0.228	0.081	3	0.027	1.163	0.337	0.075	3	0.025	1.114	0.356
	60 to 120	0.003	3	0.001	1.865	0.153	0.001	3	0	0.599	0.62	0.006	3	0.002	4.417	0.01
	120 to 180	0.002	3	0.001	1.481	0.236	0.001	3	0	0.332	0.802	0.001	3	0	1.162	0.338

		Elongation					**Refinement**				
	Time (minutes)	*Type III Sum of Squares*	*df*	*Mean Square*	*F*	*Sig.*	*Type III Sum of Squares*	*df*	*Mean Square*	*F*	*Sig.*
Time	0 to 1	0.007	1	0.007	1.449	0.237	0.24	1	0.24	0.786	0.381
	1 to 5	0.013	1	0.013	1.38	0.248	0.111	1	0.111	0.253	0.618
	5 to 15	0.004	1	0.004	0.239	0.628	1.057	1	1.057	5.798	0.021
	15 to 30	0.03	1	0.03	2.27	0.141	0.491	1	0.491	5.469	0.025
	30 to 60	0.001	1	0.001	0.146	0.705	0.073	1	0.073	1.698	0.201
	60 to 120	0	1	0	0.043	0.837	0.515	1	0.515	17.014	0
	120 to 180	0	1	0	0.061	0.807	1.992	1	1.992	3.499	0.07
Time* Sediment lithology	0 to 1	0.007	3	0.002	0.513	0.676	0.678	3	0.226	0.739	0.536
	1 to 5	0.004	3	0.001	0.133	0.94	1.781	3	0.594	1.354	0.273
	5 to 15	0.036	3	0.012	0.704	0.556	0.944	3	0.315	1.726	0.18
	15 to 30	0.063	3	0.021	1.62	0.202	0.476	3	0.159	1.767	0.171
	30 to 60	0.018	3	0.006	1.611	0.204	0.249	3	0.083	1.939	0.141
	60 to 120	0.02	3	0.007	1.363	0.27	0.119	3	0.04	1.31	0.287
	120 to 180	0.007	3	0.002	0.476	0.701	3.775	3	1.258	2.21	0.104

Appendix 13: Bootstrapped Welch's t-test illustrating the significance between flakes and handaxes.

Factor	t	df	Sig.
Weight	-5.983	16.775	**0.002***
Length	-13.512	14.244	**0.001***
Width	-10.223	13.059	**0.001***
Thickness	-10.367	16.534	**0.001***
Elongation	0.434	13.315	0.672
Refinement	3.578	10.214	**0.005***

Appendix 14: Table of planned comparisons for time and the interaction of time and artifact class.

		Weight					**Length**					**Width**				
	Time (minutes)	*Type III Sum of Squares*	*df*	*Mean Square*	*F*	*Sig.*	*Type III Sum of Squares*	*df*	*Mean Square*	*F*	*Sig.*	*Type III Sum of Squares*	*df*	*Mean Square*	*F*	*Sig.*
Time	0 to 1															
	1 to 5	0.001	1	0.001	2.293	0.147	0.004	1	0.004	13.181	0.002	0.001	1	0.001	10.761	0.004
	5 to 15	0.002	1	0.002	4.297	0.053	0.002	1	0.002	4.075	0.059	0.004	1	0.004	24.068	0
	15 to 30	0.002	1	0.002	6.336	0.022	0.001	1	0.001	11.278	0.004	0.001	1	0.001	8.702	0.009
	30 to 60	0.002	1	0.002	21.867	0	0.004	1	0.004	22.512	0	0.002	1	0.002	7.131	0.016
	60 to 120	0.002	1	0.002	33.957	0	0.009	1	0.009	18.403	0	0.009	1	0.009	23.233	0
	120 to 180	0.003	1	0.003	4.915	0.04	0.003	1	0.003	12.446	0.002	0.003	1	0.003	10.204	0.005
Time * artifact type	0 to 1															
	1 to 5	0	1	0	1.67	0.213	0.001	1	0.001	2.507	0.131	5.78E-05	1	5.78E-05	0.924	0.349
	5 to 15	0.001	1	0.001	1.445	0.245	0	1	0	0.54	0.472	0	1	0	3.124	0.094
	15 to 30	0.001	1	0.001	3.07	0.097	8.00E-07	1	8.00E-07	0.008	0.929	2.65E-05	1	2.65E-05	0.402	0.534
	30 to 60	0	1	0	4.369	0.051	0	1	0	2.576	0.126	0	1	0	1.504	0.236
	60 to 120	4.81E-05	1	4.81E-05	0.895	0.357	0	1	0	0.297	0.593	0.001	1	0.001	2.631	0.122
	120 to 180	0.001	1	0.001	1.843	0.191	0	1	0	2.09	0.165	0.001	1	0.001	1.777	0.199

		Elongation					**Refinement**				
	Time (minutes)	*Type III Sum of Squares*	*df*	*Mean Square*	*F*	*Sig.*	*Type III Sum of Squares*	*df*	*Mean Square*	*F*	*Sig.*
Time	0 to 1	0.001	1	0.001	0.573	0.459	0.036	1	0.036	7.258	0.015
	1 to 5	0.002	1	0.002	3.166	0.092	0.006	1	0.006	7.675	0.013
	5 to 15	0.001	1	0.001	0.755	0.396	0.057	1	0.057	19.865	0
	15 to 30	0	1	0	0.519	0.481	0.012	1	0.012	9.122	0.007
	30 to 60	0	1	0	0.405	0.533	0.028	1	0.028	5.379	0.032
	60 to 120	0	1	0	0.048	0.829	0.14	1	0.14	16.798	0.001
	120 to 180	1.28E-05	1	1.28E-05	0.033	0.858	0.05	1	0.05	8.556	0.009
Time * artifact type	0 to 1	0.001	1	0.001	0.597	0.45	0.022	1	0.022	4.283	0.053
	1 to 5	0.001	1	0.001	0.815	0.379	0.001	1	0.001	1.367	0.258
	5 to 15	0	1	0	0.189	0.669	0.015	1	0.015	5.088	0.037
	15 to 30	4.21E-05	1	4.21E-05	0.134	0.718	0.001	1	0.001	1.005	0.329
	30 to 60	8.00E-07	1	8.00E-07	0.001	0.98	0.009	1	0.009	1.674	0.212
	60 to 120	0.003	1	0.003	1.216	0.285	0.03	1	0.03	3.618	0.073
	120 to 180	0	1	0	0	1	0.015	1	0.015	2.475	0.133

Appendix 15: t-test comparing levels B/C and C of Caours.

	Levene's Test for Equality of Variances				
	F	*Sig.*	*t*	*df*	*Sig. (2-tailed)*
Max length	.697	.404	-.850	326	.396
Max width	.013	.908	-1.445	326	.149
Width	1.318	.252	-1.308	326	.192
Weight	.404	.525	-.878	326	.381

Appendix 16: Lithic artifact assemblage from Happisburgh 3, UK.

ID	Max length (mm)	Max width (mm)	Max Thick (mm)	Weight (g)	Edges							Ridges		
					Micro-flaking	Unifacial	Continuous	Flaking	Unifacial	Continuous	Battering	Rounding	Pitting	Battering
HSB3 2008 52	130.84	106.62	39.22	367.1	yes	no	yes	yes	no	yes	yes	yes	yes	yes
HSB3 2008 60	92.70	78.66	30.71	193.7	yes	no	yes	no	no	no	no	yes	yes	yes
HSB3 2006 1036	43.95	21.37	13.42	12.5	yes	no	yes	no	no	no	no	yes	yes	yes
HSB 2007-45	71.75	35.95	23.60	27.8	yes	no	yes	no	no	no	no	no	yes	yes
HSB 2007-42	43.65	32.83	14.84	18.6	yes	no	yes	no	no	no	yes	yes	no	yes
HSB3 2005 3001	76.79	55.33	45.94	150.3	yes	no	yes	no	no	no	yes	yes	no	yes
HSB3 2008 59	56.13	34.56	15.53	20.1	yes	no	yes	yes	no	yes	yes	yes	yes	no
HSB3 2008 58	96.66	76.63	23.51	172.8	yes	no	yes	yes	no	yes	yes	no	yes	no
HSB3 2008	46.91	36.28	10.92	17.9	yes	yes	yes	no	no	no	yes	no	yes	no
HSB 2007-72	67.99	47.62	15.56	33.5	no	no	no	yes	yes	yes	no	no	yes	no
HSB3 2008 61	45.13	38.80	16.74	20.1	yes	no	yes	yes	no	yes	no	no	yes	no
HSB3 2008 55	41.68	33.17	10.76	17.2	yes	no	yes	yes	no	yes	no	no	yes	no
HSB 2007-28	67.30	37.94	15.02	42.3	yes	no	yes	yes	yes	no	no	no	yes	no
HSB 2007-64	81.77	57.11	12.25	109.3	yes	no	yes	yes	no	no	no	no	yes	no
HSB 2007-3	86.27	72.07	16.50	125.6	yes	no	yes	no	no	no	no	no	yes	no
HSB3 2011 354	75.88	69.84	48.09	117	yes	no	yes	no	no	no	no	no	yes	no
HSB 2007-16	71.13	66.19	30.25	112	yes	no	yes	no	no	no	no	no	yes	no
HSB 2007-33	61.02	54.95	15.53	36.6	yes	no	yes	no	no	no	no	no	yes	no
HSB3 2005 3003	51.18	39.52	97.20	16.7	no	no	no	yes	no	yes	yes	yes	no	no
HSB3 2006 1032	38.69	34.95	20.16	23.1	yes	no	yes	no	no	no	yes	yes	no	no
HSB3 2006 1010	37.77	22.87	6.79	13	yes	no	yes	no	no	no	yes	yes	no	no
HSB3 2006 1004	34.74	22.03	10.27	7.7	yes	no	yes	no	no	no	yes	yes	no	no
HSB3 2006 1021	87.78	70.65	25.67	113.8	yes	no	yes	no	no	no	no	yes	no	no
HSB 2007-47	41.53	16.87		2.4	yes	no	yes	no	no	no	no	yes	no	no
HSB3 2006 1019	42.48	29.99	8.39	12.2	no	no	no	no	no	no	no	yes	no	no
HSB3 2006 1046	78.42	69.13	22.21	107.8	yes	no	yes	yes	yes	yes	yes	no	no	no
HSB3 2005 3006	84.61	52.32	35.07	122.4	no	no	no	yes	yes	yes	yes	no	no	no
HSB 2007-7	64.07	45.25	18.41	59.2	yes	yes	yes	yes	yes	no	yes	no	no	no
HSB3 2006 1035	43.90	27.08	10.54	15.5	yes	no	yes	no	no	no	yes	no	no	no
HSB3 2005 3015	23.00	10.11	4.53	1.1	yes	no	yes	no	no	no	yes	no	no	no
HSB 2007-32	77.67	50.91	20.90		yes	yes	yes	yes	yes	yes	no	no	no	no
HSB3 2011 356	16.24	9.05	6.63	3.1	yes	no	yes	yes	yes	yes	no	no	no	no
HSB3 2006 1023	36.91	26.39	10.02	9.3	no	no	no	yes	yes	yes	no	no	no	no
HSB3 2006 1038	57.89	42.50	18.19	40.3	yes	no	yes	yes	no	yes	no	no	no	no

HSB 2007-23	49.83	39.81	10.84	18.7	yes	no	yes	yes	no	yes	no	no	no	no
HSB3 2005 3002	53.51	33.03	9.67	18.3	no	no	no	yes	no	yes	no	no	no	no
HSB 2007-1	83.98	83.22	16.16	189.5	yes	no	yes	yes	no	no	no	no	no	no
HSB3 2005 3007	33.59	30.32	3.83	6.3	yes	no	yes	yes	no	no	no	no	no	no
HSB3 2006 1017	109.40	62.70	27.81	179.5	yes	no	no	yes	no	no	no	no	no	no
HSB3 2006 1007	36.10	29.25	5.18	12.4	yes	no	no	yes	no	no	no	no	no	no
HSB3 2006 1025	45.82	36.52	7.04	44.05	no	yes	yes	no	no	no	no	no	no	no
HSB3 2008 54	102.43	51.89	16.93	72.2	yes	no	yes	no	no	no	no	no	no	no
HSB 2007-27	76.73	50.27	23.21	65	yes	no	yes	no	no	no	no	no	no	no
HSB 2007-5	63.97	34.57	13.17	25.7	yes	no	yes	no	no	no	no	no	no	no
HSB3 2008 62	49.71	46.57	11.02	21.6	yes	no	yes	no	no	no	no	no	no	no
HSB 2007-2	39.55	24.66	10.09	10.2	yes	no	yes	no	no	no	no	no	no	no
HSB3 2006 1044	96.05	62.61	42.50	239.7	yes	no	yes	no	no	no	no	no	no	no
HSB 2007-4	107.39	56.31	36.94	168.6	yes	no	yes	no	no	no	no	no	no	no
HSB3 2006 1034	63.99	53.30	10.07	43.4	yes	no	yes	no	no	no	no	no	no	no
HSB 2007-31	62.39	38.76	22.04	41.9	yes	no	yes	no	no	no	no	no	no	no
HSB 2007-14	68.85	42.18	12.60	35.7	yes	no	yes	no	no	no	no	no	no	no
HSB3 2006 1049	42.70	40.28	10.74	22.9	yes	no	yes	no	no	no	no	no	no	no
HSB3 2006 1009	45.07	33.56	10.77	17.8	yes	no	yes	no	no	no	no	no	no	no
HSB 2007-30	40.78	31.27	11.40	16.6	yes	no	yes	no	no	no	no	no	no	no
HSB3 2006 1040	44.13	32.50	8.94	11.4	yes	no	yes	no	no	no	no	no	no	no
HSB3 2006 1039	36.08	32.70	11.14	10	yes	no	yes	no	no	no	no	no	no	no
HSB3 2006 1041	39.05	17.14	6.44	6.6	yes	no	yes	no	no	no	no	no	no	no
HSB 2007-39	41.00	28.54	7.65	6.3	yes	no	yes	no	no	no	no	no	no	no
HSB3 2006 1016	26.80	22.60	10.50	6	yes	no	yes	no	no	no	no	no	no	no
HSB3 2006 1048	31.94	26.68	5.45	5.5	yes	no	yes	no	no	no	no	no	no	no
HSB3 2006 1050	32.03	19.15	5.48	5.1	yes	no	yes	no	no	no	no	no	no	no
HSB 2007-49	35.63	19.00	7.51	4.7	yes	no	yes	no	no	no	no	no	no	no
HSB3 2006 1012	37.30	20.10	10.49	4.6	yes	no	yes	no	no	no	no	no	no	no
HSB3 2006 1015	27.70	19.44	7.10	3.6	yes	no	yes	no	no	no	no	no	no	no
HSB3 2006 1047	22.07	21.96	5.88	2.4	yes	no	yes	no	no	no	no	no	no	no
HSB 2007-12	32.24	18.24	3.66	2.2	yes	no	yes	no	no	no	no	no	no	no
HSB3 2005 3011	21.75	17.73	3.96	1.5	yes	no	yes	no	no	no	no	no	no	no
HSB 2007-41	52.98	38.92	14.32	23.5	yes	no	yes	no	no	no	no	no	no	no
HSB 2007-51	53.42	34.05	15.71	34.1	yes	no	no	no	no	no	no	no	no	no
HSB 2007-17	27.60	24.62	7.17	3.4	yes	no	no	no	no	no	no	no	no	no
unidentified	90.81	85.95	57.75	432.1	yes	no	no	no	no	no	no	no	no	no
HSB3 2006 1027	36.89	25.36	10.03	8.4	yes	no	no	no	no	no	no	no	no	no
HSB3 2006 1045	37.07	22.18	8.24	3.6	yes	no	no	no	no	no	no	no	no	no
HSB3 2006 1013	28.00	20.10	3.11	2.5	yes	no	no	no	no	no	no	no	no	no

HSB3 2006 1020	15.10	13.71	1.35	0.5	yes	no	no	no	no	no	no	no	no	no
HSB 2007-9	34.36	28.78	8.62	7.8	no	no	no	no	no	no	no	no	no	no
HSB 2007-6	48.03	39.20	14.81	25.1	no	no	no	no	no	no	no	no	no	no
HSB 2007-40	36.31	25.03	10.27	8.6	no	no	no	no	no	no	no	no	no	no
HSB 2007-25	38.51	21.67	6.55	6	no	no	no	no	no	no	no	no	no	no
HSB3 2006 1024	30.33	20.16	6.57	2.2	no	no	no	no	no	no	no	no	no	no
HSB3 2006 1005	17.85	13.26	4.16	1	no	no	no	no	no	no	no	no	no	no
HSB 2007-26	39.91	24.36	5.80	4.1	no	no	no	no	no	no	no	no	no	no
HSB3 2006 1002	20.19	15.61	6.84	2.3	n/a	n/a	n/a	n/a	n/a	n/a	n/a	no	no	no
HSB3 2006 1008	34.72	30.53	3.94	13.9	yes	no	yes	no	no	no	yes	n/a	no	no
HSB3 2006 1000	23.90	17.95	3.85	2	no	no	no	yes	yes	yes	no	n/a	no	no
HSB3 2008 57	66.78	31.21	13.56	23	yes	no	yes	no	no	no	no	n/a	no	no
HSB3 2006 1006	19.41	17.13	4.61	1.6	yes	no	yes	no	no	no	no	n/a	no	no
HSB3 2006 1014	27.06	15.95	4.72	1.5	no	no	no	no	no	no	no	n/a	no	no
HSB3 2006 1001	21.17	13.76	4.80	1.3	no	no	no	no	no	no	no	n/a	no	no
HSB3 2006 1003	13.51	11.64	4.20	0.8	n/a	n/a	n/a	n/a	n/a	n/a	n/a	n/a	no	no
HSB3 2006 1011	13.95	10.75	3.66	0.7	n/a	n/a	n/a	n/a	n/a	n/a	n/a	n/a	no	no
HSB3 2008 30	56.70	50.60	24.41	7.4	yes	yes	yes	yes	yes	yes	yes	n/a	n/a	n/a
HSB3 2008 56	48.70	42.40	7.02	13.7	yes	no	yes	yes	no	yes	yes	n/a	n/a	n/a
HSB 2007-11	52.50	37.86	16.87	37.4	yes	no	yes	yes	yes	no	yes	n/a	n/a	n/a
HSB3 2005 3012	21.03	20.40	2.72	1.4	no	no	no	yes	no	yes	no	n/a	n/a	n/a
HSB 2007-48	57.39	44.51	9.42	27	no	no	no	yes	yes	no	no	n/a	n/a	n/a
HSB3 2008 63	91.17	63.57	34.27	169.2	yes	no	yes	no	no	no	no	n/a	n/a	n/a
HSB3 2006 1042	69.43	50.56	15.63	40.9	yes	no	yes	no	no	no	no	n/a	n/a	n/a
HSB3 2008 53	61.24	54.83	12.77	39.2	yes	no	yes	no	no	no	no	n/a	n/a	n/a
HSB 2007-71	63.16	47.00	17.79	38	yes	no	yes	no	no	no	no	n/a	n/a	n/a
HSB3 2006 1028	30.82	26.61	10.28	6.6	yes	no	yes	no	no	no	no	n/a	n/a	n/a
HSB3 2005 3014	65.54	33.92	17.61	36.4	yes	no	yes	no	no	no	no	n/a	n/a	n/a
HSB3 2006 1026	44.01	34.59	14.03	18.5	yes	no	yes	no	no	no	no	n/a	n/a	n/a
HSB3 2005 3018	40.70	26.49	9.64	10.9	yes	no	yes	no	no	no	no	n/a	n/a	n/a
HSB3 2005 3009	35.42	19.41	7.34	6.2	yes	no	yes	no	no	no	no	n/a	n/a	n/a
HSB3 2005 3008	24.09	19.48	4.92	3.6	yes	no	yes	no	no	no	no	n/a	n/a	n/a
HSB3 2005 3004	15.15	13.56	2.35	0.6	yes	no	yes	no	no	no	no	n/a	n/a	n/a
HSB3 2005 3013	19.21	10.49	3.25	0.5	yes	no	yes	no	no	no	no	n/a	n/a	n/a
HSB3 2006 1022	84.09	51.35	11.08	49.1	yes	no	yes	no	no	no	no	n/a	n/a	n/a
HSB3 2006 1030	49.28	28.12	8.61	11.7	yes	no	yes	no	no	no	no	n/a	n/a	n/a
HSB3 2006 1029	32.40	25.77	6.00	5.4	yes	no	yes	no	no	no	no	n/a	n/a	n/a
HSB3 2006 1037	24.11	15.08	3.95	1.5	yes	no	yes	no	no	no	no	n/a	n/a	n/a
HSB 2007-24	36.96	17.37	5.20	3.3	yes	no	no	no	no	no	no	n/a	n/a	n/a
HSB 2007-22	43.68	41.57	11.33	17.3	no	no	no	no	no	no	no	n/a	n/a	n/a

HSB3 2005 3005	22.80	13.19	5.72	1.5	no	no	no	no	no	no	no	n/a	n/a	n/a
HSB3 2005 3017	26.34	13.42	1.78	1	no	no	no	no	no	no	no	n/a	n/a	n/a
HSB3 2005 3016	13.43	9.82	1.54	0.3	no	no	no	no	no	no	no	n/a	n/a	n/a
HSB3 2006 1031	23.69	11.82	4.11	1	no	no	no	no	no	115	no	n/a	n/a	n/a
HSB3 2005 10.05.2	25.60	25.58	3.86	4.62	n/a	n/a	n/a	n/a	n/a	n/a	n/a	n/a	n/a	n/a
HSB3 2006 1033	66.40	30.87	13.81	34.4	n/a	n/a	n/a	n/a	n/a	n/a	n/a	n/a	n/a	n/a
HSB3 2005 3010	25.03	21.45	6.15	2.7	n/a	n/a	n/a	n/a	n/a	n/a	n/a	n/a	n/a	n/a
Mean	48.53	34.76	13.65	38.86										
SD	24.19	18.80	12.85	68.38										
n					115	115	115	115	115	115	115	83	91	91
n=yes					91	5	82	28	12	19	19	13	16	6
%yes					79%	4%	71%	24%	10%	17%	17%	16%	18%	7%